PERSONAL RECOLLECTIONS OF JOAN OF ARC

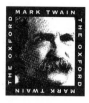

THE OXFORD MARK TWAIN
Shelley Fisher Fishkin, Editor

Personal Recollections of Joan of Arc

Mark Twain

FOREWORD

SHELLEY FISHER FISHKIN

INTRODUCTION

JUSTIN KAPLAN

AFTERWORD

SUSAN K. HARRIS

New York   Oxford

OXFORD UNIVERSITY PRESS

1996

OXFORD UNIVERSITY PRESS

Oxford New York

Athens, Auckland, Bangkok, Bogotá, Bombay
Buenos Aires, Calcutta, Cape Town, Dar es Salaam
Delhi, Florence, Hong Kong, Istanbul, Karachi
Kuala Lumpur, Madras, Madrid, Melbourne
Mexico City, Nairobi, Paris, Singapore
Taipei, Tokyo, Toronto
and associated companies in
Berlin, Ibadan

Copyright © 1996 by
Oxford University Press, Inc.
Introduction © 1996 by Justin Kaplan
Afterword © 1996 by Susan K. Harris
Illustrators and Illustrations in Mark Twain's
First American Editions © 1996 by Beverly R. David
and Ray Sapirstein
Reading the Illustrations in Joan of Arc © 1996
by Ray Sapirstein
Text design by Richard Hendel
Composition: David Thorne

Published by
Oxford University Press, Inc.
198 Madison Avenue, New York,
New York 10016

Oxford is a registered trademark of
Oxford University Press

Library of Congress
Cataloging-in-Publication Data

Twain, Mark, 1835–1910.
Personal recollections of Joan of Arc / by Mark
Twain; with an introduction by Justin Kaplan ; and
an afterword by Susan K. Harris.
p. cm. — (The Oxford Mark Twain)
Includes bibliographical references.
1. Joan, of Arc, Saint, 1412–1431—Fiction. 2. Christian
women saints—France—Fiction. I. Title. II. Series:
Twain, Mark, 1835–1910. Works. 1996.
PS1313.A1 1996
813'.4—dc20
96-16581
CIP
ISBN 0-19-510145-6 (trade ed.)
ISBN 0-19-511416-7 (lib. ed.)
ISBN 0-19-509088-8 (trade ed. set)
ISBN 0-19-511345-4 (lib. ed. set)

9 8 7 6 5 4 3 2 1

Printed in the United States of America
on acid-free paper

FRONTISPIECE
Samuel L. Clemens appears here in a photograph
taken in London, between 1895 and 1899, around
the time he published *Personal Recollections of Joan
of Arc*. (From the Mark Twain Collection of The
James S. Copley Library, La Jolla, California)

CONTENTS

EDITOR'S NOTE

The Oxford Mark Twain consists of twenty-nine volumes of facsimiles of the first American editions of Mark Twain's works, with an editor's foreword, new introductions, afterwords, notes on the texts, and essays on the illustrations in volumes with artwork. The facsimiles have been reproduced from the originals unaltered, except that blank pages in the front and back of the books have been omitted, and any seriously damaged or missing pages have been replaced by pages from other first editions (as indicated in the notes on the texts).

In the foreword, introduction, afterword, and essays on the illustrations, the titles of Mark Twain's works have been capitalized according to modern conventions, as have the names of characters (except where otherwise indicated). In the case of discrepancies between the title of a short story, essay, or sketch as it appears in the original table of contents and as it appears on its own title page, the title page has been followed. The parenthetical numbers in the introduction, afterwords, and illustration essays are page references to the facsimiles.

FOREWORD

Shelley Fisher Fishkin

Samuel Clemens entered the world and left it with Halley's Comet, little dreaming that generations hence Halley's Comet would be less famous than Mark Twain. He has been called the American Cervantes, our Homer, our Tolstoy, our Shakespeare, our Rabelais. Ernest Hemingway maintained that "all modern American literature comes from one book by Mark Twain called *Huckleberry Finn*." President Franklin Delano Roosevelt got the phrase "New Deal" from *A Connecticut Yankee in King Arthur's Court. The Gilded Age* gave an entire era its name. "The future historian of America," wrote George Bernard Shaw to Samuel Clemens, "will find your works as indispensable to him as a French historian finds the political tracts of Voltaire."[1]

There is a Mark Twain Bank in St. Louis, a Mark Twain Diner in Jackson Heights, New York, a Mark Twain Smoke Shop in Lakeland, Florida. There are Mark Twain Elementary Schools in Albuquerque, Dayton, Seattle, and Sioux Falls. Mark Twain's image peers at us from advertisements for Bass Ale (his drink of choice was Scotch), for a gas company in Tennessee, a hotel in the nation's capital, a cemetery in California.

Ubiquitous though his name and image may be, Mark Twain is in no danger of becoming a petrified icon. On the contrary: Mark Twain lives. *Huckleberry Finn* is "the most taught novel, most taught long work, and most taught piece of American literature" in American schools from junior high to the graduate level.[2] Hundreds of Twain impersonators appear in theaters, trade shows, and shopping centers in every region of the country.[3] Scholars publish hundreds of articles as well as books about Twain every year, and he is the subject of daily exchanges on the Internet. A journalist somewhere in the world finds a reason to quote Twain just about every day. Television series such as *Bonanza, Star Trek: The Next Generation,* and *Cheers* broadcast episodes that feature Mark Twain as a character. Hollywood screenwriters regularly produce movies inspired by his works, and writers of mysteries and science fiction continue to weave him into their plots.[4]

A century after the American Revolution sent shock waves throughout Europe, it took Mark Twain to explain to Europeans and to his countrymen alike what that revolution had wrought. He probed the significance of this new land and its new citizens, and identified what it was in the Old World that America abolished and rejected. The founding fathers had thought through the political dimensions of making a new society; Mark Twain took on the challenge of interpreting the social and cultural life of the United States for those outside its borders as well as for those who were living the changes he discerned.

Americans may have constructed a new society in the eighteenth century, but they articulated what they had done in voices that were largely inter-changeable with those of Englishmen until well into the nineteenth century. Mark Twain became the voice of the new land, the leading translator of what and who the "American" was — and, to a large extent, is. Frances Trollope's *Domestic Manners of the Americans,* a best-seller in England, Hector St. John de Crèvecoeur's *Letters from an American Farmer,* and Tocqueville's *Democracy in America* all tried to explain America to Europeans. But Twain did more than that: he allowed European readers to *experience* this strange "new world." And he gave his countrymen the tools to do two things they had not quite had the confidence to do before. He helped them stand before the cultural icons of the Old World unembarrassed, unashamed of America's lack of palaces and shrines, proud of its brash practicality and bold inventiveness, unafraid to reject European models of "civilization" as tainted or corrupt. And he also helped them recognize their own insularity, boorishness, arrogance, or ignorance, and laugh at it — the first step toward transcending it and becoming more "civilized," in the best European sense of the word.

Twain often strikes us as more a creature of our time than of his. He appreciated the importance and the complexity of mass tourism and public relations, fields that would come into their own in the twentieth century but were only fledgling enterprises in the nineteenth. He explored the liberating potential of humor and the dynamics of friendship, parenting, and marriage. He narrowed the gap between "popular" and "high" culture, and he meditated on the enigmas of personal and national identity. Indeed, it would be difficult to find an issue on the horizon today that Twain did not touch on somewhere in his work. Heredity versus environment? Animal rights? The boundaries of gender? The place of black voices in the cultural heritage of the United States? Twain was there.

With startling prescience and characteristic grace and wit, he zeroed in on many of the key challenges — political, social, and technological — that would face his country and the world for the next hundred years: the challenge of race relations in a society founded on both chattel slavery and ideals of equality, and the intractable problem of racism in American life; the potential of new technologies to transform our lives in ways that can be both exhilarating and terrifying — as well as unpredictable; the problem of imperialism and the difficulties entailed in getting rid of it. But he never lost sight of the most basic challenge of all: each man or woman's struggle for integrity in the face of the seductions of power, status, and material things.

Mark Twain's unerring sense of the right word and not its second cousin taught people to pay attention when he spoke, in person or in print. He said things that were smart and things that were wise, and he said them incomparably well. He defined the rhythms of our prose and the contours of our moral map. He saw our best and our worst, our extravagant promise and our stunning failures, our comic foibles and our tragic flaws. Throughout the world he is viewed as the most distinctively American of American authors — and as one of the most universal. He is assigned in classrooms in Naples, Riyadh, Belfast, and Beijing, and has been a major influence on twentieth-century writers from Argentina to Nigeria to Japan. The Oxford Mark Twain celebrates the versatility and vitality of this remarkable writer.

The Oxford Mark Twain reproduces the first American editions of Mark Twain's books published during his lifetime.[5] By encountering Twain's works in their original format — typography, layout, order of contents, and illustrations — readers today can come a few steps closer to the literary artifacts that entranced and excited readers when the books first appeared. Twain approved of and to a greater or lesser degree supervised the publication of all of this material.[6] The Mark Twain House in Hartford, Connecticut, generously loaned us its originals.[7] When more than one copy of a first American edition was available, Robert H. Hirst, general editor of the Mark Twain Project, in cooperation with Marianne Curling, curator of the Mark Twain House (and Jeffrey Kaimowitz, head of Rare Books for the Watkinson Library of Trinity College, Hartford, where the Mark Twain House collection is kept), guided our decision about which one to use.[8] As a set, the volumes also contain more than eighty essays commissioned especially for The Oxford Mark Twain, in which distinguished contributors reassess Twain's achievement as a writer and his place in the cultural conversation that he did so much to shape.

Each volume of The Oxford Mark Twain is introduced by a leading American, Canadian, or British writer who responds to Twain — often in a very personal way — as a fellow writer. Novelists, journalists, humorists, columnists, fabulists, poets, playwrights — these writers tell us what Twain taught them and what in his work continues to speak to them. Reading Twain's books, both famous and obscure, they reflect on the genesis of his art and the characteristics of his style, the themes he illuminated, and the aesthetic strategies he pioneered. Individually and collectively their contributions testify to the place Mark Twain holds in the hearts of readers of all kinds and temperaments.

Scholars whose work has shaped our view of Twain in the academy today have written afterwords to each volume, with suggestions for further reading. Their essays give us a sense of what was going on in Twain's life when he wrote the book at hand, and of how that book fits into his career. They explore how each book reflects and refracts contemporary events, and they show Twain responding to literary and social currents of the day, variously accepting, amplifying, modifying, and challenging prevailing paradigms. Sometimes they argue that works previously dismissed as quirky or eccentric departures actually address themes at the heart of Twain's work from the start. And as they bring new perspectives to Twain's composition strategies in familiar texts, several scholars see experiments in form where others saw only formlessness, method where prior critics saw only madness. In addition to elucidating the work's historical and cultural context, the afterwords provide an overview of responses to each book from its first appearance to the present.

Most of Mark Twain's books involved more than Mark Twain's words: unique illustrations. The parodic visual send-ups of "high culture" that Twain himself drew for *A Tramp Abroad*, the sketch of financial manipulator Jay Gould as a greedy and sadistic "Slave Driver" in *A Connecticut Yankee in King Arthur's Court*, and the memorable drawings of Eve in *Eve's Diary* all helped Twain's books to be sold, read, discussed, and preserved. In their essays for each volume that contains artwork, Beverly R. David and Ray Sapirstein highlight the significance of the sketches, engravings, and photographs in the first American editions of Mark Twain's works, and tell us what is known about the public response to them.

The Oxford Mark Twain invites us to read some relatively neglected works by Twain in the company of some of the most engaging literary figures of our time. Roy Blount Jr., for example, riffs in a deliciously Twain-like manner on "An Item Which the Editor Himself Could Not Understand," which may well rank as one of the least-known pieces Twain ever published. Bobbie Ann Mason celebrates the "mad energy" of Twain's most obscure comic novel, *The American Claimant*, in which the humor "hurtles beyond tall tale into simon-pure absurdity."[9] Garry Wills finds that *Christian Science* "gets us very close to the heart of American culture." Lee Smith reads "Political Economy" as a sharp and funny essay on language. Walter Mosley sees "The Stolen White Elephant," a story "reduced to a series of ridiculous telegrams related by an untrustworthy narrator caught up in an adventure that is as impossible as it is ludicrous," as a stunningly compact and economical satire of a world we still recognize as our own. Anne Bernays returns to "The Private History of a Campaign That Failed" and finds "an antiwar manifesto that is also confession, dramatic monologue, a plea for understanding and absolution, and a romp that gradually turns into atrocity even as we watch." After revisiting Captain Stormfield's heaven, Frederik Pohl finds that there "is no imaginable place more pleasant to spend eternity." Indeed, Pohl writes, "one would almost be willing to die to enter it."

While less familiar works receive fresh attention in The Oxford Mark Twain, new light is cast on the best-known works as well. Judith Martin ("Miss Manners") points out that it is by reading a court etiquette book that Twain's pauper learns how to behave as a proper prince. As important as etiquette may be in the palace, Martin notes, it is even more important in the slums.

> That etiquette is a sorer point with the ruffians in the street than with the proud dignitaries of the prince's court may surprise some readers. As in our own streets, etiquette is always a more volatile subject among those who cannot count on being treated with respect than among those who have the power to command deference.

And taking a fresh look at *Adventures of Huckleberry Finn,* Toni Morrison writes,

> much of the novel's genius lies in its quiescence, the silences that pervade it and give it a porous quality that is by turns brooding and soothing. It lies in ... the subdued images in which the repetition of a simple word, such as "lonesome," tolls like an evening bell; the moments when nothing is said, when scenes and incidents swell the heart unbearably precisely because unarticulated, and force an act of imagination almost against the will.

Engaging Mark Twain as one writer to another, several contributors to The Oxford Mark Twain offer new insights into the processes by which his books came to be. Russell Banks, for example, reads *A Tramp Abroad* as "an important revision of Twain's incomplete first draft of *Huckleberry Finn,* a second draft, if you will, which in turn made possible the third and final draft." Erica Jong suggests that *1601,* a freewheeling parody of Elizabethan manners and mores, written during the same summer Twain began *Huckleberry Finn*, served as "a warm-up for his creative process" and "primed the pump for other sorts of freedom of expression." And Justin Kaplan suggests that "one of the transcendent figures standing behind and shaping" *Joan of Arc* was Ulysses S. Grant, whose memoirs Twain had recently published, and who, like Joan, had risen unpredictably "from humble and obscure origins" to become a "military genius" endowed with "the gift of command, a natural eloquence, and an equally natural reserve."

As a number of contributors note, Twain was a man ahead of his times. *The Gilded Age* was the first "Washington novel," Ward Just tells us, because "Twain was the first to see the possibilities that had eluded so many others." Commenting on *The Tragedy of Pudd'nhead Wilson,* Sherley Anne Williams observes that "Twain's argument about the power of environment in shaping character runs directly counter to prevailing sentiment where the negro was concerned." Twain's fictional technology, wildly fanciful by the standards of his day, predicts developments we take for granted in ours. DNA cloning, fax machines, and photocopiers are all prefigured, Bobbie Ann Mason tells us, in *The American Claimant.* Cynthia Ozick points out that the "telelectrophonoscope" we meet in "From the 'London Times' of 1904" is suspiciously like what we know as "television." And Malcolm Bradbury suggests that in the "phrenophones" of "Mental Telegraphy" "the Internet was born."

Twain turns out to have been remarkably prescient about political affairs as well. Kurt Vonnegut sees in *A Connecticut Yankee* a chilling foreshadowing (or perhaps a projection from the Civil War) of "all the high-tech atrocities which followed, and which follow still." Cynthia Ozick suggests that "The Man That Corrupted Hadleyburg," along with some of the other pieces collected under that title — many of them written when Twain lived in a Vienna ruled by Karl Lueger, a demagogue Adolf Hitler would later idolize — shoot up moral flares that shed an eerie light on the insidious corruption, prejudice, and hatred that reached bitter fruition under the Third Reich. And Twain's portrait in this book of "the dissolving Austria-Hungary of the 1890s," in Ozick's view, presages not only the Sarajevo that would erupt in 1914 but also "the disintegrated components of the former Yugoslavia" and "the *fin-de-siècle* Sarajevo of our own moment."

Despite their admiration for Twain's ambitious reach and scope, contributors to The Oxford Mark Twain also recognize his limitations. Mordecai Richler, for example, thinks that "the early pages of *Innocents Abroad* suffer from being a tad broad, proffering more burlesque than inspired satire," perhaps because Twain was "trying too hard for knee-slappers." Charles Johnson notes that the Young Man in Twain's philosophical dialogue about free will and determinism (*What Is Man?*) "caves in far too soon," failing to challenge what through late-twentieth-century eyes looks like "pseudoscience" and suspect essentialism in the Old Man's arguments.

Some contributors revisit their first encounters with Twain's works, recalling what surprised or intrigued them. When David Bradley came across "Fenimore Cooper's Literary Offences" in his college library, he "did not at first realize that Twain was being his usual ironic self with all this business about the 'nineteen rules governing literary art in the domain of romantic fiction,' but by the time I figured out there was no such list outside Twain's own head, I had decided that the rules made *sense*. . . . It seemed to me they were a pretty good blueprint for writing — Negro writing included." Sherley Anne Williams remembers that part of what attracted her to *Pudd'nhead Wilson* when she first read it thirty years ago was "that Twain, writing at the end of the nineteenth century, could imagine negroes as characters, albeit white ones, who actually thought for and of themselves, whose actions were the product of their thinking rather than the spontaneous ephemera of physical instincts that stereotype assigned to blacks." Frederik Pohl recalls his first reading of *Huckleberry Finn* as "a watershed event" in his life, the first book he read as a child in which "bad people" ceased to exercise a monopoly on doing "bad things." In *Huckleberry Finn* "some seriously bad things — things like the possession and mistreatment of black slaves, like stealing and lying, even like killing other people in duels — were quite often done by people who not only thought of themselves as exemplarily moral but, by any other standards I knew how to apply, actually *were* admirable citizens." The world that

Tom and Huck lived in, Pohl writes, "was filled with complexities and con-tradictions," and resembled "the world I appeared to be living in myself."

Other contributors explore their more recent encounters with Twain, ex-plaining why they have revised their initial responses to his work. For Toni Morrison, parts of *Huckleberry Finn* that she "once took to be deliberate eva-sions, stumbles even, or a writer's impatience with his or her material," now strike her "as otherwise: as entrances, crevices, gaps, seductive invitations flashing the possibility of meaning. Unarticulated eddies that encourage div-ing into the novel's undertow — the real place where writer captures reader." One such "eddy" is the imprisonment of Jim on the Phelps farm. Instead of dismissing this portion of the book as authorial bungling, as she once did, Morrison now reads it as Twain's commentary on the 1880s, a period that "saw the collapse of civil rights for blacks," a time when "the nation, as well as Tom Sawyer, was deferring Jim's freedom in agonizing play." Morrison be-lieves that Americans in the 1880s were attempting "to bury the combustible issues Twain raised in his novel," and that those who try to kick Huck Finn out of school in the 1990s are doing the same: "The cyclical attempts to re-move the novel from classrooms extend Jim's captivity on into each genera-tion of readers."

Although imitation-Hemingway and imitation-Faulkner writing contests draw hundreds of entries annually, no one has ever tried to mount a faux-Twain competition. Why? Perhaps because Mark Twain's voice is too much a part of who we are and how we speak even today. Roy Blount Jr. suggests that it is impossible, "at least for an American writer, to parody Mark Twain. It would be like doing an impression of your father or mother: he or she is al-ready there in your voice."

Twain's style is examined and celebrated in The Oxford Mark Twain by fellow writers who themselves have struggled with the nuances of words, the structure of sentences, the subtleties of point of view, and the trickiness of opening lines. Bobbie Ann Mason observes, for example, that "Twain loved the sound of words and he knew how to string them by sound, like different shades of one color: 'The earl's barbaric eye,' 'the Usurping Earl,' 'a double-dyed humbug.'" Twain "relied on the punch of plain words" to show writers how to move beyond the "wordy romantic rubbish" so prevalent in nineteenth-century fiction, Mason says; he "was one of the first writers in America to deflower literary language." Lee Smith believes that "American writers have benefited as much from the way Mark Twain opened up the possibilities of first-person narration as we have from his use of vernacular language." (She feels that "the ghost of Mark Twain was hovering someplace in the background" when she decided to write her novel *Oral History* from the standpoint of multiple first-person narrators.) Frederick Busch maintains that "A Dog's Tale" "boasts one of the great opening sentences" of all time: "My father was a St. Bernard, my mother was a collie, but I am a Presbyterian." And Ursula Le Guin marvels at the ingenuity of the following sentence that she encounters in *Extracts from Adam's Diary.*

. . . This made her sorry for the creatures which live in there, which she calls fish, for she continues to fasten names on to things that don't need them and don't come when they are called by them, which is a matter of no consequence to her, as she is such a numskull anyway; so she got a lot of them out and brought them in last night and put them in my bed to keep warm, but I have noticed them now and then all day, and I don't see that they are any happier there than they were before, only quieter.[10]

Le Guin responds,

Now, that is a pure Mark-Twain-tour-de-force sentence, covering an immense amount of territory in an effortless, aimless ramble that seems to be heading nowhere in particular and ends up with breathtaking accuracy at the gold mine. Any sensible child would find that funny, perhaps not following all its divagations but delighted by the swing of it, by the word "numskull," by the idea of putting fish in the bed; and as that child grew older and reread it, its reward would only grow; and if that grown-up child had to write an essay on the piece and therefore earnestly studied and pored over this sentence, she would end up in unmitigated admiration of its vocabulary, syntax, pacing, sense, and rhythm, above all the beautiful timing of the last two words; and she would, and she does, still find it funny.

The fish surface again in a passage that Gore Vidal calls to our attention, from *Following the Equator*: "'The Whites always mean well when they take human fish out of the ocean and try to make them dry and warm and happy and comfortable in a chicken coop,' which is how, through civilization, they did away with many of the original inhabitants. Lack of empathy is a principal theme in Twain's meditations on race and empire."

Indeed, empathy — and its lack — is a principal theme in virtually all of Twain's work, as contributors frequently note. Nat Hentoff quotes the following thoughts from Huck in *Tom Sawyer Abroad*:

I see a bird setting on a dead limb of a high tree, singing with its head tilted back and its mouth open, and before I thought I fired, and his song stopped and he fell straight down from the limb, all limp like a rag, and I run and picked him up and he was dead, and his body was warm in my hand, and his head rolled about this way and that, like his neck was broke, and there was a little white skin over his eyes, and one little drop of blood on the side of his head; and laws! I could n't see nothing more for the tears; and I hain't never murdered no creature since that war n't doing me no harm, and I ain't going to.[11]

"The Humane Society," Hentoff writes, "has yet to say anything as powerful — and lasting."

Readers of The Oxford Mark Twain will have the pleasure of revisiting Twain's Mississippi landmarks alongside Willie Morris, whose own lower Mississippi Valley boyhood gives him a special sense of connection to Twain. Morris knows firsthand the mosquitoes described in *Life on the Mississippi* — so colossal that "two of them could whip a dog" and "four of them could hold a man down"; in Morris's own hometown they were so large during the flood season that "local wags said they wore wristwatches." Morris's Yazoo City and Twain's Hannibal shared a "rough-hewn democracy . . . complicated by all the visible textures of caste and class, . . . harmless boyhood fun and mischief right along with . . . rank hypocrisies, churchgoing sanctimonies, racial hatred, entrenched and unrepentant greed."

For the West of Mark Twain's *Roughing It*, readers will have George Plimpton as their guide. "What a group these newspapermen were!" Plimpton writes about Twain and his friends Dan De Quille and Joe Goodman in Virginia City, Nevada. "Their roisterous carryings-on bring to mind the kind of frat-house enthusiasm one associates with college humor magazines like the *Harvard Lampoon*." Malcolm Bradbury examines Twain as "a living example of what made the American so different from the European." And Hal Holbrook, who has interpreted Mark Twain on stage for some forty years, describes how Twain "played" during the civil rights movement, during the Vietnam War, during the Gulf War, and in Prague on the eve of the demise of Communism.

Why do we continue to read Mark Twain? What draws us to him? His wit? His compassion? His humor? His bravura? His humility? His understanding of who and what we are in those parts of our being that we rarely open to view? Our sense that he knows we can do better than we do? Our sense that he knows we can't? E. L. Doctorow tells us that children are attracted to *Tom Sawyer* because in this book "the young reader confirms his own hope that no matter how troubled his relations with his elders may be, beneath all their disapproval is their underlying love for him, constant and steadfast." Readers in general, Arthur Miller writes, value Twain's "insights into America's always uncertain moral life and its shifting but everlasting hypocrisies"; we appreciate the fact that he "is not using his alienation from the public illusions of his hour in order to reject the country implicitly as though he could live without it, but manifestly in order to correct it." Perhaps we keep reading Mark Twain because, in Miller's words, he "wrote much more like a father than a son. He doesn't seem to be sitting in class taunting the teacher but standing at the head of it challenging his students to acknowledge their own humanity, that is, their immemorial attraction to the untrue."

Mark Twain entered the public eye at a time when many of his countrymen considered "American culture" an oxymoron; he died four years before a world conflagration that would lead many to question whether the contradiction in terms was not "European civilization" instead. In between he worked in journalism, printing, steamboating, mining, lecturing, publishing, and editing, in virtually every region of the country. He tried his hand at humorous sketches, social satire, historical novels, children's books, poetry, drama, science fiction, mysteries, romance, philosophy, travelogue, memoir, polemic, and several genres no one had ever seen before or has ever seen since. He invented a self-pasting scrapbook, a history game, a vest strap, and a gizmo for keeping bed sheets tucked in; he invested in machines and processes designed to revolutionize typesetting and engraving, and in a food supplement called "Plasmon." Along the way he cheerfully impersonated himself and prior versions of himself for doting publics on five continents while playing out a charming rags-to-riches story followed by a devastating riches-to-rags story followed by yet another great American comeback. He had a long-running real-life engagement in a sumptuous comedy of manners, and then in a real-life tragedy not of his own design: during the last fourteen years of his life almost everyone he ever loved was taken from him by disease and death.

Mark Twain has indelibly shaped our views of who and what the United States is as a nation and of who and what we might become. He understood the nostalgia for a "simpler" past that increased as that past receded — and he saw through the nostalgia to a past that was just as complex as the present. He recognized better than we did ourselves our potential for greatness and our potential for disaster. His fictions brilliantly illuminated the world in which he lived, changing it — and us — in the process. He knew that our feet often danced to tunes that had somehow remained beyond our hearing; with perfect pitch he played them back to us.

My mother read *Tom Sawyer* to me as a bedtime story when I was eleven. I thought Huck and Tom could be a lot of fun, but I dismissed Becky Thatcher as a bore. When I was twelve I invested a nickel at a local garage sale in a book that contained short pieces by Mark Twain. That was where I met Twain's Eve. Now, *that's* more like it, I decided, pleased to meet a female character I could identify *with* instead of against. Eve had spunk. Even if she got a lot wrong, you had to give her credit for trying. "The Man That Corrupted

Hadleyburg" left me giddy with satisfaction: none of my adolescent reveries of getting even with my enemies were half as neat as the plot of the man who got back at that town. "How I Edited an Agricultural Paper" set me off in uncontrollable giggles.

People sometimes told me that I looked like Huck Finn. "It's the freckles," they'd explain — not explaining anything at all. I didn't read *Huckleberry Finn* until junior year in high school in my English class. It was the fall of 1965. I was living in a small town in Connecticut. I expected a sequel to *Tom Sawyer*. So when the teacher handed out the books and announced our assignment, my jaw dropped: "Write a paper on how Mark Twain used irony to attack racism in *Huckleberry Finn*."

The year before, the bodies of three young men who had gone to Mississippi to help blacks register to vote — James Chaney, Andrew Goodman, and Michael Schwerner — had been found in a shallow grave; a group of white segregationists (the county sheriff among them) had been arrested in connection with the murders. America's inner cities were simmering with pent-up rage that began to explode in the summer of 1965, when riots in Watts left thirty-four people dead. None of this made any sense to me. I was confused, angry, certain that there was something missing from the news stories I read each day: the why. Then I met Pap Finn. And the Phelpses.

Pap Finn, Huck tells us, "had been drunk over in town" and "was just all mud." He erupts into a drunken tirade about "a free nigger ... from Ohio — a mulatter, most as white as a white man," with "the whitest shirt on you ever see, too, and the shiniest hat; and there ain't a man in town that's got as fine clothes as what he had."

> ... they said he was a p'fessor in a college, and could talk all kinds of languages, and knowed everything. And that ain't the wust. They said he could *vote*, when he was at home. Well, that let me out. Thinks I, what is the country a-coming to? It was 'lection day, and I was just about to go and vote, myself, if I warn't too drunk to get there; but when they told me there was a State in this country where they'd let that nigger vote, I drawed out. I says I'll never vote agin. Them's the very words I said. . . . And to see the

cool way of that nigger — why, he wouldn't a give me the road if I hadn't shoved him out o' the way.[12]

Later on in the novel, when the runaway slave Jim gives up his freedom to nurse a wounded Tom Sawyer, a white doctor testifies to the stunning altruism of his actions. The Phelpses and their neighbors, all fine, upstanding, well-meaning, churchgoing folk,

> agreed that Jim had acted very well, and was deserving to have some notice took of it, and reward. So every one of them promised, right out and hearty, that they wouldn't curse him no more.
>
> Then they come out and locked him up. I hoped they was going to say he could have one or two of the chains took off, because they was rotten heavy, or could have meat and greens with his bread and water, but they didn't think of it.[13]

Why did the behavior of these people tell me more about why Watts burned than anything I had read in the daily paper? And why did a drunk Pap Finn railing against a black college professor from Ohio whose vote was as good as his own tell me more about white anxiety over black political power than anything I had seen on the evening news? Mark Twain knew that there was nothing, absolutely *nothing*, a black man could do — including selflessly sacrificing his freedom, the only thing of value he had — that would make white society see beyond the color of his skin. And Mark Twain knew that depicting racists with chilling accuracy would expose the viciousness of their world view like nothing else could. It was an insight echoed some eighty years after Mark Twain penned Pap Finn's rantings about the black professor, when Malcolm X famously asked, "Do you know what white racists call black Ph.D.'s?" and answered, "'*Nigger!*'"[14]

Mark Twain taught me things I needed to know. He taught me to understand the raw racism that lay behind what I saw on the evening news. He taught me that the most well-meaning people can be hurtful and myopic. He taught me to recognize the supreme irony of a country founded in freedom that continued to deny freedom to so many of its citizens. Every time I hear of another effort to kick Huck Finn out of school somewhere, I recall everything that Mark Twain taught *this* high school junior, and I find myself jumping into the fray.[15] I remember the black high school student who called CNN during the phone-in portion of a 1985 debate between Dr. John Wallace, a black educator spearheading efforts to ban the book, and myself. She accused Dr. Wallace of insulting her and all black high school students by suggesting they weren't smart enough to understand Mark Twain's irony. And I recall the black cameraman on the *CBS Morning News* who came up to me after he finished shooting another debate between Dr. Wallace and myself. He said he had never read the book by Mark Twain that we had been arguing about — but now he really wanted to. One thing that puzzled him, though, was why a white woman was defending it and a black man was attacking it, because as far as he could see from what we'd been saying, the book made whites look pretty bad.

As I came to understand *Huckleberry Finn* and *Pudd'nhead Wilson* as commentaries on the era now known as the nadir of American race relations, those books pointed me toward the world recorded in nineteenth-century black newspapers and periodicals and in fiction by Mark Twain's black contemporaries. My investigation of the role black voices and traditions played in shaping Mark Twain's art helped make me aware of their role in shaping all of American culture.[16] My research underlined for me the importance of changing the stories we tell about who we are to reflect the realities of what we've been.[17]

Ever since our encounter in high school English, Mark Twain has shown me the potential of American literature and American history to illuminate each other. Rarely have I found a contradiction or complexity we grapple with as a nation that Mark Twain had not puzzled over as well. He insisted on taking America seriously. And he insisted on *not* taking America seriously: "I think that there is but a single specialty with us, only one thing that can be called by the wide name 'American,'" he once wrote. "That is the national devotion to ice-water."[18]

Mark Twain threw back at us our dreams and our denial of those dreams, our greed, our goodness, our ambition, and our laziness, all rattling around together in that vast echo chamber of our talk — that sharp, spunky American talk that Mark Twain figured out how to write down without robbing it of its energy and immediacy. Talk shaped by voices that the official arbiters of "culture" deemed of no importance — voices of children, voices of slaves, voices of servants, voices of ordinary people. Mark Twain listened. And he made us listen. To the stories he told us, and to the truths they conveyed. He still has a lot to say that we need to hear.

Mark Twain lives — in our libraries, classrooms, homes, theaters, movie houses, streets, and most of all in our speech. His optimism energizes us, his despair sobers us, and his willingness to keep wrestling with the hilarious and horrendous complexities of it all keeps us coming back for more. As the twenty-first century approaches, may he continue to goad us, chasten us, delight us, berate us, and cause us to erupt in unrestrained laughter in unexpected places.

NOTES

1. Ernest Hemingway, *Green Hills of Africa* (New York: Charles Scribner's Sons, 1935), 22. George Bernard Shaw to Samuel L. Clemens, July 3, 1907, quoted in Albert Bigelow Paine, *Mark Twain: A Biography* (New York: Harper and Brothers, 1912), 3:1398.

2. Allen Carey-Webb, "Racism and *Huckleberry Finn*: Censorship, Dialogue and Change," *English Journal* 82, no. 7 (November 1993):22.

3. See Louis J. Budd, "Impersonators," in J. R. LeMaster and James D. Wilson, eds., *The Mark Twain Encyclopedia* (New York: Garland Publishing Company, 1993), 389–91.

4. See Shelley Fisher Fishkin, "Ripples and Reverberations," part 3 of *Lighting Out for the Territory: Reflections on Mark Twain and American Culture* (New York: Oxford University Press, 1996).

5. There are two exceptions. Twain published chapters from his autobiography in the *North American Review* in 1906 and 1907, but this material was not published in book form in Twain's lifetime; our volume reproduces the material as it appeared in the *North American Review*. The other exception is our final volume, *Mark Twain's Speeches*, which appeared two months after Twain's death in 1910.

An unauthorized handful of copies of *1601* was privately printed by an Alexander Gunn of Cleveland at the instigation of Twain's friend John Hay in 1880. The first American edition authorized by Mark Twain, however, was printed at the United States Military Academy at West Point in 1882; that is the edition reproduced here.

It should further be noted that four volumes — *The Stolen White Elephant and Other Detective Stories*, *Following the Equator and Anti-imperialist Essays*, *The Diaries of Adam and Eve*, and *1601, and Is Shakespeare Dead?* — bind together material originally published separately. In each case the first American edition of the material is the version that has been reproduced, always in its entirety. Because Twain constantly recycled and repackaged previously published works in his collections of short pieces, a certain amount of duplication is unavoidable. We have selected volumes with an eye toward keeping this duplication to a minimum.

Even the twenty-nine-volume Oxford Mark Twain has had to leave much out. No edition of Twain can ever claim to be "complete," for the man was too prolix, and the file drawers of both ephemera and as yet unpublished texts are deep.

6. With the possible exception of *Mark Twain's Speeches*. Some scholars suspect Twain knew about this book and may have helped shape it, although no hard evidence to that effect has yet surfaced. Twain's involvement in the production process varied greatly from book to book. For a fuller sense of authorial intention, scholars will continue to rely on the superb definitive editions of Twain's works produced by the Mark Twain Project at the University of California at Berkeley as they become available. Dense with annotation documenting textual emendation and related issues, these editions add immeasurably to our understanding of Mark Twain and the genesis of his works.

7. Except for a few titles that were not in its collection. The American Antiquarian Society in Worcester, Massachusetts, provided the first edition of *King Leopold's Soliloquy*; the Elmer Holmes Bobst Library of New York University furnished the 1906–7 volumes of the *North American Review* in which *Chapters from My Autobiography* first appeared; the Harry Ransom Humanities Research Center at the University of Texas at Austin made their copy of the West Point edition of *1601* available; and the Mark Twain Project provided the first edition of *Extract from Captain Stormfield's Visit to Heaven*.

8. The specific copy photographed for Oxford's facsimile edition is indicated in a note on the text at the end of each volume.

9. All quotations from contemporary writers in this essay are taken from their introductions to the volumes of The Oxford Mark Twain, and the quotations from Mark Twain's works are taken from the texts reproduced in The Oxford Mark Twain.

10. *The Diaries of Adam and Eve*, The Oxford Mark Twain [hereafter OMT] (New York: Oxford University Press, 1996), p. 33.

11. *Tom Sawyer Abroad*, OMT, p. 74.

12. *Adventures of Huckleberry Finn*, OMT, p. 49–50.

13. Ibid., p. 358.

14. Malcolm X, *The Autobiography of Malcolm X*, with the assistance of Alex Haley (New York: Grove Press, 1965), p. 284.

15. I do not mean to minimize the challenge of teaching this difficult novel, a challenge for which all teachers may not feel themselves prepared. Elsewhere I have developed some concrete strategies for approaching the book in the classroom, including teaching it in the context of the history of American race relations and alongside books by black writers. See Shelley Fisher Fishkin, "Teaching *Huckleberry Finn*," in James S. Leonard, ed., *Making Mark Twain Work in the Classroom* (Durham: Duke University Press, forthcoming). See also Shelley Fisher Fishkin, *Was Huck Black? Mark Twain and African-American Voices* (New York: Oxford University Press, 1993), pp. 106–8, and a curriculum kit in preparation at the Mark Twain House in Hartford, containing teaching suggestions from myself, David Bradley, Jocelyn Chadwick-Joshua, James Miller, and David E. E. Sloane.

16. See Fishkin, *Was Huck Black?* See also Fishkin, "Interrogating 'Whiteness,' Complicating 'Blackness': Remapping American Culture," in Henry Wonham, ed., *Criticism and the Color Line: Desegregating American Literary Studies* (New Brunswick: Rutgers UP, 1996, pp. 251–90 and in shortened form in *American Quarterly* 47, no. 3 (September 1995):428–66.

17. I explore the roots of my interest in Mark Twain and race at greater length in an essay entitled "Changing the Story," in Jeffrey Rubin-Dorsky and Shelley Fisher Fishkin, eds., *People of the Book: Thirty Scholars Reflect on Their Jewish Identity* (Madison: U of Wisconsin Press, 1996), pp. 47–63.

18. "What Paul Bourget Thinks of Us," *How to Tell a Story and Other Essays*, OMT, p. 197.

INTRODUCTION

Justin Kaplan

I f *Personal Recollections of Joan of Arc*, published in 1896, had been the work of another popular writer of its day — Marie Corelli, for one, or F. Marion Crawford — it would not command attention a century later. But this is a novel by Mark Twain, who once said, "I am God's fool, and all His works must be contemplated with respect." *Joan of Arc*, too, deserves respect, but for me, and I would guess for most contemporary readers, that respect comes with a certain degree of consternation. *Joan of Arc* is of less interest for its intrinsic literary quality than as a biographical crux, an event that illuminates the later life of a major American writer while not adding to his stature. It is the work of a deeply conflicted, intermittently fulfilled man and artist, a temporary resolution of the many disunities and identities of Samuel Clemens and Mark Twain.

We think of Mark Twain today as a humorist, satirist, realist, the shaper of a distinctively American prose and narrative style, and preeminently the author of *Adventures of Huckleberry Finn*. That vernacular masterpiece turns nearly every common assumption on its head and ridicules the work ethic, prayer, and pious sentiment, characterized as "tears and flapdoodle," "soul-butter and hogwash." *Joan of Arc*, on the other hand, is a sustained exercise in romantic medievalism, patriotic impulse, and devotional fervor, all framed in peculiarly conventional terms. Its prose is genteel, nerveless, transatlantic rather than native. This serial-sentence description of springtime in Rouen is just the sort of thing that at other times in his life Mark Twain would have mocked:

... the chill was departed out of the air, the wild flowers were springing in the glades and glens, the birds were piping in the woods, all nature was brilliant with sunshine, all spirits were renewed and refreshed, all hearts glad, the world was alive with hope and cheer, the plain beyond the Seine stretched away soft and rich and green, the river was limpid and lovely, the leafy islands were dainty to see, and flung still daintier reflections of themselves upon the shining water. (405)

For all the storybook spectacle and action that fill its pages, *Joan of Arc* is surprisingly slack. The secondary characters are either mouthpieces or stereotypes out of juvenile adventure stories, and the heroine, a peasant girl from the tiny village of Domrémy, is dematerialized and idealized beyond all credence. The narrator of Joan's story, Sieur Louis de Conte, says at the outset that he had come to recognize her "for what she was — the most noble life that was ever born into this world save only One" (2).

William Dean Howells, Mark Twain's friend, fervent admirer, and perceptive critic, was disappointed by the book, as he freely confessed in his review in *Harper's Weekly*.

It would be impossible for anyone who was not a prig to keep to the archaic attitude and parlance which the author attempts here and there; and I wish he had frankly refused to attempt it at all. I wish his personal recollections of Joan could have been written by some Southwestern American, translated to Domrémy by some mighty magic of imagination. . . . My suffering begins when he does the supposed medieval thing. Then I suspect that his armor is of tin, that the castles and rocks are pasteboard, that the mob of citizens and soldiers who fill the air with their two-up-and-two-down combats, and the well-known muffled roar of their voices, has been hired in at so much a night, and that Joan is sometimes in an awful temper behind the scenes.

Nonetheless, as Howells wrote at the end of his review, "this curious book of the arch-humorist of the century . . . has a vitalizing force." Mark Twain not only had "the facts" of Joan's life on his side but played on them the light of his imagination.

Joan "is rare, she is exquisite, she is all that is lovely," said Mark Twain's authorized biographer, Albert Bigelow Paine, who went on to write his own book about her. "Considered from every point of view, *Joan of Arc* is Mark Twain's supreme literary expression, the loftiest, the most delicate, the most luminous example of his work. . . . It is bathed in the atmosphere of romance, but it is the ultimate of realism, too; not hard, sordid, ugly realism, but noble, spiritual, divine realism." But compared to Mark Twain's other excursions into medievalism, *Joan of Arc* lacks the plot ingenuities of *The Prince and the Pauper*, the comic and anarchic furies of *A Connecticut Yankee*, and the questioning of reality that shaped the posthumously published *No. 44, The Mysterious Stranger*. *Joan of Arc* was an act of piety, a surrender to the historical given of a virgin-martyr whom Mark Twain infused with the intolerant idealism of his own favorite daughter, Susy Clemens. He said that he wrote the book out of love, "not for lucre." "I like the *Joan of Arc* best of all my books," he noted in 1908. "It *is* the best; I know it perfectly well."

With readers of the 1890s, Mark Twain paid a price for being the most celebrated American humorist — "humorist" taken in its most superficial and eventually demeaning sense of genial entertainer, benevolent storyteller, and harmless, consistently quotable household sage. The trouble with humor, as he discovered early in his career when he chose the pseudonym Mark Twain and decided his vocation was "to excite the *laughter* of God's creatures," was that few people took it seriously or recognized its power to explode nonsense and lies. Paine recalled that the first reviewers and readers of *Joan of Arc* were beguiled by its "lofty charm." But, knowing it was written by Mark Twain, they expected at least a lurking joke, were disappointed, and in the end decided that he "had gone out of his proper field." This was just as the author himself had anticipated: in writing *Joan of Arc* he had tried once again — as he had in *The Prince and the Pauper* — to escape what critic James M. Cox has called "the fate of humor."

"I shall never be accepted seriously over my own signature," he said when he began the book. "I shall write it anonymously" — as if Mark Twain and his laughter had never existed. But for a man who so mastered the craft of publicity that he was to describe himself as "the most conspicuous person on the planet," anonymity was no more possible than levitation. The nominal narrator, Joan's page and secretary, Sieur Louis de Conte, even bears the initials of Samuel Langhorne Clemens.[*] From time to time, through the muffling folds of Sieur Louis' wide-eyed and reverent narration, one hears the Old Adam, sardonic, amused, and briefly liberated, yielding to burlesque ("the forbidden thing," Paine cautioned, "which was so likely to be Mark Twain's undoing"). He mocks the gullibility and vanity of "the damned human race" and the cruelty of institutionalized religion: "It seemed to me that I could hear the bones snap and the flesh tear apart, and I did not see how that body of annointed servants of the merciful Jesus could sit there and look so placid and indifferent" (411). He celebrates the art of the tall story: "It was more stirring and interesting to hear him tell about a battle the tenth time than it was the first time, because he did not tell it twice the same way, but always made a new battle of it and a better one, with more casualties on the enemy's side each time, and more general wreck and disaster all around, and more widows and orphans and suffering in the neighborhood where it happened" (116).

Distant as it is from *Huckleberry Finn* in ethos, mood, and style, *Joan of Arc*, a story set in Europe in the fifteenth century, sprang from the ingrained contradictions of a well-traveled American writer who lived into the opening years of the twentieth. Mark Twain's aversion to France and the French, for example, was almost pathological. "When all other interests fail," he wrote in 1895, a Frenchman "can turn in and see if he can't find out who his father was." The French were "the connecting link between man and monkey," practiced unspeakable "bestialities," and in point of civilized behavior ranked lower than the Comanches, who at least "do not fight among themselves, whereas a favorite pastime with the French, from time immemorial, has been

---

[*] "Page" echoes the name of Mark Twain's nemesis, the inventor James W. Paige, and also recalls a passage early in *A Connecticut Yankee*: "He arrived, looked me over with a smiling and impudent curiosity; said he had come for me, and informed me that he was a page. 'Go 'long,' I said; 'you ain't more than a paragraph.'"

the burning and slaughtering of each other," a major component of the action in *Joan of Arc*. "No weapon has drunk such rivers of French blood as the French sword. No hatred has been so implacable as the Frenchman's hatred of his brother. No other creature's religion has wrought such marvels of murderous atrocity as the meek and lowly religion of the Frenchman."

Mark Twain chose as the heroine of what he hoped to be his crowning work, the one that would raise him to the Olympus of mainstream literature and exorcise the curse of humor, an avatar of French nationalism and militarism. "With Joan of Arc," he wrote in his novel, "love of country was more than a sentiment — it was a passion. She was the Genius of Patriotism — she was Patriotism embodied, concreted, made flesh, and palpable to the touch and visible to the eye" (461). Elsewhere in the writings of Mark Twain "the holy fire of patriotism" (as he called it in "The War Prayer") is dealt with skeptically if not with downright contempt. It was a matter of training, self-approval, and conformity to public opinion, like any other "corn-pone opinion."

"When the personages of a tale deal in conversation," he wrote in his devastating critique of James Fenimore Cooper, "the talk shall sound like human talk, and be talk such as human beings would be likely to talk in the given circumstances." Even allowing for stylized period rhetoric, this is not the sort of "talk" that runs through *Joan of Arc*. "Black news is come," announces a peasant boy called the Sunflower. "A treaty has been made at Troyes between France and the English and Burgundians. By it France is betrayed and delivered over, tied hand and foot, to the enemy. It is the work of the Duke of Burgundy and that she-devil the Queen of France" (31). The discussion that follows this announcement combines history primer with newspaper editorial.

Despite such violations of essential stance and manner, Mark Twain's choice of Joan of Arc as the subject of a novel was not simply an accommodation to middlebrow standards and to Joan's popularity, nor was it a reheating of old soup, as one could say of *The American Claimant, Tom Sawyer Abroad*, and other books he turned out during the same period in order to stave off bankruptcy. During the early 1890s, as he was writing *Joan of Arc*,

Mark Twain's business affairs took a decisive downward turn. His two major investments were failing: one was the Paige typesetter, a mechanical wonder he had poured money into for years; the other, the publishing house of Charles L. Webster, which in better days had brought out *Huckleberry Finn* and U.S. Grant's *Personal Memoirs*. ("There are two times in a man's life when he should not speculate," Mark Twain was to say, "when he can't afford it, and when he can.")

In 1891, to cut personal expenses, he closed the Hartford house that had been the emblem of his literary success and domestic happiness and moved his family to Europe. Sieur Louis begins writing his account of Joan in 1492, the Columbian year Mark Twain commemorated in a maxim that more than hints at his own mood four centuries later: "It was wonderful to find America, but it would have been more wonderful to miss it." "The billows of hell have been rolling over me," he said in 1893, in the thick of his entrepreneurial nightmares. "Get me out of business!" he wrote to his partner in the publishing house. "And I will be yours forever gratefully." Finally, with "just barely enough head left on my shoulders to protect me from being used as a convenience for the dogs," he declared bankruptcy in 1894. The next year he set off on a marathon round-the-world lecture tour that rescued him from poverty and also furnished material for the last of his travel books, *Following the Equator*. His comfort, refuge, and escape from his troubles during this period was the writing of *Joan of Arc*.

The historian Henry Adams saw western culture as the clash of the mechanical Dynamo, symbol of modern technology and capitalism, and the Virgin, symbol of medieval faith and thought. Betrayed by the Dynamo, Mark Twain threw himself at the feet of the Virgin.

According to Paine, Mark Twain's passion for Joan of Arc's story went back to his early years in Hannibal when he found a vagrant page from a book about her blowing along the sidewalk. "There arose within him a deep compassion for the gentle Maid of Orléans," Paine writes, "a burning resentment toward her captors, a powerful and indestructible interest in her sad history." Much later, by his own liberal estimate, Mark Twain was to spend twelve years studying her history and about two years (1892–1894) writing it. Here was a story that had everything a historical novelist with a taste for pageantry, spectacle, and "effects" could desire: forced marches, sieges, panoplied combat, a royal court, gorgeous ceremony, intrigue, betrayal, inquisitorial trials, imprisonment, and finally, a fiery death at the stake (a scene he had already rehearsed to opposite effect, in chapter 6 of *A Connecticut Yankee*). And the story's remarkable heroine, endowed by the Holy Spirit with beauty, charisma, courage, eloquence, and the gift of leadership, was an illiterate peasant girl who stood as an equal with kings, nobles, and princes of the church.

The historical Joan claimed to have heard "voices" — of Saints Michael, Catherine, and Margaret — telling her to lead the liberation of France from the Burgundians and English. Dressed and armed as a man, she took command of French troops, raised the siege of Orléans, and, her ultimate triumph, saw the dauphin crowned Charles VII in the Cathedral of Rheims. Betrayed to the English, she stood trial before an ecclesiastical court for heresy and witchcraft, was found guilty, and at the age of nineteen was burned at the stake as a relapsed heretic. By the 1890s, when Mark Twain began intensive work on the book, Joan was a timely as well as compelling heroine. A surge of popular adoration culminated in her beatification in 1909, a year before Mark Twain died. She was canonized Saint Joan in 1920.

Shakespeare (*The First Part of King Henry the Sixth*) had portrayed her as a witch; Voltaire (*La Pucelle*), as a dupe of her "voices," an object lesson in superstition and credulity; Schiller (*Die Jungfrau von Orleans*), as a romantic heroine. In George Bernard Shaw's *Saint Joan* (1923) she became an archetypal Protestant who set her own conscience against established authority. (Shaw described Mark Twain's Joan as "an unimpeachable American school teacher in armor.") Writing in the voice of Joan's page and secretary, Mark Twain portrays her as a "wonderful child" whose spirit had no peer in "its purity from all alloy of self-seeking, self-interest, personal ambition. . . . This cannot be said of any other person whose name appears in profane history" (461). Given such a high degree of idealization, and lacking any opposing edge, Mark Twain's Joan is more a character out of a school pageant than she is a flesh-and-blood historical figure who led armies into battle and endured imprisonment and martyrdom. "Can it be true," the narrator imagines his contemporaries asking, that "this little creature, this girl, this child with the good face, the sweet face, the beautiful face, the dear and bonny face, . . . has carried fortresses by storm, charged at the head of victorious armies, blown the might of England out of her path with a breath, and fought a long campaign . . . against the massed brains and learning of France" (429).

It is this sort of question that Mark Twain's novel isn't able to answer, because Joan, for all her commanding presence and battle victories, remains, as she was at the beginning of the book, a "wise little child," "this fragile girl," "that wonderful child, that sublime personality," abstracted beyond any recognition that she could ever become a mature woman. This was a vital part of her appeal for Mark Twain. He saw Joan as a paragon of purity and selflessness who would never have to meet the confounding test of sexuality. "The higher life absorbed her and suppressed her physical (sexual) development," he wrote in one of his notes for the book; he was commenting on the recorded testimony of some Domrémy women that the Maid of Orléans had never menstruated. In the novel he seems to be uncomfortably circling around this piece of information. "She was such a vision of young bloom and beauty and grace, and such an incarnation of pluck and life and go! She was growing more and more ideally beautiful every day, as was plain to be seen — and these were days of development; for she was well past seventeen, now — in fact she was getting close upon seventeen and a half — indeed, just a little woman, as you may say" (224). This "little woman" would never grow up to be the much heralded but also dreaded "New Woman" of Mark Twain's era.

With the exception of the slave Roxy in his *Pudd'nhead Wilson*, Mark Twain's work is conspicuously lacking in mature, sexualized female characters. Tom's calf love, Becky Thatcher, has her counterpart in the pre-nubile "Angelfish" with whom Mark Twain surrounded himself during the last five years of his life. In Joan he created a stereotype of nonsexual young womanhood so pure, single-mindedly devout, and unashamedly sentimentalized that even Susy Clemens, at times his harshest critic, was thoroughly proud and pleased. To hear him read aloud from the manuscript was "uplifting and revealing." It promised to be "his loveliest book," she told her sister Clara, "perhaps even more sweet and beautiful than *The Prince and the Pauper*" (which he had dedicated to his "good-mannered and agreeable children").

Physically and in point of character, taste, love of birds and animals, and natural eloquence, Susy was his model for Joan. He had always hoped she would be a writer. "Every now and then in her vivacious talk," he recalled, "she threw out phrases of such admirable grace and force, such precision of form, that they thrilled through one's consciousness like the passage of the electric spark." "Eloquence was a native gift of Joan of Arc," Mark Twain's narrator says. "It came from her lips without effort and without preparation. Her words were as sublime as her deeds, as sublime as her character; they had their source in a great heart and were coined in a great brain" (382).

Mark Twain favored Susy over his other two daughters to the point of obsession and almost went into mourning when in 1890 she left home for college at Bryn Mawr. He cast about for pretexts to visit her there, and would even have delivered Susy's laundry from home if he had been allowed to, his wife said. It was clear that Livy, who had wanted her to go away to college in the first place, thought it best that father and daughter should be separated. His grief over Susy's death in August 1896, three months after *Joan of Arc* came out in volume form, brought him to the brink of insanity. "It is one of the mysteries of our nature," he reflected ten years later, "that a man, all unprepared, can receive a thunder-stroke like that and live."

Susy had been more than his model for Joan: she was the presiding spirit of the entire book and its demon as well. When nearly eleven she drew up a list of famous men that delighted him: "Longfellow, Papa (Mark Twain), Columbus, Teneson, Ferdinad." She knew her father was a great man, but she was not at all sure a humorist was any better than a clown, and she wanted him to be a great man in some other way, while he in turn had a guilty sense that he had failed her just by being Mark Twain. "In a great many directions he has greater ability than in the gifts which have made him famous," she wrote about him. She wanted him to be a moral philosopher, and the author not of *Huckleberry Finn*, which she had come to dislike for its coarse realism, low characters, and scorn for polite behavior, but of *The Prince and the Pauper*, "unquestionably the best book" he had ever written up to then, and later *Joan of Arc*, "perhaps even more sweet and beautiful."

Susy sometimes expressed her aspirations for her father in open resentment of that deliberate and brilliant literary creation, Mark Twain. "How I hate that name! I should never like to hear it again! My father should not be satisfied with it! He should show himself the great writer that he is, not merely a funny man! Funny! That's all the people see in him — a maker of funny speeches!" More even than her mother, whom Van Wyck Brooks blamed for Mark Twain's surrender to conventional values, Susy Clemens demanded purity, gentility, high sentiment — the criteria of the late-nineteenth-century female reading audience characterized by one of Mark Twain's contemporaries as "the iron Madonna who strangles in her fond embrace the American novelist."

He had first submitted to the Madonna's embraces three decades earlier when he left California in pursuit of fame, love, and acceptance and began giving lasting hostages to the social order. Bound for Europe and the eventual writing of his first great success, *The Innocents Abroad*, he had invited Mary Mason Fairbanks, a "most refined, intelligent, and cultivated lady," to scan his writing for vulgarities and vernacularisms, for lack of charity and too much irreverence. Without hypocrisy but with a certain willing suspension of identity he yielded to her influence, just as he yielded to the influence of Olivia Langdon, the Elmira heiress he courted and won. He quoted Scripture in his love letters to Livy and apparently gave his prospective father-in-law the notion that he would combine his newfound religious faith with his broad experience of the world and write a life of Christ. Less than a year after he married Livy he said in half-jest, "I would deprive myself of sugar in my coffee if she wished it, or quit wearing socks if she thought them immoral."

In malleability and direction he was the antithesis of Walt Whitman, also sea-changed in his thirties. The journalist Walter Whitman Jr. became the poet Walt of the open road; Samuel L. Clemens inhabited a mansion at Nook Farm, in Hartford, Connecticut, that was one of the wonders of the region. Within five years of his marriage, Sam Clemens, sagebrush Bohemian, had been transformed, at least part time, into an upper-class paterfamilias, a man of property, and a writer surrounded by women and seeking their approval. Divided in goal and sensibility, his work during the 1890s followed two divergent paths. One led toward *Pudd'nhead Wilson*, as a novelist his last long look at America, the other, toward *Joan of Arc*.

If Susy is one of the transcendent figures standing behind and shaping this novel, another is Ulysses S. Grant, general in chief of the Union armies, supreme war hero, and two-term president of the United States. His two-volume *Personal Memoirs* was the most spectacular commercial triumph of Mark Twain's doomed publishing house. Grant's life, like Joan's (and like Mark Twain's as well), had been a saga of the unpredictable, of the bestowal and possession of unsuspected powers. The reluctant soldier, sent to West Point against his will, rose from business failure, firewood salesman, and village drunk to become a global hero. Like Joan, Grant came from humble and obscure origins, had been elected by history and the Holy Spirit, and was endowed with charisma, military genius, the gift of command, a natural eloquence, and an equally natural reserve that made him seem expressionless, an iron visage of determination and authority. General William Tecumseh Sherman, a nonbeliever, said that he fought under Grant with "the faith a Christian has in his Savior."

Comparably, Joan is graced with a "strange deep light in her eye which we named the battle-light" (184). "This new light in the eye and this new bearing," Sieur Louis says, "were born of the authority and leadership which had this day been vested in her by the decree of God, and they asserted that authority as plainly as speech could have done it, yet without ostentation or bravado. This calm consciousness of command, and calm outward expression of it, remained with her thenceforth until her mission was accomplished" (59). The "little country-maid of seventeen" had become "General of the Armies of France, with a prince of the blood for subordinate," but "yesterday she was not even a sergeant, not even a corporal, not even a private.... It made me dizzy to think of these things, they were so out of the common order, and seemed so impossible" (132). "Her spirits were high, and her bearing martial" (69), Sieur Louis says of "this untaught country damsel, unused to dictating anything at all to anybody" (156). "I caught the infection and felt a great impulse stirring in me that was like what one feels when he hears the roll of the drums and the tramp of marching men" (69).

Mark Twain's accounts of Joan's eloquence and bearing echo what he had written in 1886, after Matthew Arnold mocked Grant for writing "an English without charm or high breeding." Citing Grant's most famous utterances — "Unconditional and immediate surrender," "I propose to fight it out on this line if it takes all summer," "Let us have peace" — Mark Twain wrote a thundering rejoinder that voiced his own conscious and defiant nativism.

> There is that about the sun which makes us forget his spots: and when we think of General Grant our pulses quicken and his grammar vanishes: we only remember that this is the simple soldier, who, all untaught of the silken phrase-makers, linked words together with an art surpassing the art of the schools and put into them a something which will still bring to American ears, as long as America shall last, the roll of his vanished drums and the tread of his marching hosts.

One could suggest another parallel figure to Joan among Mark Twain's contemporaries: Mary Baker Eddy, founder of the Church of Christ, Scientist. Mark Twain loathed her, although for a brief time he subscribed to Mrs. Eddy's "mind-cure" and believed that Christian Science, as a business venture, at any rate, was "the Standard Oil of the future." Like Joan, she was perceived by many as a threat to domestic order and the conventional role of women. (She had "a hunger for power," Mark Twain said, "such as has never been seen in the world before.") Like Joan, who almost instantly mobilized the armies of France, Mrs. Eddy oversaw the growth of Christian Science from a local sect to a world religion within a remarkably short time. Like Joan, who heard the voices of saints guiding her on her mission, Mrs. Eddy claimed to have her own direct line — to Jesus of Nazareth and the Virgin Mary — and could be seen as divinely inspired, cunningly manipulative, or completely delusional. That such parallels exist, although Mark Twain would have rejected them out of hand, is one of the many anomalies of his *Joan of Arc*.

PERSONAL
RECOLLECTIONS OF
JOAN OF ARC

PERSONAL
RECOLLECTIONS
OF
JOAN
OF
ARC

MARK TWAIN

PERSONAL RECOLLECTIONS

OF

JOAN OF ARC

BY

THE SIEUR LOUIS DE CONTE

(HER PAGE AND SECRETARY)

FREELY TRANSLATED
OUT OF THE ANCIENT FRENCH INTO MODERN ENGLISH
FROM THE ORIGINAL UNPUBLISHED MANUSCRIPT
IN THE NATIONAL ARCHIVES OF FRANCE

BY

JEAN FRANÇOIS ALDEN

ILLUSTRATED

FROM ORIGINAL DRAWINGS BY

F. V. DU MOND

AND FROM REPRODUCTIONS OF
OLD PAINTINGS AND STATUES

NEW YORK
HARPER & BROTHERS PUBLISHERS
1896

JEHANNE D'ARC
DIT
LA
PUCELLE D'ORLEANS

Consider this unique and imposing distinction. Since the writing of human history began, Joan of Arc is the only person, of either sex, who has ever held supreme command of the military forces of a nation *at the age of seventeen.*

LOUIS KOSSUTH.

Authorities examined in verification of the truthfulness of this narrative:

J. E. J. QUICHERAT, *Condamnation et Réhabilitation de Jeanne d'Arc.*

J. FABRE, *Procès de Condamnation de Jeanne d'Arc.*

H. A. WALLON, *Jeanne d'Arc.*

M. SEPET, *Jeanne d'Arc.*

J. MICHELET, *Jeanne d'Arc.*

BERRIAT DE SAINT-PRIX, *La Famille de Jeanne d'Arc.*

La Comtesse A. DE CHABANNES, *La Vierge Lorraine.*

Monseigneur RICARD, *Jeanne d'Arc la Vénérable.*

Lord RONALD GOWER, F.S.A., *Joan of Arc.*

JOHN O'HAGAN, *Joan of Arc.*

JANET TUCKEY, *Joan of Arc the Maid.*

TRANSLATOR'S PREFACE

———

To arrive at a just estimate of a renowned man's character one must judge it by the standards of his time, not ours. Judged by the standards of one century, the noblest characters of an earlier one lose much of their lustre; judged by the standards of to-day, there is probably no illustrious man of four or five centuries ago whose character could meet the test at all points. But the character of Joan of Arc is unique. It can be measured by the standards of all times without misgiving or apprehension as to the result. Judged by any of them, judged by all of them, it is still flawless, it is still ideally perfect; it still occupies the loftiest place possible to human attainment, a loftier one than has been reached by any other mere mortal.

When we reflect that her century was the brutalest, the wickedest, the rottenest in history since the darkest ages, we are lost in wonder at the miracle of such a product from such a soil. The contrast between her and her century is the contrast between day and night. She was truthful when lying was the common speech of men; she was honest when honesty was become a lost virtue; she was a keeper of promises when the keeping of a promise was expected of no one; she gave her great mind to great thoughts and great purposes when other great minds wasted themselves upon pretty fancies or upon poor ambitions; she was modest and fine and delicate when to be loud and coarse might be said to be universal; she was full of pity when a merciless cruelty was the rule; she was steadfast when stability was unknown, and honorable in an age which had forgotten what honor was; she was a rock of convictions in a time when men believed in nothing and scoffed at all things; she was unfailingly true in an age that was false to the core; she maintained her personal dignity unimpaired in an age of fawnings and servilities; she was of a dauntless courage when hope and courage had perished in the hearts of her nation; she was spotlessly pure in mind and body when society in the highest places was foul in both—she was all these things in an age when crime was the common business of lords and princes, and when the highest personages in Christendom were able to astonish even that infamous era and make it stand aghast at the spectacle of their atrocious lives black with unimaginable treacheries, butcheries, and bestialities.

She was perhaps the only entirely unselfish person whose name has a place in profane history. No vestige or suggestion of self-seeking can be found in any word or deed of hers. When she had rescued her King from his vagabondage, and set his crown upon his head, she was offered rewards and honors, but she refused them all, and would take nothing. All she would take for herself—if the King would grant it—was leave to go back to her village home, and tend her sheep again, and feel her mother's arms about her, and be her housemaid and helper. The selfishness of this unspoiled general of victorious armies, companion of princes, and idol of an applauding and grateful nation, reached but that far and no farther.

The work wrought by Joan of Arc may fairly be regarded as ranking any recorded in history, when one considers the conditions under which it was undertaken, the obstacles in the way, and the means at her disposal. Cæsar carried conquest far, but he did it with the trained and confident veterans of Rome, and was a trained soldier himself; and Napoleon swept away the disciplined armies of Europe, but he also was a trained soldier, and he began his work with patriot battalions inflamed and inspired by the miracle-working new breath of Liberty breathed upon them by the Revolution — eager young apprentices to the splendid trade of war, not old and broken men-at-arms, despairing survivors of an age-long accumulation of monotonous defeats; but Joan of Arc, a mere child in years, ignorant, unlettered, a poor village girl unknown and without influence, found a great nation lying in chains, helpless and hopeless under an alien domination, its treasury bankrupt, its soldiers disheartened and dispersed, all spirit torpid, all courage dead in the hearts of the people through long years of foreign and domestic outrage and oppression, their King cowed, resigned to its fate, and preparing to fly the country; and she laid her hand upon this nation, this corpse, and it rose and followed her. She led it from victory to victory, she turned back the tide of the Hundred Years' War, she fatally crippled the English power, and died with the earned title of DELIVERER OF FRANCE, *which she bears to this day.*

And for all reward, the French King whom she had crowned stood supine and indifferent while French priests took the noble child, the most innocent, the most lovely, the most adorable the ages have produced, and burned her alive at the stake.

A PECULIARITY OF JOAN OF ARC'S HISTORY

The details of the life of Joan of Arc form a biography which is unique among the world's biographies in one respect: *It is the only story of a human life which comes to us under oath,* the only one which comes to us from the witness-stand. The official records of the Great Trial of 1431, and of the Process of Rehabilitation of a quarter of a century later, are still preserved in the National Archives of France, and they furnish with remarkable fulness the facts of her life. The history of no other life of that remote time is known with either the certainty or the comprehensiveness that attaches to hers.

The Sieur Louis de Conte is faithful to her official history in his Personal Recollections, and thus far his trustworthiness is unimpeachable; but his mass of added particulars must depend for credit upon his own word alone.

THE TRANSLATOR.

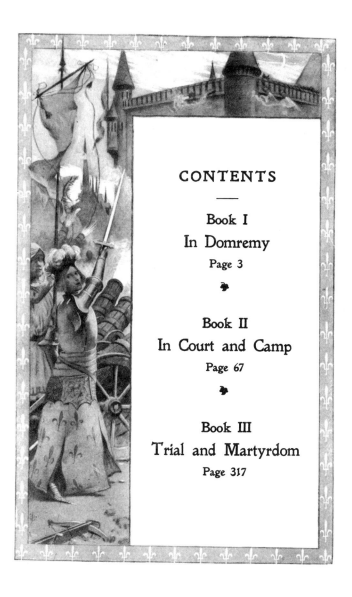

CONTENTS

ILLUSTRATIONS

PERSONAL RECOLLECTIONS OF JOAN OF ARC

THE SIEUR LOUIS DE CONTE

TO HIS GREAT-GREAT-GRAND NEPHEWS AND NIECES

THIS is the year 1492. I am eighty-two years of age. The things I am going to tell you are things which I saw myself as a child and as a youth.

In all the tales and songs and histories of Joan of Arc which you and the rest of the world read and sing and study in the books wrought in the late invented art of printing, mention is made of me, the Sieur Louis de Conte—I was her page and secretary. I was with her from the beginning until the end.

I was reared in the same village with her. I played with her every day, when we were little children together, just as you play with your mates. Now that we perceive how great she was; now that her name fills the whole world, it seems strange that what I am saying is true; for it is as if a perishable paltry candle should speak of the eternal sun riding in the heavens and say, " He was gossip and housemate to me when we were candles together." And yet it is true, just as I say. I was her playmate, and I fought at her side in the wars; to this day I carry in my mind, fine and clear, the picture of that dear little figure, with breast bent to the flying horse's neck, charging at the head of the armies of France, her hair streaming back, her silver mail ploughing steadily deeper and deeper into the thick of the battle, sometimes nearly drowned from sight by tossing heads of horses, uplifted sword-arms, wind-blown plumes, and intercepting shields. I was with her to the end; and when that black day came whose accusing shadow will lie always upon the memory of the mitred French slaves of England who were her assassins, and upon France who stood idle and essayed no rescue, my hand was the last she touched in life.

As the years and the decades drifted by, and the spectacle of the marvellous child's meteor-flight across the war-firmament of France and its extinction in the smoke-clouds of the stake receded deeper and deeper into the past and grew ever more strange and wonderful and divine and pathetic, I came to comprehend and recognize her at last for what she was— the most noble life that was ever born into this world save only One.

Book 1

IN DOMREMY

CHAPTER I

I, the Sieur Louis de Conte, was born in Neufchâteau, the 6th of January, 1410; that is to say, exactly two years before Joan of Arc was born in Domremy. My family had fled to those distant regions from the neighborhood of Paris in the first years of the century. In politics they were Armagnacs— patriots: they were for our own French King, crazy and impotent as he was. The Burgundian party, who were for the English, had stripped them, and done it well. They took everything but my father's small nobility, and when he reached Neufchâteau he reached it in poverty and with a broken spirit. But the political atmosphere there was the sort he liked, and that was something. He came to a region of comparative quiet; he left behind him a region peopled with furies, madmen, devils, where slaughter was a daily pastime and no man's life safe for a moment. In Paris, mobs roared through the streets nightly, sacking, burning, killing, unmolested, uninterrupted. The sun rose upon wrecked and smoking buildings, and upon mutilated corpses lying here, there, and yonder about the streets, just as they fell, and stripped naked by thieves, the unholy gleaners after the mob. None had the courage to gather these dead for burial; they were left there to rot and create plagues.

And plagues they did create. Epidemics swept away the people like flies, and the burials were conducted secretly and by night; for public funerals were not allowed, lest the revelation of the magnitude of the plague's work unman the people and plunge them into despair. Then came, finally, the bitterest winter which had visited France in five hundred years. Famine, pestilence, slaughter, ice, snow—Paris had all these at once. The dead lay in heaps about the streets, and *wolves entered the city in daylight and devoured them.*

Ah, France had fallen low—so low! For more than three quarters of a century the English fangs had been bedded in her flesh, and so cowed had her armies become by ceaseless rout and defeat that it was said and accepted that the mere sight of an English army was sufficient to put a French one to flight.

When I was five years old the prodigious disaster of Agincourt fell upon France; and although the English king went home to enjoy his glory, he left the country prostrate and a prey to roving bands of Free Companions in the service of the Burgundian party, and one of these bands came raiding through Neufchâteau one night, and by the light of our burning roof-thatch I saw all that were dear to me in this world (save an elder brother, your ancestor, left behind with the Court) butchered while they begged for mercy, and heard the butchers laugh at their prayers and mimic their pleadings. I was overlooked, and escaped without hurt. When the savages were gone I crept out and cried the night away watching the burning houses; and I was all alone, except for the company of the dead and the wounded, for the rest had taken flight and hidden themselves.

I was sent to Domremy, to the priest, whose house-keeper became a loving mother to me. The priest in the course of time taught me to read and write, and he and I were the only persons in the village who possessed this learning.

At the time that the house of this good priest, Guillaume Fronte, became my home, I was six years old. We lived close by the village church, and the small garden of Joan's parents was behind the church. As to that family, there were Jacques d'Arc the father, his wife Isabel Romée; three sons —Jacques, ten years old, Pierre, eight, and Jean, seven; Joan, four, and her baby sister Catherine, about a year old. I had these children for playmates from the beginning. I had some other playmates besides—particularly four boys: Pierre Morel, Etienne Roze, Noël Rainguesson, and Edmond

Aubrey, whose father was maire at that time; also two girls, about Joan's age, who by-and-by became her favorites; one was named Haumette, the other was called Little Mengette. These girls were common peasant children, like Joan herself. When they grew up, both married common laborers. Their estate was lowly enough, you see; yet a time came, many years after, when no passing stranger, howsoever great he might be, failed to go and pay his reverence to those two humble old women who had been honored in their youth by the friendship of Joan of Arc.

These were all good children, just of the ordinary peasant type; not bright, of course—you would not expect that—but good-hearted and companionable, obedient to their parents and the priest; and as they grew up they became properly stocked with narrownesses and prejudices got at second hand from their elders, and adopted without reserve; and without examination also—which goes without saying. Their religion was inherited, their politics the same. John Huss and his sort might find fault with the Church, in Domremy it disturbed nobody's faith; and when the split came, when I was fourteen, and we had three Popes at once, nobody in Domremy was worried about how to choose among them—the Pope of Rome was the right one, a Pope outside of Rome was no Pope at all. Every human creature in the village was an Armagnac—a patriot—and if we children hotly hated nothing else in the world, we did certainly hate the English and Burgundian name and polity in that way.

CHAPTER II

OUR Domremy was like any other humble little hamlet of that remote time and region. It was a maze of crooked, narrow lanes and alleys shaded and sheltered by the overhanging thatch roofs of the barn-like houses. The houses were dimly lighted by wooden-shuttered windows—that is, holes in the walls which served for windows. The floors were of dirt, and there was very little furniture. Sheep and cattle grazing was the main industry; all the young folks tended flocks.

The situation was beautiful. From one edge of the village a flowery plain extended in a wide sweep to the river—the Meuse; from the rear edge of the village a grassy slope rose gradually, and at the top was the great oak forest—a forest that was deep and gloomy and dense, and full of interest for us children, for many murders had been done in it by outlaws in old times, and in still earlier times prodigious dragons that spouted fire and poisonous vapors from their nostrils had their homes in there. In fact, one was still living in there in our own time. It was as long as a tree, and had a body as big around as a tierce, and scales like overlapping great tiles, and deep ruby eyes as large as a cavalier's hat, and an anchor-fluke on its tail as big as I don't know what, but very big, even unusually so for a dragon, as everybody said who knew about dragons. It was thought that this dragon was of a brilliant blue color, with gold mottlings, but no one had ever seen it, therefore this was not known to be so, it was only an opinion. It was not my opinion; I think there is no sense in forming an opinion when there is no evidence to form it on. If you build a person without any bones in him he may look fair enough to the eye, but he will be limber and cannot stand up; and I consider that *evidence* is the bones of an opinion. But I will take up this matter more at large at another time, and try to make the justness of my position appear. As to that dragon, I always held the belief that its color was gold and without blue, for that has always been the color of dragons. That this dragon lay but a little way within the wood at one time is shown by the fact that Pierre Morel was in there one day and smelt it, and recognized it by the smell. It gives one a horrid idea of how near to us the deadliest danger can be and we not suspect it.

In the earliest times a hundred knights from many remote places in the earth would have gone in there one after another, to kill the dragon and get the reward, but in our time that method had gone out, and the priest had become the one that abolished dragons. Père Guillaume Fronte did it in this case. He had a procession, with candles and incense and banners, and marched around the edge of the wood and exorcised the dragon, and it was never heard of again, although it was the opinion of many that the smell never wholly passed away. Not that any had ever smelt the smell again, for none had; it was only an opinion, like that other—and lacked bones, you see. I know that the creature was there before the exorcism, but whether it was there afterwards or not is a thing which I cannot be so positive about.

In a noble open space carpeted with grass on the high ground towards Vaucouleurs stood a most majestic beech-tree with wide-reaching arms and a grand spread of shade, and by it a limpid spring of cold water; and on summer days the children went there—oh, every summer for more than five hundred years—went there and sang and danced around the tree for hours together, refreshing themselves at the spring from time to time, and it was most lovely and enjoyable. Also they made wreaths of flowers and hung them upon the tree and about the spring to please the fairies that lived there; for they liked that, being idle innocent little creatures, as all fairies are, and fond of anything delicate and pretty like wild flowers put together in that way. And in return for this attention the fairies did any friendly thing they could for the children, such as keeping the spring always full and clear and cold, and driving away serpents and insects that sting; and so there was never any unkindness between the fairies and the children during more than five hundred years—tradition said a thousand—but only the warmest affection and the most perfect trust and confidence; and whenever a child died the fairies mourned just as that child's playmates did, and the sign of it was there to see: for before the dawn on the day of the funeral they hung a little immortelle over the place where that child was used to sit under the tree. I know this to be true by my own eyes; it is not hearsay. And the reason it was known that the fairies did it was this—that it was made all of black flowers of a sort not known in France anywhere.

Now from time immemorial all children reared in Domremy were called the Children of the Tree; and they loved that name, for it carried with it a mystic privilege not granted to any others of the children of this world. Which was this: whenever one of these came to die, then beyond the vague and formless images drifting through his darkening mind rose soft and rich and fair a vision of the Tree—if all was well with his soul. That was what some said. Others said the vision came in two ways: once as a warning, one or two years in advance of death, when the soul was the captive of sin, and then the Tree appeared in its desolate winter aspect—then that soul was smitten with an awful fear. If repentance came, and purity of life, the vision came again, this time summer-clad and beautiful; but if it were otherwise with that soul the vision was withheld, and it passed from life knowing its doom. Still others said that the vision came but once, and then only to the sinless dying forlorn in distant lands and pitifully longing for some last dear reminder of their home. And what reminder of it could go to their hearts like the picture of the Tree that was the darling of their love and the

THE FAIRY TREE

comrade of their joys and comforter of their small griefs all through the divine days of their vanished youth ?

Now the several traditions were as I have said, some believing one and some another. One of them I knew to be the truth, and that was the last one. I do not say anything against the others; I think they were true, but I only *know* that the last one was; and it is my thought that if one keep to the things he knows, and not trouble about the things which he cannot be sure about, he will have the steadier mind for it —and there is profit in that. I know that when the Children of the Tree die in a far land, then—if they be at peace with God—they turn their longing eyes toward home, and there, far-shining, as through a rift in a cloud that curtains heaven, they see the soft picture of the Fairy Tree, clothed in a dream of golden light; and they see the bloomy mead sloping away to the river, and to their perishing nostrils is blown faint and sweet the fragrance of the flowers of home. And then the vision fades and passes—but *they* know, *they* know! and by their transfigured faces you know also, you who stand looking on; yes, you know the message that has come, and that it has come from heaven.

Joan and I believed alike about this matter. But Pierre Morel, and Jacques d'Arc, and many others believed that the vision appeared twice—to a sinner. In fact they and many others said they *knew* it. Probably because their fathers had known it and had told them; for one gets most things at second hand in this world.

Now one thing that does make it quite likely that there were really two apparitions of the Tree is this fact: From the most ancient times if one saw a villager of ours with his face ash-white and rigid with a ghastly fright, it was common for every one to whisper to his neighbor, "Ah, he is in sin, and has got his warning." And the neighbor would shudder at the thought and whisper back, "Yes, poor soul, he has seen the Tree."

Such evidences as these have their weight; they are not to be put aside with a wave of the hand. A thing that is backed by the cumulative experience of centuries naturally gets nearer and nearer to being proof all the time; and if this continue and continue, it will some day become authority—and authority is a bedded rock, and will abide.

In my long life I have seen several cases where the Tree appeared announcing a death which was still far away; but in none of these was the person in a state of sin. No; the apparition was in these cases only a special grace; in place of deferring the tidings of that soul's redemption till the day of death, the apparition brought them long before, and with them peace—peace that might no more be disturbed—the eternal peace of God. I myself, old and broken, wait with serenity; for I have seen the vision of the Tree. I have seen it, and am content.

Always, from the remotest times, when the children joined hands and danced around the Fairy Tree they sang a song which was the Tree's Song, the Song of *L'Arbre Fée de Bourlemont*. They sang it to a quaint sweet air—a solacing sweet air which has gone murmuring through my dreaming spirit all my life when I was weary and troubled, resting me and carrying me through night and distance home again. No stranger can know or feel what that song has been, through the drifting centuries, to exiled Children of the Tree, homeless and heavy of heart in countries foreign to their speech and ways. You will think it a simple thing, that song, and poor perchance; but if you will remember what it was to us, and what it brought before our eyes when it floated through our memories, then you will respect it. And you will understand how the water wells up in our eyes and makes all things dim, and our voices break and we cannot sing the last lines:

> "And when in exile wand'ring we
> Shall fainting yearn for glimpse of thee,
> O rise upon our sight!"

And you will remember that Joan of Arc sang this song with us around the Tree when she was a little child, and always loved it. And that hallows it, yes, you will grant that:

L'ARBRE FÉE DE BOURLEMONT

SONG OF THE CHILDREN

Now what has kept your leaves so green,
   Arbre Fée de Bourlemont?
The children's tears! They brought each grief,
   And you did comfort them and cheer
   Their bruisèd hearts, and steal a tear
    That healèd rose a leaf.

And what has built you up so strong,
   Arbre Fée de Bourlemont?
The children's love! They've loved you long:
   Ten hundred years, in sooth,
They've nourished you with praise and song,
And warmed your heart and kept it young—
    A thousand years of youth!

Bide alway green in our young hearts,
   Arbre Fée de Bourlemont!
And we shall alway youthful be,
   Not heeding Time his flight;
And when in exile wand'ring we
Shall fainting yearn for glimpse of thee,
    O rise upon our sight!

The fairies were still there when we were children, but
we never saw them; because, a hundred years before that,
the priest of Domremy had held a religious function under
the tree and denounced them as being blood kin of the
Fiend and barred out from redemption; and then he warned
them never to show themselves again, nor hang any more
immortelles, on pain of perpetual banishment from that parish.

All the children pleaded for the fairies, and said they were
their good friends and dear to them and never did them any
harm, but the priest would not listen, and said it was sin and
shame to have such friends. The children mourned and
could not be comforted; and they made an agreement among
themselves that they would always continue to hang flower-

wreaths on the tree as a perpetual sign to the fairies that they were still loved and remembered, though lost to sight.

But late one night a great misfortune befell. Edmond Aubrey's mother passed by the Tree, and the fairies were stealing a dance, not thinking anybody was by; and they were so busy, and so intoxicated with the wild happiness of it, and with the bumpers of dew sharpened up with honey which they had been drinking, that they noticed nothing; so Dame Aubrey stood there astonished and admiring, and saw the little fantastic atoms holding hands, as many as three hundred of them, tearing around in a great ring half as big as an ordinary bedroom, and leaning away back and spreading their mouths with laughter and song, which she could hear quite distinctly, and kicking their legs up as much as three inches from the ground in perfect abandon and hilarity—oh, the very maddest and witchingest dance the woman ever saw.

But in about a minute or two minutes the poor little ruined creatures discovered her. They burst out in one heart-breaking squeak of grief and terror and fled every which way, with their wee hazel-nut fists in their eyes and crying; and so disappeared.

The heartless woman—no, the foolish woman; she was not heartless, but only thoughtless—went straight home and told the neighbors all about it, whilst we, the small friends of the fairies, were asleep and not witting the calamity that was come upon us, and all unconscious that we ought to be up and trying to stop these fatal tongues. In the morning everybody knew, and the disaster was complete, for where everybody knows a thing the priest knows it, of course. We all flocked to Père Fronte, crying and begging—and he had to cry, too, seeing our sorrow, for he had a most kind and gentle nature; and he did not want to banish the fairies, and said so; but said he had no choice, for it had been decreed that if they ever revealed themselves to man again, they must go. This all happened at the worst time possible, for Joan of Arc was ill of a fever and out of her head, and what could we do who had not her gifts of reasoning and persuasion? We flew in a swarm to her bed and cried out, "Joan, wake! Wake, there is no moment to lose! Come and plead for the fairies—come and save them; only you can do it."

But her mind was wandering, she did not know what we said nor what we meant; so we went away knowing all was lost. Yes, all was lost, forever lost; the faithful friends of the children for five hundred years must go, and never come back any more.

It was a bitter day for us, that day that Père Fronte held the function under the tree and banished the fairies. We could not wear mourning that any could have noticed, it would not have been allowed; so we had to be content with some poor small rag of black tied upon our garments where it made no show; but in our hearts we wore mourning big and noble and occupying all the room, for our *hearts* were ours; they could not get at them to prevent that.

The great tree—*l'Arbre Fée de Bourlemont* was its beautiful name—was never afterward quite as much to us as it had been before, but it was always dear; is dear to me yet when I go there, now, once a year in my old age, to sit under it and bring back the lost playmates of my youth and group them about me and look upon their faces through my tears and break my heart, oh, my God! No, the place was not quite the same afterwards. In one or two ways it could not be; for, the fairies' protection being gone, the spring lost much of its freshness and coldness, and more than two-thirds of its volume, and the banished serpents and stinging insects returned, and multiplied, and became a torment and have remained so to this day.

When that wise little child, Joan, got well, we realized how much her illness had cost us; for we found that we had been right in believing she could save the fairies. She burst into a great storm of anger, for so little a creature, and went straight to Père Fronte, and stood up before him where he sat, and made reverence and said:

"The fairies were to go if they showed themselves to people again, is it not so?"

"Yes, that was it, dear."

"If a man comes prying into a person's room at midnight when that person is half naked, will you be so unjust as to say that that person is showing himself to that man ?"

"Well—no." The good priest looked a little troubled and uneasy when he said it.

"Is a sin a sin anyway, even if one did not intend to commit it ?"

Père Fronte threw up his hands and cried out—

"Oh, my poor little child, I see all my fault," and he drew her to his side and put his arm around her and tried to make his peace with her, but her temper was up so high that she could not get it down right away, but buried her head against his breast and broke out crying and said :

"Then the fairies committed no sin, for there was no intention to commit one, they not knowing that any one was by; and because they were little creatures and could not speak for themselves and say the law was against the intention, not against the innocent act, and because they had no friend to think that simple thing for them and say it, they have been sent away from their home forever, and it was wrong, *wrong* to do it !"

The good father hugged her yet closer to his side and said:

"Oh, out of the mouths of babes and sucklings the heedless and unthinking are condemned : would God I could bring the little creatures back, for your sake. And mine, yes, and mine; for I have been unjust. There, there, don't cry—nobody could be sorrier than your poor old friend—don't cry, dear."

"But I can't stop right away, I've *got* to. And it is no little matter, this thing that you have done. Is being sorry penance enough for such an act ?"

Père Fronte turned away his face, for it would have hurt her to see him laugh, and said :

"Oh, thou remorseless but most just accuser, no, it is not. I will put on sackcloth and ashes ; there—are you satisfied ?"

Joan's sobs began to diminish, and she presently looked

up at the old man through her tears, and said, in her simple way:

"Yes, that will do—if it will clear you."

Père Fronte would have been moved to laugh again, perhaps, if he had not remembered in time that he had made a contract, and not a very agreeable one. It must be fulfilled. So he got up and went to the fireplace, Joan watching him with deep interest, and took a shovelful of cold ashes, and was going to empty them on his old gray head when a better idea came to him, and he said:

"Would you mind helping me, dear?"

"How, father?"

He got down on his knees and bent his head low, and said:

"Take the ashes and put them on my head for me."

The matter ended there, of course. The victory was with the priest. One can imagine how the idea of such a profanation would strike Joan or any other child in the village. She ran and dropped upon her knees by his side and said:

"Oh, it is dreadful. I didn't know that that was what one meant by sackcloth and ashes—do please get up, father."

"But I can't until I am forgiven. Do you forgive me?"

"I? Oh, you have done nothing to me, father; it is *yourself* that must forgive yourself for wronging those poor things. Please get up, father, won't you?"

"But I am worse off now than I was before. I thought I was earning *your* forgiveness, but if it is my own, I can't be lenient; it would not become me. Now what can I do? Find me some way out of this with your wise little head."

The Père would not stir, for all Joan's pleadings. She was about to cry again; then she had an idea, and seized the shovel and deluged her own head with the ashes, stammering out through her chokings and suffocations—

"There—now it is done. Oh, please get up, father."

The old man, both touched and amused, gathered her to his breast and said—

"Oh, you incomparable child! It's a humble martyrdom,

and not of a sort presentable in a picture, but the right and true spirit is in it; that I testify."

Then he brushed the ashes out of her hair, and helped her scour her face and neck and properly tidy herself up. He was in fine spirits now, and ready for further argument, so he took his seat and drew Joan to his side again, and said:

"Joan, you were used to make wreaths there at the Fairy Tree with the other children; is it not so?"

That was the way he always started out when he was going to corner me up and catch me in something—just that gentle, indifferent way that fools a person so, and leads him into the trap, he never noticing which way he is travelling until he is in and the door shut on him. He enjoyed that. I knew he was going to drop corn along in front of Joan now. Joan answered:

"Yes, father."

"Did you hang them on the tree?"

"No, father."

"Didn't hang them there?"

"No."

"Why didn't you?"

"I—well, I didn't wish to."

"Didn't wish to?"

"No, father."

"What did you do with them?"

"I hung them in the church."

"Why didn't you want to hang them in the tree?"

"Because it was said that the fairies were of kin to the Fiend, and that it was sinful to show them honor."

"Did you believe it was wrong to honor them so?"

"Yes. I thought it must be wrong."

"Then if it was wrong to honor them in that way, and if they were of kin to the Fiend, they could be dangerous company for you and the other children, couldn't they?"

"I suppose so—yes, I think so."

He studied a minute, and I judged he was going to spring his trap, and he did. He said:

"Then the matter stands like this. They were banned creatures, of fearful origin; they could be dangerous company for the children. Now give me a rational reason, dear, if you can think of any, why you call it a wrong to drive them into banishment, and why you would have saved them from it. In a word, what loss have you suffered by it?"

How stupid of him to go and throw his case away like that! I could have boxed his ears for vexation if he had been a boy. He was going along all right until he ruined everything by winding up in that foolish and fatal way. What had *she* lost by it! Was he never going to find out what kind of a child Joan of Arc was? Was he never going to learn that things which merely concerned her own gain or loss she cared nothing about? Could he never get the simple fact into his head that the sure way and the only way to rouse her up and set her on fire was to show her where some *other* person was going to suffer wrong or hurt or loss? Why, he had gone and set a trap for himself—that was all he had accomplished.

The minute those words were out of his mouth her temper was up, the indignant tears rose in her eyes, and she burst out on him with an energy and passion which astonished him, but didn't astonish me, for I knew he had fired a mine when he touched off his ill-chosen climax.

"Oh, father, how can you talk like that? Who owns France?"

"God and the King."

"Not Satan?"

"Satan, my child? This is the footstool of the Most High—Satan owns no handful of its soil."

"Then who gave those poor creatures their home? God. Who protected them in it all those centuries? God. Who allowed them to dance and play there all those centuries and found no fault with it? God. Who disapproved of God's approval and put a threat upon them? A man. Who caught them again in harmless sports that God allowed and a man forbade, and carried out that threat, and drove the poor things away from the home the good God gave them in His mercy and His pity, and sent down His rain and dew and sunshine upon it five hundred years in token of His peace? It was *their* home—theirs, by the grace of God and His good heart, and no man had a right to rob them of it. And they were the gentlest, truest friends that children ever had, and did them sweet and loving service all these five long centuries, and never any hurt or harm; and the children loved them, and now they mourn for them, and there is no healing for their grief. And what had the children done that they should suffer this cruel stroke? The poor fairies *could* have been dangerous company for the children? Yes, but never had been; and *could* is no argument. Kinsman of the Fiend? What of it? Kinsmen of the Fiend have *rights*, and these had; and children have rights, and these had; and if I had been here I would have spoken—I would have begged for the children and the fiends, and stayed your hand and saved them all. But now—oh, now, all is lost; everything is lost, and there is no help more!"

Then she finished with a blast at that idea that fairy kinsmen of the Fiend ought to be shunned and denied human sympathy and friendship because salvation was barred against them. She said that for that very *reason* people ought to pity them, and do every humane and loving thing they could to make them forget the hard fate that had been put upon them by accident of birth and no fault of their own. "Poor little creatures!" she said. "What can a person's heart be made of that can pity a Christian's child and yet can't pity a devil's child, that a thousand times more *needs* it!"

She had torn loose from Père Fronte, and was crying, with her knuckles in her eyes, and stamping her small feet in a fury; and now she burst out of the place and was gone before we could gather our senses together out of this storm of words and this whirlwind of passion.

The Père had got upon his feet, toward the last, and now he stood there passing his hand back and forth across his forehead like a person who is dazed and troubled; then he turned and wandered toward the door of his little workroom, and as he passed through it I heard him murmur sorrowfully :

"Ah me, poor children, poor fiends, they *have* rights, and she said true—I never thought of that. God forgive me, I am to blame."

When I heard that, I knew I was right in the thought that he had set a trap for himself. It was so, and he had walked into it, you see. I seemed to feel encouraged, and wondered if mayhap I might get him into one; but upon reflection my heart went down, for this was not my gift.

CHAPTER III

Speaking of this matter reminds me of many incidents, many things that I could tell, but I think I will not try to do it now. It will be more to my present humor to call back a little glimpse of the simple and colorless good times we used to have in our village homes in those peaceful days—especially in the winter. In the summer we children were out on the breezy uplands with the flocks from dawn till night, and then there was noisy frolicking and all that; but winter was the cosey time, winter was the snug time. Often we gathered in old Jacques d'Arc's big dirt-floored apartment, with a great fire going, and played games, and sang songs, and told fortunes, and listened to the old villagers tell tales and histories and lies and one thing and another till twelve o'clock at night.

One winter's night we were gathered there—it was the winter that for years afterward they called the hard winter—and that particular night was a sharp one. It blew a gale outside, and the screaming of the wind was a stirring sound, and I think I may say it was beautiful, for I think it *is* great and fine and beautiful to hear the wind rage and storm and blow its clarions like that, when you are inside and comfortable. And we were. We had a roaring fire, and the pleasant *spit-spit* of the snow and sleet falling in it down the chimney, and the yarning and laughing and singing went on at a noble rate till about ten o'clock, and then we had a supper of hot porridge and beans, and meal cakes with butter, and appetites to match.

Little Joan sat on a box apart, and had her bowl and bread on another one, and her pets around her, helping. She had

more than was usual of them or economical, because all the
outcast cats came and took up with her, and homeless or un-
lovable animals of other kinds heard about it and came, and
these spread the matter to the other creatures, and they came
also ; and as the birds and the other timid wild things of the
woods were not afraid of her, but always had an idea she was
a friend when they came across her, and generally struck up
an acquaintance with her to get invited to the house, she always had samples of those breeds in stock. She was hospi-
table to them all, for an animal was an animal to her, and
dear by mere reason of being an animal, no matter about
its sort or social station ; and as she would allow of no cages,
no collars, no fetters, but left the creatures free to come and
go as they liked, that contented them, and they came ; but
they didn't go, to any extent, and so they were a marvellous
nuisance, and made Jacques d'Arc swear a good deal ; but his
wife said God gave the child the instinct, and knew what He
was doing when He did it, therefore it must have its course ;
it would be no sound prudence to meddle with His affairs
when no invitation had been extended. So the pets were left
in peace, and here they were, as I have said, rabbits, birds,
squirrels, cats, and other reptiles, all around the child, and full
of interest in her supper, and helping what they could. There
was a very small squirrel on her shoulder, sitting up, as those
creatures do, and turning a rocky fragment of prehistoric
chestnut-cake over and over in its knotty hands, and hunting
for the less indurated places, and giving its elevated bushy
tail a flirt and its pointed ears a toss when it found one—
signifying thankfulness and surprise—and then it filed that
place off with those two slender front teeth which a squirrel
carries for that purpose and not for ornament, for ornamental
they never could be, as any will admit that have noticed them.

Everything was going fine and breezy and hilarious, but
then there came an interruption, for somebody hammered on
the door. It was one of those ragged road-stragglers—the
eternal wars kept the country full of them. He came in, all
over snow, and stamped his feet and shook and brushed him-

self, and shut the door, and took off his limp ruin of a hat and slapped it once or twice against his leg to knock off its fleece of snow, and then glanced around on the company with a pleased look upon his thin face, and a most yearning and famished one in his eye when it fell upon the victuals, and then he gave us a humble and conciliatory salutation, and said it was a blessed thing to have a fire like that on such a night, and a roof overhead like this, and that rich food to eat, and loving friends to talk with—ah, yes, this was true, and God help the homeless, and such as must trudge the roads in this weather.

Nobody said anything. The embarrassed poor creature stood there and appealed to one face after the other with his eyes, and found no welcome in any, the smile on his own face flickering and fading and perishing, meanwhile; then he dropped his gaze, the muscles of his face began to twitch, and he put up his hand to cover this womanish sign of weakness.

"Sit down!"

This thunder-blast was from old Jacques d'Arc, and Joan was the object of it. The stranger was startled, and took his hand away, and there was Joan standing before him offering him her bowl of porridge. The man said,

"God Almighty bless you, my darling!" and then the tears came, and ran down his cheeks, but he was afraid to take the bowl.

"Do you hear me? Sit down, I say!"

There could not be a child more easy to persuade than Joan, but this was not the way. Her father had not the art; neither could he learn it. Joan said,

"Father, he is hungry; I can see it."

"Let him go work for food, then. We are being eaten out of house and home by his like, and I have said I would endure it no more, and will keep my word. He has the face of a rascal anyhow, and a villain. Sit *down*, I tell you!"

"I know not if he is a rascal or no, but he is hungry, father, and shall have my porridge—I do not need it."

"If you don't obey me I'll— Rascals are not entitled to help

JOAN'S VISION

from honest people, and no bite nor sup shall they have in this house. *Joan!*"

She set her bowl down on the box and came over and stood before her scowling father, and said :

" Father, if you will not let me, then it must be as you say ; but I would that you would think—then you would see that it is not right to punish one part of him for what the other part has done; for it is that poor stranger's head that does the evil things, but it is not his head that is hungry, it is his stomach, and it has done no harm to anybody, but is without blame, and innocent, not having any *way* to do a wrong, even if it was minded to it. Please let—"

"What an idea ! It is the most idiotic speech I ever heard."

But Aubrey, the maire, broke in, he being fond of an argument, and having a pretty gift in that regard, as all acknowledged. Rising in his place and leaning his knuckles upon the table and looking about him with easy dignity, after the manner of such as be orators, he began, smooth and persuasive :

" I will differ with you there, gossip, and will undertake to show the company"—here he looked around upon us and nodded his head in a confident way—" that there is a grain of sense in what the child has said ; for look you, it is of a certainty most true and demonstrable that it is a man's head that is master and supreme ruler over his whole body. Is that granted ? Will any deny it ?" He glanced around again ; everybody indicated assent. " Very well, then; that being the case, no part of the body is responsible for the result when it carries out an order delivered to it by the head ; ergo, the head is alone responsible for crimes done by a man's hands or feet or stomach—do you get the idea ? am I right thus far ?" Everybody said yes, and said it with enthusiasm, and some said, one to another, that the maire was in great form to-night and at his very best—which pleased the maire exceedingly and made his eyes sparkle with pleasure, for he overheard these things ; so he went on in the same fertile and brilliant way. "Now, then, we will consider what the term responsibility means, and how it affects the case in point. Responsibility makes a man responsible for only those things for which he is properly responsible" — and he waved his spoon around in a wide sweep to indicate the comprehensive nature of that class of responsibilities which render people responsible, and several exclaimed, admiringly, " He is right! —he has put that whole tangled thing into a nutshell—it is wonderful!" After a little pause to give the interest opportunity to gather and grow, he went on: "Very good. Let us suppose the case of a pair of tongs that falls upon a man's foot, causing a cruel hurt. Will you claim that the tongs are punishable for that? The question is answered: I see by your faces that you would call such a claim absurd. Now, why is it absurd? It is absurd because, there being no reasoning faculty — that is to say, no faculty of personal command—in a pair of tongs, personal responsibility for the acts of the tongs is wholly absent from the tongs; and therefore, responsibility being absent, punishment cannot ensue. Am I right?" A hearty burst of applause was his answer. "Now, then, we arrive at a man's stomach. Consider how exactly, how marvellously, indeed, its situation corresponds to that of a pair of tongs. Listen—and take careful note, I beg you. Can a man's stomach plan a murder? No. Can it plan a theft? No. Can it plan an incendiary fire? No. Now answer me — *can a pair of tongs?*" (There were admiring shouts of "No!" and "The cases are just exact!" and "Don't he do it splendid!") "Now, then, friends and neighbors, a stomach which cannot plan a crime cannot be a principal in the commission of it—that is plain, as you see. The matter is narrowed down by that much; we will narrow it further. Can a stomach, of its own motion, assist at a crime? The answer is no, because command is absent, the reasoning faculty is absent, volition is absent — as in the case of the tongs. We perceive, now, do we not, that the stomach is totally irresponsible for crimes committed, either in whole or in part, by it?" He got a rousing cheer for response. " Then what do we arrive at as our verdict? Clearly this: that there is no such thing in this world as a guilty stomach; that in the body of the veriest rascal resides a pure and innocent stomach; that, whatever its owner may do, *it* at least should be sacred in our eyes; and that while God gives us minds to think just and charitable and honorable thoughts, it should be and *is* our privilege, as well as our duty, not only to feed the hungry stomach that resides in a rascal, having pity for its sorrow and its need, but to do it gladly, gratefully, in recognition of its sturdy and loyal maintenance of its purity and innocence in the midst of temptation and in company so repugnant to its better feelings. I am done."

Well, you never saw such an effect! They rose—the whole house rose—and clapped, and cheered, and praised him to the skies; and one after another, still clapping and shouting, they crowded forward, some with moisture in their eyes, and wrung his hands, and said such glorious things to him that he was clear overcome with pride and happiness, and couldn't say a word, for his voice would have broken, sure. It was splendid to see; and everybody said he had never come up to that speech in his life before, and never could do it again. Eloquence *is* a power, there is no question of that. Even old Jacques d'Arc was carried away, for once in his life, and shouted out—

"It's all right, Joan—give him the porridge!"

She was embarrassed, and did not seem to know what to say, and so didn't say anything. It was because she had given the man the porridge long ago, and he had already eaten it all up. When she was asked why she had not waited until a decision was arrived at, she said the man's stomach was very hungry, and it would not have been wise to wait, since she could not tell what the decision would be. Now that was a good and thoughtful idea for a child.

The man was not a rascal at all. He was a very good fellow, only he was out of luck, and surely that was no crime at that time in France. Now that his stomach was proved to be innocent, it was allowed to make itself at home; and as soon as it was well filled and needed nothing moie, the man unwound his tongue and turned it loose, and it was really a noble one to go. He had been in the wars for years, and the things he told, and the way he told them, fired everybody's patriotism away up high, and set all hearts to thumping and all pulses to leaping; then, before anybody rightly knew how the change was made, he was leading us a sublime march through the ancient glories of France, and in fancy we saw the titanic forms of the twelve paladins rise out of the mists of the past and face their fate; we heard the tread of the innumerable hosts sweeping down to shut them in; we saw this human tide flow and ebb, ebb and flow, and waste away before that little band of heroes; we saw each detail pass before us of that most stupendous, most disastrous, yet most adored and glorious day in French legendary history; here and there and yonder, across that vast field of the dead and dying, we saw this and that and the other paladin dealing his prodigious blows with weary arm and failing strength, and one by one we saw them fall, till only one remained—he that was without peer, he whose name gives name to the Song of Songs, the song which no Frenchman can hear and keep his feelings down and his pride of country cool; then, grandest and pitifulest scene of all, we saw his own pathetic death; and our stillness, as we sat with parted lips and breathless, hanging upon this man's words, gave us a sense of the awful stillness that reigned in that field of slaughter when that last surviving soul had passed.

And now, in this solemn hush, the stranger gave Joan a pat or two on the head and said:

"Little maid—whom God keep!—you have brought me from death to life this night; now listen: here is your reward," and at that supreme time for such a heart-melting, soul-rousing surprise, without another word he lifted up the most noble and pathetic voice that was ever heard, and began to pour out the great Song of Roland!

Think of that, with a French audience all stirred up and ready. Oh, where was your spoken eloquence now! what was it to this! How fine he looked, how stately, how inspired, as he stood there with that mighty chant welling from his lips and his heart, his whole body transfigured, and his rags along with it.

Everybody rose and stood, while he sang, and their faces glowed and their eyes burned; and the tears came and flowed down their cheeks, and their forms began to sway unconsciously to the swing of the song, and their bosoms to heave and pant; and moanings broke out, and deep ejaculations; and when the last verse was reached, and Roland lay dying, all alone, with his face to the field and to his slain, lying there in heaps and winrows, and took off and held up his gauntlet to God with his failing hand, and breathed his beautiful prayer with his paling lips, all burst out in sobs and wailings. But when the final great note died out and the song was done, they all flung themselves in a body at the singer, stark mad with love of him and love of France and pride in her great deeds and old renown, and smothered him with their embracings; but Joan was there first, hugged close to his breast, and covering his face with idolatrous kisses.

The storm raged on outside, but that was no matter; this was the stranger's home now, for as long as he might please.

CHAPTER IV

ALL children have nicknames, and we had ours. We got one apiece early, and they stuck to us; but Joan was richer in this matter, for as time went on she earned a second, and then a third, and so on, and we gave them to her. First and last she had as many as half a dozen. Several of these she never lost. Peasant girls are bashful naturally; but she surpassed the rule so far, and colored so easily, and was so easily embarrassed in the presence of strangers, that we nicknamed her the Bashful. We were all patriots, but she was called *the* Patriot, because our warmest feeling for our country was cold beside hers. Also she was called the Beautiful; and this was not merely because of the extraordinary beauty of her face and form, but because of the loveliness of her character. These names she kept, and one other—the Brave.

We grew along up, in that plodding and peaceful region, and got to be good-sized boys and girls—big enough, in fact, to begin to know as much about the wars raging perpetually to the west and north of us as our elders, and also to feel as stirred up over the occasional news from those red fields as they did. I remember certain of these days very clearly. One Tuesday a crowd of us were romping and singing around the Fairy Tree, and hanging garlands on it in memory of our lost little fairy friends, when little Mengette cried out:

"Look! What is that?"

When one exclaims like that, in a way that shows astonishment and apprehension, he gets attention. All the panting breasts and flushed faces flocked together, and all the eager eyes were turned in one direction—down the slope, toward the village.

"It's a black flag."

"A black flag! No—is it?"

"You can see for yourself that it is nothing else."

"It *is* a black flag, sure! Now, has any ever seen the like of that before?"

"What can it mean?"

"Mean? It means something dreadful—what else?"

"That is nothing to the point; anybody knows that without the telling. But *what?*—that is the question."

"It is a chance that he that bears it can answer as well as any that are here, if you can contain yourself till he come."

"He runs well. Who is it?"

Some named one, some another; but presently all saw that it was Étienne Roze, called the Sunflower, because he had yellow hair and a round, pock-marked face. His ancestors had been Germans some centuries ago. He came straining up the slope, now and then projecting his flag-stick aloft and giving his black symbol of woe a wave in the air, whilst all eyes watched him, all tongues discussed him, and every heart beat faster and faster with impatience to know his news. At last he sprang among us, and struck his flag-stick into the ground, saying:

"There! Stand there and represent France while I get my breath. She needs no other flag, now."

All the giddy chatter stopped. It was as if one had announced a death. In that chilly hush there was no sound audible but the panting of the breath-blown boy. When he was presently able to speak, he said:

"Black news is come. A treaty has been made at Troyes between France and the English and Burgundians. By it France is betrayed and delivered over, tied hand and foot, to the enemy. It is the work of the Duke of Burgundy and that she-devil the Queen of France. It marries Henry of England to Catharine of France—"

"Is not this a lie? Marries the daughter of France to the Butcher of Agincourt? It is not to be believed. You have not heard aright."

"If you cannot believe that, Jacques d'Arc, then you have a difficult task indeed before you, for worse is to come. Any child that is born of that marriage—if even a girl—is to inherit the thrones of both England and France, and this double ownership is to remain with its posterity forever!"

"Now *that* is certainly a lie, for it runs counter to our Salic law, and so is not legal and cannot have effect," said Edmond Aubrey, called the Paladin, because of the armies he was always going to eat up some day. He would have said more, but he was drowned out by the clamors of the others, who all burst into a fury over this feature of the treaty, all talking at once and nobody hearing anybody, until presently Haumette persuaded them to be still, saying:

"It is not fair to break him up so in his tale; pray let him go on. You find fault with his history because it seems to be lies. That were reason for satisfaction—*that* kind of lies—not discontent. Tell the rest, Étienne."

"There is but this to tell: Our King, Charles VI., is to reign until he dies, then Henry V. of England is to be Regent of France until a child of his shall be old enough to—"

"*That* man is to reign over us—the Butcher? It is lies! all lies!" cried the Paladin. "Besides, look you — what becomes of our Dauphin? What says the treaty about him?"

"Nothing. It takes away his throne and makes him an outcast."

Then everybody shouted at once and said the news was a lie; and all began to get cheerful again, saying, "Our King would have to sign the treaty to make it good; and that he would not do, seeing how it serves his own son."

But the Sunflower said: "I will ask you this: Would the *Queen* sign a treaty disinheriting her son?"

"That viper? Certainly. Nobody is talking of her. Nobody expects better of her. There is no villany she will stick at, if it feed her spite; and she hates her son. Her signing it is of no consequence. The King must sign."

"I will ask you another thing. What is the King's condition? Mad, isn't he?"

"Yes, and his people love him all the more for it. It brings him near to them by his sufferings; and pitying him makes them love him."

"You say right, Jacques d'Arc. Well, what would you of one that is mad? Does he know what he does? No. Does he do what others make him do? Yes. Now, then, I tell you he has signed the treaty."

"Who made him do it?"

"You know, without my telling. The Queen."

Then there was another uproar—everybody talking at once, and all heaping execrations upon the Queen's head. Finally Jacques d'Arc said:

"But many reports come that are not true. Nothing so shameful as this has ever come before, nothing that cuts so deep, nothing that has dragged France so low; therefore there is hope that this tale is but another idle rumor. Where did you get it?"

The color went out of his sister Joan's face. She dreaded the answer; and her instinct was right.

"The curé of Maxey brought it."

There was a general gasp. We knew him, you see, for a trusty man.

"Did he believe it?"

The hearts almost stopped beating. Then came the answer:

"He did. And that is not all. He said he *knew* it to be true."

Some of the girls began to sob; the boys were struck silent. The distress in Joan's face was like that which one sees in the face of a dumb animal that has received a mortal hurt. The animal bears it, making no complaint; she bore it also, saying no word. Her brother Jacques put his hand on her head and caressed her hair to indicate his sympathy, and she gathered the hand to her lips and kissed it for thanks, not saying anything. Presently the reaction came, and the boys began to talk. Noël Rainguesson said:

"Oh, are we never going to be men! We do grow along

so slowly, and France never needed soldiers as she needs them now, to wipe out this black insult."

"I hate youth !" said Pierre Morel, called the Dragon-fly because his eyes stuck out so. "You've always got to wait, and wait, and wait—and here are the great wars wasting away for a hundred years, and you never get a chance. If I could only be a soldier now !"

"As for me, I'm not going to wait much longer," said the Paladin; "and when I do start you'll hear from me, I promise you that. There are some who, in storming a castle, prefer to be in the rear; but as for me, give me the front or none; I will have none in front of me but the officers."

Even the girls got the war spirit, and Marie Dupont said—

"I would I were a man; I would start this minute !" and looked very proud of herself, and glanced about for applause.

"So would I," said Cécile Letellier, sniffing the air like a war-horse that smells the battle; "I warrant you I would not turn back from the field though all England were in front of me."

"Pooh !" said the Paladin; "girls can brag, but that's all they are good for. Let a thousand of them come face to face with a handful of soldiers once, if you want to see what running is like. Here's little Joan—next *she'll* be threatening to go for a soldier !"

The idea was so funny, and got such a good laugh, that the Paladin gave it another trial, and said : "Why, you can just see her !—see her plunge into battle like any old veteran. Yes, indeed; and not a poor shabby common soldier like us, but an officer—an officer, mind you, with armor on, and the bars of a steel helmet to blush behind and hide her embarrassment when she finds an army in front of her that she hasn't been introduced to. An officer? Why, she'll be a captain ! A captain, I tell you, with a hundred men at her back—or maybe girls. Oh, no common-soldier business for her ! And, dear me, when she starts for that other army, you'll think there's a hurricane blowing it away !"

Well, he kept it up like that till he made their sides ache with laughing; which was quite natural, for certainly it was a very funny idea—at that time—I mean, the idea of that gentle little creature, that wouldn't hurt a fly, and couldn't bear the sight of blood, and was so girlish and shrinking in all ways, rushing into battle with a gang of soldiers at her back. Poor thing, she sat there confused and ashamed to be so laughed at; and yet at that very minute there was something about to happen which would change the aspect of things, and make those young people see that when it comes to laughing, the person that laughs last has the best chance. For just then a face which we all knew and all feared projected itself from behind the Fairy Tree, and the thought that shot through us all was, crazy Benoist has gotten loose from his cage, and we are as good as dead! This ragged and hairy and horrible creature glided out from behind the tree, and raised an axe as he came. We all broke and fled, this way and that, the girls screaming and crying. No, not all; all but Joan. She stood up and faced the man, and remained so. As we reached the wood that borders the grassy clearing and jumped into its shelter, two or three of us glanced back to see if Benoist was gaining on us, and that is what we saw—Joan standing, and the maniac gliding stealthily toward her with his axe lifted. The sight was sickening. We stood where we were, trembling and not able to move. I did not want to see the murder done, and yet I could not take my eyes away. Now I saw Joan step forward to meet the man, though I believed my eyes must be deceiving me. Then I saw him stop. He threatened her with his axe, as if to warn her not to come further, but she paid no heed, but went steadily on, until she was right in front of him—right under his axe. Then she stopped, and seemed to begin to talk with him. It made me sick, yes, giddy, and everything swam around me, and I could not see anything for a time—whether long or brief I do not know. When this passed and I looked again, Joan was walking by the man's side toward the village, holding him by his hand. The axe was in her other hand.

One by one the boys and girls crept out, and we stood there gazing, open-mouthed, till those two entered the village and were hid from sight. It was then that we named her the Brave.

We left the black flag there to continue its mournful office, for we had other matter to think of now. We started for the village on a run, to give warning, and get Joan out of her peril; though for one, after seeing what I had seen, it seemed to me that while Joan had the axe the man's chance was not the best of the two. When we arrived the danger was past, the madman was in custody. All the people were flocking to the little square in front of the church to talk and exclaim and wonder over the event, and it even made the town forget the black news of the treaty for two or three hours.

All the women kept hugging and kissing Joan, and praising her, and crying, and the men patted her on the head and said they wished she was a man, they would send her to the wars and never doubt but that she would strike some blows that would be heard of. She had to tear herself away and go and hide, this glory was so trying to her diffidence.

Of course the people began to ask us for the particulars. I was so ashamed that I made an excuse to the first comer, and got privately away and went back to the Fairy Tree, to get relief from the embarrassment of those questionings. There I found Joan, but she was there to get relief from the embarrassment of glory. One by one the others shirked the inquirers and joined us in our refuge. Then we gathered around Joan, and asked her how she had dared to do that thing. She was very modest about it, and said :

"You make a great thing of it, but you mistake; it was not a great matter. It was not as if I had been a stranger to the man. I know him, and have known him long; and he knows me, and likes me. I have fed him through the bars of his cage many times ; and last December when they chopped off two of his fingers to remind him to stop seizing and wounding people passing by, I dressed his hand every day till it was well again."

"That is all well enough," said Little Mengette, "but he is

a madman, dear, and so his likings and his gratitude and friendliness go for nothing when his rage is up. You did a perilous thing."

"Of course you did," said the Sunflower. "Didn't he threaten to kill you with the axe?"

"Yes."

"Didn't he threaten you more than once?"

"Yes."

"Didn't you feel afraid?"

"No—at least not much—very little."

"Why didn't you?"

She thought a moment, then said, quite simply—

"I don't know."

It made everybody laugh. Then the Sunflower said it was like a lamb trying to think out how it had come to eat a wolf, but had to give it up.

Cécile Letellier asked, "Why didn't you run when we did?"

"Because it was necessary to get him to his cage; else he would kill some one. Then he would come to the like harm himself."

It is noticeable that this remark, which implies that Joan was entirely forgetful of herself and her own danger, and had thought and wrought for the preservation of other people alone, was not challenged, or criticised, or commented upon by anybody there, but was taken by all as matter of course and true. It shows how clearly her character was defined, and how well it was known and established.

There was silence for a time, and perhaps we were all thinking of the same thing—namely, what a poor figure we had cut in that adventure as contrasted with Joan's performance. I tried to think up some good way of explaining why I had run away and left a little girl at the mercy of a maniac armed with an axe, but all of the explanations that offered themselves to me seemed so cheap and shabby that I gave the matter up and remained still. But others were less wise. Noël Rainguesson fidgeted a while, then broke out with a remark which showed what his mind had been running on:

"The fact is, I was taken by surprise. That is the reason. If I had had a moment to think, I would no more have thought of running than I would think of running from a baby. For, after all, what is Théophile Benoist, that I should seem to be afraid of him? Pooh! the idea of being afraid of that poor thing! I only wish he would come along now—I'd show you!"

"So do I!" cried Pierre Morel. "If I wouldn't make him climb this tree quicker than—well, you'd see what I would do! Taking a person by surprise, that way—why, I never meant to run; not in earnest, I mean. I never thought of running in earnest; I only wanted to have some fun, and when I saw Joan standing there, and him threatening her, it was all I could do to restrain myself from going there and just tearing the livers and lights out of him. I *wanted* to do it bad enough, and if it was to do over again, I *would!* If ever he comes fooling around me again, I'll—"

"Oh, hush!" said the Paladin, breaking in with an air of disdain; "the way you people talk, a person would think there's something heroic about standing up and facing down that poor remnant of a man. Why, it's nothing! There's small glory to be got in facing *him* down, I should say. Why, I wouldn't want any better fun than to face down a hundred like him. If he was to come along here now, I would walk up to him just as I am now—I wouldn't care if he had a thousand axes—and say—"

And so he went on and on, telling the brave things he would say and the wonders he would do; and the others put in a word from time to time, describing over again the gory marvels they would do if ever that madman ventured to cross their path again, for next time they would be ready for him, and would soon teach him that if he thought he could surprise them twice because he had surprised them once, he would find himself very seriously mistaken, that's all.

And so, in the end, they all got back their self-respect; yes, and even added somewhat to it; indeed, when the sitting broke up they had a finer opinion of themselves than they had ever had before.

CHAPTER V

THEY were peaceful and pleasant, those young and smoothly flowing days of ours; that is, that was the case as a rule, we being remote from the seat of war, but at intervals roving bands approached near enough for us to see the flush in the sky at night which marked where they were burning some farmstead or village, and we all knew, or at least felt, that some day they would come yet nearer, and we should have our turn. This dull dread lay upon out spirits like a physical weight. It was greatly augmented a couple of years after the Treaty of Troyes.

It was truly a dismal year for France. One day we had been over to have one of our occasional pitched battles with those hated Burgundian boys of the village of Maxey, and had been whipped, and were arriving on our side of the river after dark, bruised and weary, when we heard the bell ringing the tocsin. We ran all the way, and when we got to the square we found it crowded with the excited villagers, and weirdly lighted by smoking and flaring torches.

On the steps of the church stood a stranger, a Burgundian priest, who was telling the people news which made them weep, and rave, and rage, and curse, by turns. He said our old mad King was dead, and that now we and France and the crown were the property of an English baby lying in his cradle in London. And he urged us to give that child our allegiance, and be its faithful servants and well-wishers; and said we should now have a strong and stable government at last, and that in a little time the English armies would start on their last march, and it would be a brief one, for all that it would need to do would be to conquer what odds and ends of our country yet remained under that rare and almost forgotten rag, the banner of France.

The people stormed and raged at him, and you could see dozens of them stretch their fists above the sea of torch-lighted faces and shake them at him; and it was all a wild picture, and stirring to look at; and the priest was a first-rate part of it, too, for he stood there in the strong glare and looked down on those angry people in the blandest and most indifferent way, so that while you wanted to burn him at the stake, you still admired the aggravating coolness of him. And his winding up was the coolest thing of all. For he told them how, at the funeral of our old King, the French King-at-Arms had broken his staff of office over the coffin of "Charles VI. and his dynasty," at the same time saying, in a loud voice, "God grant long life to Henry, King of France and England, our sovereign lord!" and then he asked them to join him in a hearty Amen to *that!*

The people were white with wrath, and it tied their tongues for the moment, and they could not speak. But Joan was standing close by, and she looked up in his face, and said in her sober, earnest way—

"I would I might see thy head struck from thy body!"—then, after a pause, and crossing herself—"if it were the will of God."

This is worth remembering, and I will tell you why: it is the only harsh speech Joan ever uttered in her life. When I shall have revealed to you the storms she went through, and the wrongs and persecutions, then you will see that it was wonderful that she said but one bitter thing while she lived.

From the day that that dreary news came we had one scare after another, the marauders coming almost to our doors every now and then; so that we lived in ever-increasing apprehension, and yet were somehow mercifully spared from actual attack. But at last our turn did really come. This was in the spring of '28. The Burgundians swarmed in with a great noise, in the middle of a dark night, and we had to jump up and fly for our lives. We took the road to Neufchâteau, and rushed along in the wildest disorder, everybody trying to get ahead, and thus the movements of all were impeded ; but Joan had a cool head—the only cool head there—and she took command and brought order out of that chaos. She did her work quickly and with decision and despatch, and soon turned the panic flight into a quite steady-going march. You will grant that for so young a person, and a girl at that, this was a good piece of work.

She was sixteen now, shapely and graceful, and of a beauty so extraordinary that I might allow myself any extravagance of language in describing it and yet have no fear of going beyond the truth. There was in her face a sweetness and serenity and purity that justly reflected her spiritual nature. She was deeply religious, and this is a thing which sometimes gives a melancholy cast to a person's countenance, but it was not so in her case. Her religion made her inwardly content and joyous ; and if she was troubled at times, and showed the pain of it in her face and bearing, it came of distress for her country ; no part of it was chargeable to her religion.

A considerable part of our village was destroyed, and when it became safe for us to venture back there we realized what other people had been suffering in all the various quarters of France for many years—yes, decades of years. For the first time we saw wrecked and smoke-blackened homes, and in the lanes and alleys carcasses of dumb creatures that had been slaughtered in pure wantonness—among them calves and lambs that had been pets of the children ; and it was pity to see the children lament over them.

And then, the taxes, the taxes ! Everybody thought of that. That burden would fall heavy, now, in the commune's crippled condition, and all faces grew long with the thought of it. Joan said—

"Paying taxes with naught to pay them with is what the rest of France has been doing these many years, but we never knew the bitterness of that before. We shall know it now."

And so she went on talking about it and growing more

and more troubled about it, until one could see that it was filling all her mind.

At last we came upon a dreadful object. It was the madman — hacked and stabbed to death in his iron cage in the corner of the square. It was a bloody and dreadful sight. Hardly any of us young people had ever seen a man before who had lost his life by violence; so this cadaver had an awful fascination for us; we could not take our eyes from it. I mean, it had that sort of fascination for all of us but one. That one was Joan. She turned away in horror, and could not be persuaded to go near it again. There—it is a striking reminder that we are but creatures of use and custom; yes, and it is a reminder, too, of how harshly and unfairly fate deals with us sometimes. For it was so ordered that the very ones among us who were most fascinated with mutilated and bloody death were to live their lives in peace, while that other, who had a native and deep horror of it, must presently go forth and have it as a familiar spectacle every day on the field of battle.

You may well believe that we had plenty of matter for talk, now, since the raiding of our village seemed by long odds the greatest event that had really ever occurred in the world; for although these dull peasants may have *thought* they recognized the bigness of some of the previous occurrences that had filtered from the world's history dimly into their minds, the truth is that they hadn't. One biting little fact, visible to their eyes of flesh and felt in their own personal vitals, became at once more prodigious to them than the grandest remote episode in the world's history which they had got at second-hand and by hearsay. It amuses me now when I recall how our elders talked then. They fumed and fretted in a fine fashion.

"Ah yes," said old Jacques d'Arc, "things are come to a pretty pass indeed! The King must be informed of this. It is time that he cease from idleness and dreaming, and get at his proper business." He meant our young disinherited King, the hunted refugee, Charles VII.

IN THE FOREST

" You say well," said the maire. " He should be informed, and that at once. It is an outrage that such things should be permitted. Why, we are not safe in our beds, and he taking his ease yonder. It shall be made known, indeed it shall—all France shall hear of it !"

To hear them talk, one would have imagined that all the previous ten thousand sackings and burnings in France had been but fables, and this one the only fact. It is always the way : words will answer as long as it is only a person's neighbor who is in trouble, but when that person gets into trouble himself, it is time that the King rise up and *do* something.

The big event filled us young people with talk, too. We let it flow in a steady stream while we tended the flocks. We were beginning to feel pretty important, now, for I was eighteen and the other youths were from one to four years older—young men, in fact. One day the Paladin was arrogantly criticising the patriot generals of France and said—

" Look at Dunois, Bastard of Orleans—call *him* a general ! Just put me in his place once—never mind what I would do, it is not for me to say, I have no stomach for talk, my way is to *act* and let others do the talking—but just put me in his place once, that's all ! And look at Saintrailles—pooh ! and that blustering La Hire, now what a general *that* is !"

It shocked everybody to hear these great names so flippantly handled, for to us these renowned soldiers were almost gods. In their far-off splendor they rose upon our imaginations dim and huge, shadowy and awful, and it was a fearful thing to hear them spoken of as if they were mere men, and their acts open to comment and criticism. The color rose in Joan's face, and she said—

" I know not how any can be so hardy as to use such words regarding these sublime men, who are the very pillars of the French State, supporting it with their strength and preserving it at daily cost of their blood. As for me, I could count myself honored past all deserving if I might be allowed but the privilege of looking upon them once — at a distance, I mean, for it would not become one of my degree to approach them too near."

The Paladin was disconcerted for a moment, seeing by the faces around him that Joan had put into words what the others felt, then he pulled his complacency together and fell to fault-finding again. Joan's brother Jean said—

"If you don't like what our generals do, why don't you go to the great wars yourself and better their work? You are always talking about going to the wars, but you don't go."

"Look you," said the Paladin, "it is easy to say that. Now I will tell you why I remain chafing here in a bloodless tranquillity which my reputation teaches you is repulsive to my nature. I do not go because I am not a gentleman. That is the whole reason. What can one private soldier do in a contest like this? Nothing. He is not permitted to rise from the ranks. If I were a gentleman would I remain here? Not one moment. I can save France—ah, you may laugh, but I know what is in me, I know what is hid under this peasant cap. I can save France, and I stand ready to do it, but not under these present conditions. If they want me, let them send for me; otherwise, let them take the consequences; I shall not budge but as an officer."

"Alas, poor France—France is lost!" said Pierre d'Arc.

"Since you sniff so at others, why don't you go to the wars yourself, Pierre d'Arc?"

"Oh, I haven't been sent for, either. I am no more a gentleman than you. Yet I will go ; I promise to go. I promise to go as a private under your orders—when you are sent for."

They all laughed, and the Dragon-fly said—

"So soon ? Then you need to begin to get ready; you might be called for in five years—who knows? Yes, in my opinion you'll march for the wars in five years."

"He will go sooner," said Joan. She said it in a low voice and musingly, but several heard it.

"How do you know that, Joan ?" said the Dragon-fly, with a surprised look. But Jean d'Arc broke in and said—

"I want to go myself, but as I am rather young yet, I also will wait, and march when the Paladin is sent for."

"No," said Joan, "he will go with Pierre."

She said it as one who talks to himself aloud without knowing it, and none heard it but me. I glanced at her and saw that her knitting-needles were idle in her hands, and that her face had a dreamy and absent look in it. There were fleeting movements of her lips as if she might be occasionally saying parts of sentences to herself. But there was no sound, for I was the nearest person to her and I heard nothing. But I set my ears open, for those two speeches had affected me uncannily, I being superstitious and easily troubled by any little thing of a strange and unusual sort.

Noël Rainguesson said—

"There is one way to let France have a chance for her salvation. We've got *one* gentleman in the commune, at any rate. Why can't the Scholar change name and condition with the Paladin? Then he can be an officer. France will send for him then, and he will sweep these English and Burgundian armies into the sea like flies."

I was the Scholar. That was my nickname, because I could read and write. There was a chorus of approval, and the Sunflower said—

"That is the very thing—it settles every difficulty. The Sieur de Conte will easily agree to that. Yes, he will march at the back of Captain Paladin and die early, covered with common-soldier glory."

"He will march with Jean and Pierre, and live till these wars are forgotten," Joan muttered; "and at the eleventh hour Noël and the Paladin will join these, but not of their own desire." The voice was so low that I was not perfectly sure that these were the words, but they seemed to be. It makes one feel creepy to hear such things.

"Come, now," Noël continued, "it's all arranged; there's nothing to do but organize under the Paladin's banner and go forth and rescue France. You'll all join?"

All said yes, except Jacques d'Arc, who said—

"I'll ask you to excuse me. It is pleasant to talk war, and I am with you there, and I've always thought I should go soldiering about this time, but the look of our wrecked village and that carved-up and bloody madman have taught me that I am not made for such work and such sights. I could never be at home in that trade. Face swords and the big guns and death? It isn't in me. No, no; count me out. And besides, I'm the eldest son, and deputy prop and protector of the family. Since you are going to carry Jean and Pierre to the wars, somebody must be left behind to take care of our Joan and her sister. I shall stay at home, and grow old in peace and tranquillity."

"He will stay at home, but not grow old," murmured Joan.

The talk rattled on in the gay and careless fashion privileged to youth, and we got the Paladin to map out his campaigns and fight his battles and win his victories and extinguish the English and put our King upon his throne and set his crown upon his head. Then we asked him what he was going to answer when the King should require him to name his reward. The Paladin had it all arranged in his head, and brought it out promptly:

"He shall give me a dukedom, name me premier peer, and make me Hereditary Lord High Constable of France."

"And marry you to a princess—you're not going to leave that out, are you?"

The Paladin colored a trifle, and said, brusquely—

"He may keep his princesses—I can marry more to my taste."

Meaning Joan, though nobody suspected it at that time. If any had, the Paladin would have been finely ridiculed for his vanity. There was no fit mate in that village for Joan of Arc. Every one would have said that.

In turn, each person present was required to say what reward he would demand of the King if he could change places with the Paladin and do the wonders the Paladin was going to do. The answers were given in fun, and each of us tried to outdo his predecessors in the extravagance of the reward he would claim; but when it came to Joan's turn and they rallied her out of her dreams and asked her to testify, they had to explain to her what the question was, for her thought had been absent, and she had heard none of this latter part of our talk. She supposed they wanted a serious answer, and she gave it. She sat considering some moments, then she said—

"If the Dauphin out of his grace and nobleness should say to me, 'Now that I am rich and am come to my own again, choose and have,' I should kneel and ask him to give command that our village should nevermore be taxed."

It was so simple and out of her heart that it touched us and we did not laugh, but fell to thinking. We did not laugh; but there came a day when we remembered that speech with a mournful pride, and were glad that we had not laughed, perceiving then how honest her words had been, and seeing how faithfully she made them good when the time came, asking just that boon of the King and refusing to take even any least thing for herself.

CHAPTER VI

ALL through her childhood and up to the middle of her fourteenth year, Joan had been the most light-hearted creature and the merriest in the village, with a hop-skip-and-jump gait and a happy and catching laugh ; and this disposition, supplemented by her warm and sympathetic nature and frank and winning ways, had made her everybody's pet. She had been a hot patriot all this time, and sometimes the war news had sobered her spirits and wrung her heart and made her acquainted with tears, but always when these interruptions had run their course her spirits rose and she was her old self again.

But now for a whole year and a half she had been mainly grave ; not melancholy, but given to thought, abstraction, dreams. She was carrying France upon her heart, and she found the burden not light. I knew that this was her trouble, but others attributed her abstraction to religious ecstasy, for she did not share her thinkings with the village at large, yet gave me glimpses of them, and so I knew, better than the rest, what was absorbing her interest. Many a time the idea crossed my mind that she had a secret—a secret which she was keeping wholly to herself, as well from me as from the others. This idea had come to me because several times she had cut a sentence in two and changed the subject when apparently she was on the verge of a revelation of some sort. I was to find this secret out, but not just yet.

The day after the conversation which I have been reporting we were together in the pastures and fell to talking about France, as usual. For her sake I had always talked hopefully before, but that was mere lying, for really there was not anything to hang a rag of hope for France upon. Now it was such a pain to lie to her, and cost me such shame to offer this treachery to one so snow-pure from lying and treachery, and even from suspicion of such basenesses in others, as she was, that I was resolved to face about, now, and begin over again, and never insult her more with deception. I started on the new policy by saying—still opening up with a small lie, of course, for habit is habit, and not to be flung out of the window by any man, but coaxed down-stairs a step at a time—

"Joan, I have been thinking the thing all over, last night, and have concluded that we have been in the wrong all this time; that the case of France is desperate; that it has been desperate ever since Agincourt; and that to-day it is more than desperate, it is hopeless."

I did not look her in the face while I was saying it; it could not be expected of a person. To break her heart, to crush her hope with a so frankly brutal speech as that, without one charitable soft place in it—it seemed a shameful thing, and it was. But when it was out, the weight gone, and my conscience rising to the surface, I glanced at her face to see the result.

There was none to see. At least none that I was expecting. There was a barely perceptible suggestion of wonder in her serious eyes, but that was all; and she said, in her simple and placid way—

"The case of France hopeless? Why should you think that? Tell me."

It is a most pleasant thing to find that what you thought would inflict a hurt upon one whom you honor, has not done it. I was relieved, now, and could say all my say without any furtivenesses and without embarrassment. So I began:

"Let us put sentiment and patriotic illusions aside, and look the facts in the face. What do they say? They speak as plainly as the figures in a merchant's account-book. One has only to add the two columns up to see that the French house is bankrupt, that one-half of its property is already in the English sheriff's hands and the other half in nobody's —

4

except those of irresponsible raiders and robbers confessing allegiance to nobody. Our King is shut up with his favorites and fools in inglorious idleness and poverty in a narrow little patch of the kingdom—a sort of back lot, as one may say— and has no authority there or anywhere else, hasn't a farthing to his name, nor a regiment of soldiers; he is not fighting, he is not intending to fight, he means to make no further resistance; in truth there is but one thing that he is intending to do—give the whole thing up, pitch his crown into the sewer, and run away to Scotland. There are the facts. Are they correct?"

"Yes, they are correct."

"Then it is as I have said: one needs but to add them together in order to realize what they mean."

She asked, in an ordinary, level tone—

"What—that the case of France is hopeless?"

"Necessarily. In face of these facts, doubt of it is impossible."

"How can you say that? How can you *feel* like that?"

"How can I? How could I think or feel in any other way, in the circumstances? Joan, with these fatal figures before you, have you really any hope for France—really and actually?"

"Hope—oh, more than that! France will win her freedom and keep it. Do not doubt it."

It seemed to me that her clear intellect must surely be clouded to-day. It must be so, or she would see that those figures *could* mean only the one thing. Perhaps if I marshalled them again she would see. So I said:

"Joan, your heart, which worships France, is beguiling your head. You are not perceiving the importance of these figures. Here—I want to make a picture of them, here on the ground with a stick. Now, this rough outline is France. Through its middle, east and west, I draw a river."

"Yes; the Loire."

"Now, then, this whole northern half of the country is in the tight grip of the English."

"Yes."

"And this whole southern half is really in nobody's hands at all—as our King confesses by meditating desertion and flight to a foreign land. England has armies here; opposition is dead; she can assume full possession whenever she may choose. In very truth, *all* France is gone, France is already lost, France has ceased to exist. What was France is now but a British province. Is this true?"

Her voice was low, and just touched with emotion, but distinct:

"Yes, it is true."

"Very well. Now add this clinching fact, and surely the sum is complete: When have French soldiers won a victory? Scotch soldiers, under the French flag, have won a barren fight or two a few years back, but I am speaking of French ones. Since eight thousand Englishmen nearly annihilated sixty thousand Frenchmen a dozen years ago at Agincourt, French courage has been paralyzed. And so it is a common saying, to-day, that if you confront fifty French soldiers with five English ones, the French will run."

"It is a pity, but even these things are true."

"Then certainly the day for hoping is *past*."

I believed the case would be clear to her now. I thought it could not fail to be clear to her, and that she would say, herself, that there was no longer any ground for hope. But I was mistaken; and disappointed also. She said, without any doubt in her tone:

"France will rise again. You shall see."

"Rise?—with this burden of English armies on her back!"

"She will cast it off; she will trample it under foot!" This with spirit.

"Without soldiers to fight with?"

"The drums will summon them. They will answer, and they will march."

"March to the rear, as usual?"

"No; to the front—ever to the front—always to the front! You shall see."

"And the pauper King?"

"He will mount his throne—he will wear his crown."

"Well, of a truth this makes one's head dizzy. Why, if I could believe that in thirty years from now the English domination would be broken and the French monarch's head find itself hooped with a real crown of sovereignty—"

"Both will have happened before two years are sped."

"Indeed? and who is going to perform all these sublime impossibilities?"

"God."

It was a reverent low note, but it rang clear.

What could have put those strange ideas in her head? This question kept running in my mind during two or three days. It was inevitable that I should think of madness. What other way was there to account for such things? Grieving and brooding over the woes of France had weakened that strong mind, and filled it with fantastic phantoms—yes, that must be it.

But I watched her, and tested her, and it was not so. Her eye was clear and sane, her ways were natural, her speech direct and to the point. No, there was nothing the matter with her mind; it was still the soundest in the village and the best. She went on thinking for others, planning for others, sacrificing herself for others, just as always before. She went on ministering to her sick and to her poor, and still stood ready to give the wayfarer her bed and content herself with the floor. There was a secret somewhere, but madness was not the key to it. This was plain.

Now the key did presently come into my hands, and the way that it happened was this. You have heard all the world talk of this matter which I am about to speak of, but you have not heard an eye-witness talk of it before.

I was coming from over the ridge, one day—it was the 15th of May, '28—and when I got to the edge of the oak forest and was about to step out of it upon the turfy open space in which the haunted beech-tree stood, I happened to cast a glance from cover, first—then I took a step backward, and stood in the shelter and concealment of the foliage. For I had caught sight of Joan, and thought I would devise some sort of playful surprise for her. Think of it—that trivial conceit was neighbor, with but a scarcely measurable interval of time between, to an event destined to endure forever in histories and songs.

The day was overcast, and all that grassy space wherein the Tree stood lay in a soft rich shadow. Joan sat on a natural seat formed by gnarled great roots of the Tree. Her hands lay loosely, one reposing in the other, in her lap. Her head was bent a little toward the ground, and her air was that of one who is lost in thought, steeped in dreams, and not conscious of herself or of the world. And now I saw a most strange thing, for I saw a *white* shadow come slowly gliding along the grass toward the Tree. It was of grand proportions —a robed form, with wings—and the whiteness of this shadow was not like any other whiteness that we know of, except it be the whiteness of the lightnings, but even the lightnings are not so intense as it was, for one can look at them without hurt, whereas this brilliancy was so blinding that it pained my eyes and brought the water into them. I uncovered my head, perceiving that I was in the presence of something not of this world. My breath grew faint and difficult, because of the terror and the awe that possessed me.

Another strange thing. The wood had been silent—smitten with that deep stillness which comes when a storm-cloud darkens a forest, and the wild creatures lose heart and are afraid; but now all the birds burst forth in song, and the joy, the rapture, the ecstasy of it was beyond belief; and was so eloquent and so moving, withal, that it was plain it was an act of worship. With the first note of those birds Joan cast herself upon her knees, and bent her head low and crossed her hands upon her breast.

She had not seen the shadow yet. Had the song of the birds told her it was coming? It had that look to me.

Then the like of this must have happened before. Yes, there might be no doubt of that.

The shadow approached Joan slowly; the extremity of it reached her, flowed over her, clothed her in its awful splendor. In that immortal light her face, only humanly beautiful before, became divine; flooded with that transforming glory her mean peasant habit was become like to the raiment of the sun-clothed children of God as we see them thronging the terraces of the Throne in our dreams and imaginings.

Presently she rose and stood, with her head still bowed a little, and with her arms down and the ends of her fingers lightly laced together in front of her; and standing so, all drenched with that wonderful light, and yet apparently not knowing it, she seemed to listen—but I heard nothing. After a little she raised her head, and looked up as one might look up toward the face of a giant, and then clasped her hands and lifted them high, imploringly, and began to plead. I heard some of the words. I heard her say—

"But I am so young! oh, so young to leave my mother and my home, and go out into the strange world to undertake a thing so great! Ah, how can I talk with men, be comrade with men?—soldiers! It would give me over to insult, and rude usage, and contempt. How can I go to the great wars, and lead armies?—I a girl, and ignorant of such things, knowing nothing of arms, nor how to mount a horse, nor ride it. . . . Yet—if it is commanded—"

Her voice sank a little, and was broken by sobs, and I made out no more of her words. Then I came to myself. I reflected that I had been intruding upon a mystery of God—and what might my punishment be? I was afraid, and went deeper into the wood. Then I carved a mark in the bark of a tree, saying to myself, it may be that I am dreaming and have not seen this vision at all. I will come again, when I know that I am awake and not dreaming, and see if this mark is still here; then I shall know.

CHAPTER VII

I HEARD my name called. It was Joan's voice. It startled me, for how could she know I was there? I said to myself, it is part of the dream; it is all dream—voice, vision and all; the fairies have done this. So I crossed myself and pronounced the name of God, to break the enchantment. I knew I was awake now and free from the spell, for no spell can withstand this exorcism. Then I heard my name called again, and I stepped at once from under cover, and there indeed was Joan, but not looking as she had looked in the dream. For she was not crying, now, but was looking as she had used to look a year and a half before, when her heart was light and her spirits high. Her old-time energy and fire were back, and a something like exaltation showed itself in her face and bearing. It was almost as if she had been in a trance all that time and had come awake again. Really, it was just as if she had been away and lost, and was come back to us at last; and I was so glad that I felt like running to call everybody and have them flock around her and give her welcome. I ran to her excited, and said—

"Ah, Joan, I've got such a wonderful thing to tell you about! You would never imagine it. I've had a dream, and in the dream I saw you right here where you are standing now, and—"

But she put up her hand and said—

"It was not a dream."

It gave me a shock, and I began to feel afraid again.

"Not a dream?" I said, "how can you know about it, Joan?"

"Are you dreaming now?"

"I—I suppose not. I think I am not."

"Indeed you are not. I know you are not. And you were not dreaming when you cut the mark in the tree."

I felt myself turning cold with fright, for now I knew of a certainty that I had not been dreaming, but had really been in the presence of a dread something not of this world. Then I remembered that my sinful feet were upon holy ground— the ground where that celestial shadow had rested. I moved quickly away, smitten to the bones with fear. Joan followed, and said—

"Do not be afraid; indeed there is no need. Come with me. We will sit by the spring and I will tell you all my secret."

When she was ready to begin, I checked her and said—

"First tell me this. You could not see me in the wood; how did you know I cut a mark in the tree?"

"Wait a little; I will soon come to that; then you will see."

"But tell me one thing now; what was that awful shadow that I saw?"

"I will tell you, but do not be disturbed; you are not in danger. It was the shadow of an archangel—Michael, the chief and lord of the armies of heaven."

I could but cross myself and tremble for having polluted that ground with my feet.

"You were not afraid, Joan? Did you see his face—did you see his form?"

"Yes; I was not afraid, because this was not the first time. I was afraid the first time."

"When was that, Joan?"

"It is nearly three years ago, now."

"So long? Have you seen him many times?"

"Yes, many times."

"It is this, then, that has changed you; it was this that made you thoughtful and not as you were before. I see it now. Why did you not tell us about it?"

"It was not permitted. It is permitted now, and soon I

shall tell all. But only you, now. It must remain a secret a few days still."

"Has none seen that white shadow before but me?"

"No one. It has fallen upon me before when you and others were present, but none could see it. To-day it has been otherwise, and I was told why; but it will not be visible again to any."

"It was a sign to me, then—and a sign with a meaning of some kind?"

"Yes, but I may not speak of that."

" Strange—that that dazzling light could rest upon an object before one's eyes and not be visible."

"With it comes speech, also. Several saints come, attended by myriads of angels, and they speak to me; I hear their voices, but others do not. They are very dear to me—my Voices; that is what I call them to myself."

"Joan, what do they tell you?"

"All manner of things—about France I mean."

"What things have they been used to tell you?"

She sighed, and said—

"Disasters—only disasters, and misfortunes, and humiliations. There was naught else to foretell."

" They spoke of them to you *beforehand?*"

"Yes. So that I knew what was going to happen before it happened. It made me grave—as you saw. It could not be otherwise. But always there was a word of hope, too. More than that: France was to be rescued, and made great and free again. But how and by whom—that was not told. Not until to-day." As she said those last words a sudden deep glow shone in her eyes, which I was to see there many times in after-days when the bugles sounded the charge and learn to call it the battle-light. Her breast heaved, and the color rose in her face. " But to-day I know. God has chosen the meanest of His creatures for this work; and by His command, and in His protection, and by His strength, not mine, I am to lead His armies, and win back France, and set the crown upon the head of His servant that is Dauphin and shall be King."

I was amazed, and said—

"You, Joan? You, a child, lead armies?"

"Yes. For one little moment or two the thought crushed me; for it is as you say—I am only a child; a child and ignorant—ignorant of everything that pertains to war, and not fitted for the rough life of camps and the companionship of soldiers. But those weak moments passed; they will not come again. I am enlisted, I will not turn back, God helping me, till the English grip is loosed from the throat of France. My Voices have never told me lies, they have not lied to-day. They say I am to go to Robert de Baudricourt, governor of Vaucouleurs, and he will give me men-at-arms for escort and send me to the King. A year from now a blow will be struck which will be the beginning of the end, and the end will follow swiftly."

"Where will it be struck?"

"My Voices have not said; nor what will happen this present year, before it is struck. It is appointed me to strike it, that is all I know; and follow it with others, sharp and swift, undoing in ten weeks England's long years of costly labor, and setting the crown upon the Dauphin's head—for such is God's will; my Voices have said it, and shall I doubt it? No; it will be as they have said, for they say only that which is true."

These were tremendous sayings. They were impossibilities to my reason, but to my heart they rang true; and so, while my reason doubted, my heart believed—believed, and held fast to its belief from that day. Presently I said—

"Joan, I believe the things which you have said, and now I am glad that I am to march with you to the great wars—that is, if it is with you I am to march when I go."

She looked surprised, and said—

"It is true that you will be with me when I go to the wars, but how did you know?"

"I shall march with you, and so also will Jean and Pierre, but not Jacques."

"All true—it is so ordered, as was revealed to me lately, but I did not know until to-day that the marching would be

JOAN BEFORE THE GOVERNOR

with me, or that I should march at all. How did you know these things?"

I told her when it was that she had said them. But she did not remember about it. So then I knew that she had been asleep, or in a trance or an ecstasy of some kind, at that time. She bade me keep these and the other revelations to myself for the present, and I said I would, and kept the faith I promised.

None who met Joan that day failed to notice the change that had come over her. She moved and spoke with energy and decision; there was a strange new fire in her eye, and also a something wholly new and remarkable in her carriage and in the set of her head. This new light in the eye and this new bearing were born of the authority and leadership which had this day been vested in her by the decree of God, and they asserted that authority as plainly as speeeh could have done it, yet without ostentation or bravado. This calm consciousness of command, and calm unconscious outward expression of it, remained with her thenceforth until her mission was accomplished.

Like the other villagers, she had always accorded me the deference due my rank; but now, without word said on either side, she and I changed places: she gave orders, not suggestions, I received them with the deference due a superior, and obeyed them without comment. In the evening she said to me—

"I leave before dawn. No one will know it but you. I go to speak with the governor of Vaucouleurs as commanded, who will despise me and treat me rudely, and perhaps refuse my prayer at this time. I go first to Burey, to persuade my uncle Laxart to go with me, it not being meet that I go alone. I may need you in Vaucouleurs; for if the governor will not receive me I will dictate a letter to him, and so must have some one by me who knows the art of how to write and spell the words. You will go from here to-morrow in the afternoon, and remain in Vaucouleurs until I need you."

I said I would obey, and she went her way. You see how clear a head she had, and what a just and level judgment. She did not order me to go with her; no, she would not subject her good name to gossiping remark. She knew that the governor, being a noble, would grant me, another noble, audience; but no, you see, she would not have that, either. A poor peasant girl presenting a petition through a young nobleman — how would that look? She always protected her modesty from hurt; and so, for reward, she carried her good name unsmirched to the end. I knew what I must do, now, if I would have her approval: go to Vaucouleurs, keep out of her sight, and be ready when wanted.

I went the next afternoon, and took an obscure lodging; the next day I called at the castle and paid my respects to the governor, who invited me to dine with him at noon of the following day. He was an ideal soldier of the time; tall, brawny, gray-headed, rough, full of strange oaths acquired here and there and yonder in the wars and treasured as if they were decorations. He had been used to the camp all his life, and to his notion war was God's best gift to man. He had his steel cuirass on, and wore boots that came above his knees, and was equipped with a huge sword; and when I looked at this martial figure, and heard the marvellous oaths, and guessed how little of poetry and sentiment might be looked for in this quarter, I hoped the little peasant girl would not get the privilege of confronting this battery, but would have to content herself with the dictated letter.

I came again to the castle the next day at noon, and was conducted to the great dining-hall and seated by the side of the governor at a small table which was raised a couple of steps higher than the general table. At the small table sat several other guests besides myself, and at the general table sat the chief officers of the garrison. At the entrance door stood a guard of halberdiers, in morion and breastplate.

As for talk, there was but one topic, of course—the desperate situation of France. There was a rumor, some one said, that Salisbury was making preparations to march against Orleans. It raised a turmoil of excited conversation, and opinions fell thick and fast. Some believed he would march at once, others that he could not accomplish the investment before fall, others that the siege would be long, and bravely contested ; but upon one thing all voices agreed : that Orleans must eventually fall, and with it France. With that, the prolonged discussion ended, and there was silence. Every man seemed to sink himself in his own thoughts, and to forget where he was. This sudden and profound stillness where before had been so much animation, was impressive and solemn. Now came a servant and whispered something to the governor, who said—

"Would talk with *me* ?"

"Yes, your Excellency."

"H'm ! A strange idea, certainly. Bring them in."

It was Joan and her uncle Laxart. At the spectacle of the great people the courage oozed out of the poor old peasant and he stopped midway and would come no further, but remained there with his red nightcap crushed in his hands and bowing humbly here, there, and everywhere, stupefied with embarrassment and fear. But Joan came steadily forward, erect and self-possessed, and stood before the governor. She recognized me, but in no way indicated it. There was a buzz of admiration, even the governor contributing to it, for I heard him mutter, "By God's grace, it is a beautiful creature !" He inspected her critically a moment or two, then said—

"Well, what is your errand, my child ?"

"My message is to you, Robert de Baudricourt, governor of Vaucouleurs, and it is this : that you will send and tell the Dauphin to wait and not give battle to his enemies, for God will presently send him help."

This strange speech amazed the company, and many murmured, "The poor young thing is demented." The governor scowled, and said—

"What nonsense is this ? The King—or the Dauphin, as you call him—needs no message of *that* sort. He will wait, give yourself no uneasiness as to that. What further do you desire to say to me?"

"This. To beg that you will give me an escort of men-at-arms and send me to the Dauphin."

"What for?"

"That he may make me his general, for it is appointed that I shall drive the English out of France, and set the crown upon his head."

"What—you? Why, you are but a child!"

"Yet am I appointed to do it, nevertheless."

"Indeed? And when will all this happen?"

"Next year he will be crowned, and after will remain master of France."

There was a great and general burst of laughter, and when it had subsided the governor said—

"Who has sent you with these extravagant messages?"

"My Lord."

"What Lord?"

"The King of Heaven."

Many murmured, "Ah, poor thing, poor thing!" and others, "Ah, her mind is but a wreck!" The governor hailed Laxart, and said—

"Harkye!—take this mad child home and whip her soundly. That is the best cure for her ailment."

As Joan was moving away she turned and said, with simplicity—

"You refuse me the soldiers, I know not why, for it is my Lord that has commanded you. Yes, it is He that has made the command; therefore must I come again, and yet again; then I shall have the men-at-arms."

There was a great deal of wondering talk, after she was gone; and the guards and servants passed the talk to the town, the town passed it to the country; Domremy was already buzzing with it when we got back.

CHAPTER VIII

HUMAN nature is the same everywhere: it deifies success, it has nothing but scorn for defeat. The village considered that Joan had disgraced it with her grotesque performance and its ridiculous failure; so all the tongues were busy with the matter, and as bilious and bitter as they were busy; insomuch that if the tongues had been teeth she would not have survived her persecutions. Those persons who did not scold, did what was worse and harder to bear; for they ridiculed her, and mocked at her, and ceased neither day nor night from their witticisms and jeerings and laughter. Haumette and Little Mengette and I stood by her, but the storm was too strong for her other friends, and they avoided her, being ashamed to be seen with her because she was so unpopular, and because of the sting of the taunts that assailed them on her account. She shed tears in secret, but none in public. In public she carried herself with serenity, and showed no distress, nor any resentment — conduct which should have softened the feeling against her, but it did not. Her father was so incensed that he could not talk in measured terms about her wild project of going to the wars like a man. He had dreamed of her doing such a thing, some time before, and now he remembered that dream with apprehension and anger, and said that rather than see her unsex herself and go away with the armies, he would require her brothers to drown her; and that if they should refuse, he would do it with his own hands.

But none of these things shook her purpose in the least. Her parents kept a strict watch upon her to keep her from leaving the village, but she said her time was not yet; that when the time to go was come she should know it, and then the keepers would watch in vain.

The summer wasted along; and when it was seen that her purpose continued steadfast, the parents were glad of a chance which finally offered itself for bringing her projects to an end through marriage. The Paladin had the effrontery to pretend that she had engaged herself to him several years before, and now he claimed a ratification of the engagement.

She said his statement was not true, and refused to marry him. She was cited to appear before the ecclesiastical court at Toul to answer for her perversity; when she declined to have counsel, and elected to conduct her case herself, her parents and all her ill-wishers rejoiced, and looked upon her as already defeated. And that was natural enough; for who would expect that an ignorant peasant girl of sixteen would be otherwise than frightened and tongue-tied when standing for the first time in presence of the practised doctors of the law, and surrounded by the cold solemnities of a court? Yet all these people were mistaken. They flocked to Toul to see and enjoy this fright and embarrassment and defeat, and they had their trouble for their pains. She was modest, tranquil, and quite at her ease. She called no witnesses, saying she would content herself with examining the witnesses for the prosecution. When they had testified, she rose and reviewed their testimony in a few words, pronounced it vague, confused, and of no force, then she placed the Paladin again on the stand and began to search him. His previous testimony went rag by rag to ruin under her ingenious hands, until at last he stood bare, so to speak, he that had come so richly clothed in fraud and falsehood. His counsel began an argument, but the court declined to hear it, and threw out the case, adding a few words of grave compliment for Joan, and referring to her as "this marvellous child."

After this victory, with this high praise from so imposing a source added, the fickle village turned again, and gave Joan countenance, compliment, and peace. Her mother took her back to her heart, and even her father relented and said he was proud of her. But the time hung heavy on her hands, nevertheless, for the siege of Orleans was begun, the clouds lowered darker and darker over France, and still her Voices said wait, and gave her no direct commands. The winter set in, and wore tediously along ; but at last there was a change.

Book II

IN COURT AND CAMP

CHAPTER I

The 5th of January, 1429, Joan came to me with her uncle Laxart, and said—

"The time is come. My Voices are not vague, now, but clear, and they have told me what to do. In two months I shall be with the Dauphin."

Her spirits were high, and her bearing martial. I caught the infection and felt a great impulse stirring in me that was like what one feels when he hears the roll of the drums and the tramp of marching men.

"I believe it," I said.

"I also believe it," said Laxart. "If she had told me before, that she was commanded of God to rescue France, I should not have believed; I should have let her seek the governor by her own ways and held myself clear of meddling in the matter, not doubting she was mad. But I have seen her stand before those nobles and mighty men unafraid, and say her say; and she had not been able to do that but by the help of God. That, I know. Therefore with all humbleness I am at her command, to do with me as she will."

"My uncle is very good to me," Joan said. "I sent and asked him to come and persuade my mother to let him take me home with him to tend his wife, who is not well. It is arranged, and we go at dawn to-morrow. From his house I shall go soon to Vaucouleurs, and wait and strive until my prayer is granted. Who were the two cavaliers who sat to your left at the governor's table that day?"

"One was the Sieur Jean de Novelonpont de Metz, the other the Sieur Bertrand de Poulengy."

"Good metal—good metal, both. I marked them for men of mine. . . . What is it I see in your face? Doubt?"

I was teaching myself to speak the truth to her, not trimming it or polishing it; so I said—

"They considered you out of your head, and said so. It is true they pitied you for being in such misfortune, but still they held you to be mad."

This did not seem to trouble her in any way or wound her. She only said—

"The wise change their minds when they perceive that they have been in error. These will. They will march with me. I shall see them presently. . . . You seem to doubt again? Do you doubt?"

"N-no. Not now. I was remembering that it was a year ago, and that they did not belong there, but only chanced to stop a day on their journey."

"They will come again. But as to matters now in hand; I came to leave with you some instructions. You will follow me in a few days. Order your affairs, for you will be absent long."

"Will Jean and Pierre go with me?"

"No; they would refuse now, but presently they will come, and with them they will bring my parents' blessing, and likewise their consent that I take up my mission. I shall be stronger, then—stronger for that; for lack of it I am weak, now." She paused a little while, and the tears gathered in her eyes; then she went on: "I would say good-by to Little Mengette. Bring her outside the village at dawn; she must go with me a little of the way—"

"And Haumette?"

She broke down and began to cry, saying—

"No, oh, no—she is too dear to me, I could not bear it, knowing I should never look upon her face again."

Next morning I brought Mengette, and we four walked along the road in the cold dawn till the village was far behind; then the two girls said their good-byes, clinging about each other's neck, and pouring out their grief in loving words

THE GOVERNOR KEEPS HIS PROMISE TO JOAN

and tears, a pitiful sight to see. And Joan took one long look back upon the distant village, and the Fairy Tree, and the oak forest, and the flowery plain, and the river, as if she was trying to print these scenes on her memory so that they would abide there always and not fade, for she knew she would not see them any more in this life; then she turned, and went from us, sobbing bitterly. It was her birthday and mine. She was seventeen years old.

CHAPTER II

AFTER a few days, Laxart took Joan to Vaucouleurs, and found lodging and guardianship for her with Catherine Royer, a wheelwright's wife, an honest and good woman. Joan went to mass regularly, she helped do the house-work, earning her keep in that way, and if any wished to talk with her about her mission—and many did—she talked freely, making no concealments regarding the matter now. I was soon housed near by, and witnessed the effects which followed. At once the tidings spread that a young girl was come who was appointed of God to save France. The common people flocked in crowds to look at her and speak with her, and her fair young loveliness won the half of their belief, and her deep earnestness and transparent sincerity won the other half. The well-to-do remained away and scoffed, but that is their way.

Next, a prophecy of Merlin's, more than eight hundred years old, was called to mind, which said that in a far future time France would be lost by a woman and restored by a woman. France was now, for the first time, lost—and by a woman, Isabel of Bavaria, her base Queen; doubtless this fair and pure young girl was commissioned of Heaven to complete the prophecy.

This gave the growing interest a new and powerful impulse; the excitement rose higher and higher, and hope and faith along with it; and so from Vaucouleurs wave after wave of this inspiring enthusiasm flowed out over the land, far and wide, invading all the villages and refreshing and revivifying the perishing children of France; and from these villages came people who wanted to see for themselves, hear for themselves; and they did see and hear, and believe. They filled the town; they more than filled it; inns and lodgings were packed, and yet half of the inflow had to go without shelter. And still they came, winter as it was, for when a man's soul is starving, what does he care for meat and roof so he can but get that nobler hunger fed? Day after day, and still day after day, the great tide rose. Domremy was dazed, amazed, stupefied, and said to itself, "Was this world-wonder in our familiar midst all these years and we too dull to see it?" Jean and Pierre went out from the village stared at and envied like the great and fortunate of the earth, and their progress to Vaucouleurs was like a triumph, all the country-side flocking to see and salute the brothers of one with whom angels had spoken face to face, and into whose hands by command of God they had delivered the destinies of France.

The brothers brought the parents' blessing and Godspeed to Joan, and their promise to bring it to her in person later; and so, with this culminating happiness in her heart and the high hope it inspired, she went and confronted the governor again. But he was no more tractable than he had been before. He refused to send her to the King. She was disappointed, but in no degree discouraged. She said—

"I must still come to you until I get the men-at-arms; for so it is commanded, and I may not disobey. I must go to the Dauphin, though I go on my knees."

I and the two brothers were with Joan daily, to see the people that came and hear what they said; and one day, sure enough, the Sieur Jean de Metz came. He talked with her in a petting and playful way, as one talks with children, and said—

"What are you doing here, my little maid? Will they drive the King out of France, and shall we all turn English?"

She answered him in her tranquil, serious way—

"I am come to bid Robert de Baudricourt take or send me to the King, but he does not heed my words."

"Ah, you have an admirable persistence, truly; a whole year has not turned you from your wish. I saw you when you came before."

Joan said, as tranquilly as before—

"It is not a wish, it is a purpose. He will grant it. I can wait."

"Ah, perhaps it will not be wise to make too sure of *that*, my child. These governors are stubborn people to deal with. In case he shall not grant your prayer—"

"He will grant it. He must. It is not matter of choice."

The gentleman's playful mood began to disappear—one could see that, by his face. Joan's earnestness was affecting him. It always happened that people who began in jest with her, ended by being in earnest. They soon began to perceive depths in her that they had not suspected ; and then her manifest sincerity and the rocklike steadfastness of her convictions were forces which cowed levity, and it could not maintain its self-respect in their presence. The Sieur de Metz was thoughtful for a moment or two, then he began, quite soberly—

"Is it necessary that you go to the King soon ?—that is, I mean—"

"Before Mid-Lent, even though I wear away my legs to the knees !"

She said it with that sort of repressed fieriness that means so much when a person's heart is in a thing. You could see the response in that nobleman's face ; you could see his eye light up ; there was sympathy there. He said, most earnestly—

"God knows I think you should have the men-at-arms, and that somewhat would come of it. What is it that you would do ? What is your hope and purpose ?"

"To rescue France. And it is appointed that I shall do it. For no one else in the world, neither kings, nor dukes, nor any other, can recover the kingdom of France, and there is no help but in me."

The words had a pleading and pathetic sound, and they touched that good nobleman. I saw it plainly. Joan dropped her voice a little, and said : "But indeed I would rather spin with my poor mother, for this is not my calling ; but I must go and do it, for it is my Lord's will."

"Who is your Lord?"

"He is God."

Then the Sieur de Metz, following the impressive old feudal fashion, knelt and laid his hands within Joan's, in sign of fealty, and made oath that by God's help he himself would take her to the King.

The next day came the Sieur Bertrand de Poulengy, and he also pledged his oath and knightly honor to abide with her and follow whithersoever she might lead.

This day, too, toward evening, a great rumor went flying abroad through the town—namely, that the very governor himself was going to visit the young girl in her humble lodgings. So in the morning the streets and lanes were packed with people waiting to see if this strange thing would indeed happen. And happen it did. The governor rode in state, attended by his guards, and the news of it went everywhere, and made a great sensation, and modified the scoffings of the people of quality and raised Joan's credit higher than ever.

The governor had made up his mind to one thing: Joan was either a witch or a saint, and he meant to find out which it was. So he brought a priest with him to exorcise the devil that was in her in case there was one there. The priest performed his office, but found no devil. He merely hurt Joan's feelings and offended her piety without need, for he had already confessed her before this, and should have known, if he knew anything, that devils cannot abide the confessional, but utter cries of anguish and the most profane and furious cursings whenever they are confronted with that holy office.

The governor went away troubled and full of thought, and not knowing what to do. And while he pondered and studied, several days went by and the 14th of February was come. Then Joan went to the castle and said—

"In God's name, Robert de Baudricourt, you are too slow about sending me, and have caused damage thereby, for this day the Dauphin's cause has lost a battle near Orleans, and will suffer yet greater injury if you do not send me to him soon."

The governor was perplexed by this speech, and said—

" To-day, child, *to-day?* How can you know what has happened in that region to-day? It would take eight or ten days for the word to come."

" My Voices have brought the word to me, and it is true. A battle was lost to-day, and you are in fault to delay me so."

The governor walked the floor a while, talking within himself, but letting a great oath fall outside now and then; and finally he said—

" Harkye! go in peace, and wait. If it shall turn out as you say, I will give you the letter and send you to the King, and not otherwise."

Joan said with fervor—

" Now God be thanked, these waiting days are almost done. In nine days you will fetch me the letter."

Already the people of Vaucouleurs had given her a horse and had armed and equipped her as a soldier. She got no chance to try the horse and see if she could ride it, for her great first duty was to abide at her post and lift up the hopes and spirits of all who would come to talk with her, and prepare them to help in the rescue and regeneration of the kingdom. This occupied every waking moment she had. But it was no matter. There was nothing she could not learn—and in the briefest time, too. Her horse would find this out in the first hour. Meantime the brothers and I took the horse in turn and began to learn to ride. And we had teaching in the use of the sword and other arms, also.

On the 20th Joan called her small army together—the two knights and her two brothers and me—for a private council of war. No, it was not a council, that is not the right name, for she did not consult with us, she merely gave us orders. She mapped out the course she would travel toward the King, and did it like a person perfectly versed in geography; and this itinerary of daily marches was so arranged as to avoid here and there peculiarly dangerous regions by flank movements—which showed that she knew her political geography as intimately as she knew her physical geography; yet she had never had a day's schooling, of course, and was without education. I was astonished, but thought her Voices must have taught her. But upon reflection I saw that this was not so. By her references to what this and that and the other person had told her, I perceived that she had been diligently questioning those crowds of visiting strangers, and that out of them she had patiently dug all this mass of invaluable knowledge. The two knights were filled with wonder at her good sense and sagacity.

She commanded us to make preparations to travel by night and sleep by day in concealment, as almost the whole of our long journey would be through the enemy's country.

Also, she commanded that we should keep the date of our departure a secret, since she meant to get away unobserved. Otherwise we should be sent off with a grand demonstration which would advertise us to the enemy, and we should be ambushed and captured somewhere. Finally she said—

"Nothing remains, now, but that I confide to you the date of our departure, so that you may make all needful preparation in time, leaving nothing to be done in haste and badly at the last moment. We march the 23d, at eleven of the clock at night."

Then we were dismissed. The two knights were startled—yes, and troubled; and the Sieur Bertrand said—

" Even if the governor shall really furnish the letter and the escort, he still may not do it in time to meet the date she has chosen. Then how can she venture to name that date? It is a great risk—a great risk to select and decide upon the date, in this state of uncertainty."

I said—

"Since she has named the 23d, we may trust her. The Voices have told her, I think. We shall do best to obey."

We did obey. Joan's parents were notified to come before the 23d, but prudence forbade that they be told why this limit was named.

All day, the 23d, she glanced up wistfully whenever new

bodies of strangers entered the house, but her parents did not appear. Still she was not discouraged, but hoped on. But when night fell, at last, her hopes perished, and the tears came ; however, she dashed them away, and said—

"It was to be so, no doubt ; no doubt it was so ordered ; I must bear it, and will."

De Metz tried to comfort her by saying—

"The governor sends no word ; it may be that they will come to-morrow, and—"

He got no further, for she interrupted him, saying—

"To what good end ? We start at eleven to-night."

And it was so. At ten the governor came, with his guard and torch-bearers, and delivered to her a mounted escort of men-at-arms, with horses and equipments for me and for the brothers, and gave Joan a letter to the King. Then he took off his sword, and belted it about her waist with his own hands, and said—

"You said true, child. The battle *was* lost, on the day you said. So I have kept my word. Now go—come of it what may."

Joan gave him thanks, and he went his way.

The lost battle was the famous disaster that is called in history the Battle of the Herrings.

All the lights in the house were at once put out, and a little while after, when the streets had become dark and still, we crept stealthily through them and out at the western gate and rode away under whip and spur.

CHAPTER III

WE were twenty-five strong, and well equipped. We rode in double file, Joan and her brothers in the centre of the column, with Jean de Metz at the head of it and the Sieur Bertrand at its extreme rear. The knights were so placed to prevent desertions—for the present. In two or three hours we should be in the enemy's country, and then none would venture to desert. By-and-by we began to hear groans and sobs and execrations from different points along the line, and upon inquiry found that six of our men were peasants who had never ridden a horse before, and were finding it very difficult to stay in their saddles, and moreover were now beginning to suffer considerable bodily torture. They had been seized by the governor at the last moment and pressed into the service to make up the tale, and he had placed a veteran alongside of each with orders to help him stick to the saddle, and kill him if he tried to desert.

These poor devils had kept quiet as long as they could, but their physical miseries were become so sharp by this time that they were obliged to give them vent. But we were within the enemy's country now, so there was no help for them, they must continue the march, though Joan said that if they chose to take the risk they might depart. They preferred to stay with us. We modified our pace now, and moved cautiously, and the new men were warned to keep their sorrows to themselves and not get the command into danger with their curses and lamentations.

Toward dawn we rode deep into a forest, and soon all but the sentries were sound asleep in spite of the cold ground and the frosty air.

I woke at noon out of such a solid and stupefying sleep that at first my wits were all astray, and I did not know where I was nor what had been happening. Then my senses cleared, and I remembered. As I lay there thinking over the strange events of the past month or two the thought came into my mind, greatly surprising me, that one of Joan's prophecies had failed; for where were Noël and the Paladin, who were to join us at the eleventh hour? By this time, you see, I had gotten used to expecting everything Joan said to come true. So, being disturbed and troubled by these thoughts, I opened my eyes. Well, there stood the Paladin leaning against a tree and looking down on me! How often that happens: you think of a person, or speak of a person, and there he stands before you, and you not dreaming he is near. It looks as if his *being* near is really the thing that makes you think of him, and not just an accident, as people imagine. Well, be that as it may, there was the Paladin, anyway, looking down in my face and waiting for me to wake. I was ever so glad to see him, and jumped up and shook him by the hand, and led him a little way from the camp—he limping like a cripple —and told him to sit down, and said—

" Now, where have you dropped down from? And how did you happen to light in this place? And what do the soldier-clothes mean? Tell me all about it."

He answered—

" I marched with you last night."

" No!" (To myself I said, " The prophecy has not all failed—half of it has come true.")

" Yes, I did. I hurried up from Domremy to join, and was within a half a minute of being too late. In fact, I was too late, but I begged so hard that the governor was touched by my brave devotion to my country's cause — those are the words he used—and so he yielded, and allowed me to come."

I thought to myself, this is a lie, he is one of those six the governor recruited by force at the last moment; I know it, for Joan's prophecy said he would join at the eleventh hour, but not by his own desire. Then I said aloud—

"I am glad you came ; it is a noble cause, and one should not sit at home in times like these."

"Sit at home ! I could no more do it than the thunder-stone could stay hid in the clouds when the storm calls it."

"That is the right talk. It sounds like you."

That pleased him.

"I'm glad you know me. Some don't. But they will, presently. They will know me well enough before I get done with this war."

"That is what I think. I believe that wherever danger confronts you you will make yourself conspicuous."

He was charmed with this speech, and it swelled him up like a bladder. He said—

"If I know myself—and I think I do—my performances in this campaign will give you occasion more than once to remember those words."

"I were a fool to doubt it. That, I know."

"I shall not be at my best, being but a common soldier; still, the country will hear of me. If I were where I belong; if I were in the place of La Hire, or Saintrailles, or the Bastard of Orleans—well, I say nothing, I am not of the talking kind, like Noël Rainguesson and his sort, I thank God. But it will be *something*, I take it—a novelty in this world, I should say—to raise the fame of a private soldier above theirs, and extinguish the glory of their names with its shadow."

"Why, look here, my friend," I said, "do you know that you have hit out a most remarkable idea there? Do you realize the gigantic proportions of it? For look you: to be a general of vast renown, what is that? Nothing—history is clogged and confused with them; one cannot keep their names in his memory, there are so many. But a common soldier of supreme renown—why, he would stand alone ! He would be the one moon in a firmament of mustard-seed stars; his name would outlast the human race ! My friend, who gave you that idea ?"

He was ready to burst with happiness, but he suppressed

6

betrayal of it as well as he could. He simply waved the compliment aside with his hand and said, with complacency—

"It is nothing. I have them often—ideas like that—and even greater ones. I do not consider this one much."

"You astonish me; you do indeed. So it is really your own?"

"Quite. And there is plenty more where it came from"— tapping his head with his finger, and taking occasion at the same time to cant his morion over his right ear, which gave him a very self-satisfied air—"I do not need to borrow my ideas, like Noël Rainguesson."

"Speaking of Noël, when did you see him last?"

"Half an hour ago. He is sleeping yonder like a corpse. Rode with us last night."

I felt a great upleap in my heart, and said to myself, now I am at rest and glad; I will never doubt her prophecies again. Then I said aloud—

"It gives me joy. It makes me proud of our village. There is no keeping our lion-hearts at home in these great times, I see that."

"Lion-heart! Who—that baby? Why, he begged like a dog to be let off. Cried, and said he wanted to go to his mother. Him a lion-heart!—that tumble-bug!"

"Dear me, why I supposed he volunteered, of course. Didn't he?"

"Oh yes, volunteered the way people do to the headsman. Why, when he found I was coming up from Domremy to volunteer, he asked me to let him come along in my protection, and see the crowds and the excitement. Well, we arrived and saw the torches filing out at the Castle, and ran there, and the governor had him seized, along with four more, and he begged to be let off, and I begged for his place, and at last the governor allowed me to join, but wouldn't let Noël off, because he was disgusted with him he was such a cry-baby. Yes, and much good *he'll* do the King's service: he'll eat for six and run for sixteen. I hate a pigmy with half a heart and nine stomachs!"

THE PALADIN'S APPEARANCE IN CAMP

"Why, this is very surprising news to me, and I am sorry and disappointed to hear it. I thought he was a very manly fellow."

The Paladin gave me an outraged look, and said:

"I don't see how you can talk like that, I'm sure I don't. I don't see how you could have got such a notion. I don't dislike him, and I'm not saying these things out of prejudice, for I don't allow myself to have prejudices against people. I like him, and have always comraded with him from the cradle, but he must allow me to speak my mind about his faults, and I am willing he shall speak his about mine, if I have any. And true enough, maybe I have; but I reckon they'll bear inspection—I have that idea, anyway. A manly fellow! You should have heard him whine and wail and swear, last night, because the saddle hurt him. Why didn't the saddle hurt me? Pooh —I was as much at home in it as if I had been born there. And yet it was the first time I was ever on a horse. All those old soldiers admired my riding; they said they had never seen anything like it. But him—why, they had to hold him on, all the time."

An odor as of breakfast came stealing through the wood; the Paladin unconsciously inflated his nostrils in lustful response, and got up and limped painfully away, saying he must go and look to his horse.

At bottom he was all right and a good-hearted giant, without any harm in him, for it is no harm to bark, if one stops there and does not bite, and it is no harm to be an ass, if one is content to bray and not kick. If this vast structure of brawn and muscle and vanity and foolishness seemed to have a libellous tongue, what of it? There was no malice behind it; and besides, the defect was not of his own creation; it was the work of Noël Rainguesson, who had nurtured it, fostered it, built it up and perfected it, for the entertainment he got out of it. His careless light heart had to have somebody to nag and chaff and make fun of, the Paladin had only needed development in order to meet its requirements, consequently the development was taken in hand and diligently attended to and looked after, gnat-and-bull fashion, for years, to the neglect and damage of far more important concerns. The result was an unqualified success. Noël prized the society of the Paladin above everybody else's ; the Paladin preferred anybody's to Noël's. The big fellow was often seen with the little fellow, but it was for the same reason that the bull is often seen with the gnat.

With the first opportunity, I had a talk with Noël. I welcomed him to our expedition, and said—

" It was fine and brave of you to volunteer, Noël."

His eye twinkled, and he answered—

" Yes, it was rather fine I think. Still, the credit doesn't all belong to me ; I had help."

" Who helped you ?'

" The governor."

" How?"

" Well, I'll tell you the whole thing. I came up from Domremy to see the crowds and the general show, for I hadn't ever had any experience of such things, of course, and this was a great opportunity ; but I hadn't any mind to volunteer. I overtook the Paladin on the road and let him have my company the rest of the way, although he did not want it and said so ; and while we were gawking and blinking in the glare of the governor's torches they seized us and four more and added us to the escort, and that is really how I came to volunteer. But after all, I wasn't sorry, remembering how dull life would have been in the village without the Paladin."

" How did he feel about it? Was he satisfied?"

" I think he was glad."

" Why ?"

" Because he said he wasn't. He was taken by surprise, you see, and it is not likely that he could tell the truth without preparation. Not that he would have prepared, if he had had the chance, for I do not think he would. I am not charging him with that. In the same space of time that he could prepare to speak the truth, he could also prepare to lie ; besides, his judgment would be cool then, and would warn him against fooling with new methods in an emergency. No, I am sure he was glad, because he said he wasn't."

" Do you think he was very glad ?"

" Yes, I know he was. He begged like a slave, and bawled for his mother. He said his health was delicate, and he didn't know how to ride a horse, and knew he couldn't outlive the first march. But really he wasn't looking as delicate as he was feeling. There was a cask of wine there, a proper lift for four men. The governor's temper got afire, and he delivered an oath at him that knocked up the dust where it struck the ground, and told him to shoulder that cask or he would carve him to cutlets and send him home in a basket. The Paladin did it, and that secured his promotion to a privacy in the escort without any further debate."

" Yes, you seem to make it quite plain that he was glad to join—that is, if your premises are right that you start from. How did he stand the march last night ?"

" About as I did. If he made the more noise, it was the privilege of his bulk. We stayed in our saddles because we had help. We are equally lame to-day, and if he likes to sit down, let him ; I prefer to stand."

CHAPTER IV

WE were called to quarters and subjected to a searching inspection by Joan. Then she made a short little talk in which she said that even the rude business of war could be conducted better without profanity and other brutalities of speech than with them, and that she should strictly require us to remember and apply this admonition. She ordered half an hour's horsemanship-drill for the novices then, and appointed one of the veterans to conduct it. It was a ridiculous exhibition, but we learned something, and Joan was satisfied and complimented us. She did not take any instruction herself or go through the evolutions and manœuvres, but merely sat her horse like a martial little statue and looked on. That was sufficient for her, you see. She would not miss or forget a detail of the lesson, she would take it all in with her eye and her mind, and apply it afterward with as much certainty and confidence as if she had already practised it.

We now made three night-marches of twelve or thirteen leagues each, riding in peace and undisturbed, being taken for a roving band of Free Companions. Country folk were glad to have that sort of people go by without stopping. Still, they were very wearing marches, and not comfortable, for the bridges were few and the streams many, and as we had to ford them we found the water dismally cold, and afterward had to bed ourselves, still wet, on the frosty or snowy ground, and get warm as we might and sleep if we could, for it would not have been prudent to build fires. Our energies languished under these hardships and deadly fatigues, but Joan's did not. Her step kept its spring and firmness and her eye its fire. We could only wonder at this, we could not explain it.

But if we had had hard times before, I know not what to call the five nights that now followed, for the marches were as fatiguing, the baths as cold, and we were ambuscaded seven times in addition, and lost two novices and three veterans in the resulting fights. The news had leaked out and gone abroad that the inspired Virgin of Vaucouleurs was making for the King with an escort, and all the roads were being watched now.

These five nights disheartened the command a good deal. This was aggravated by a discovery which Noël made, and which he promptly made known at headquarters. Some of the men had been trying to understand why Joan continued to be alert, vigorous, and confident while the strongest men in the company were fagged with the heavy marches and exposure and were become morose and irritable. There, it shows you how men can have eyes and yet not see. All their lives those men had seen their own womenfolks hitched up with a cow and dragging the plough in the fields while the men did the driving. They had also seen other evidences that women have far more endurance and patience and fortitude than men—but what good had their seeing these things been to them? None. It had taught them nothing. They were still surprised to see a girl of seventeen bear the fatigues of war better than trained veterans of the army. Moreover, they did not reflect that a great soul, with a great purpose, can make a weak body strong and keep it so ; and here was the greatest soul in the universe ; but how could they know that, those dumb creatures ? No, they knew nothing, and their reasonings were of a piece with their ignorance. They argued and discussed among themselves, with Noël listening, and arrived at the decision that Joan was a witch, and had her strange pluck and strength from Satan ; so they made a plan to watch for a safe opportunity and take her life.

To have secret plottings of this sort going on in our midst was a very serious business, of course, and the knights asked

Joan's permission to hang the plotters, but she refused without hesitancy. She said:

"Neither these men nor any others can take my life before my mission is accomplished, therefore why should I have their blood upon my hands? I will inform them of this, and also admonish them. Call them before me."

When they came she made that statement to them in a plain matter-of-fact way, and just as if the thought never entered her mind that any one could doubt it after she had given her word that it was true. The men were evidently amazed and impressed to hear her say such a thing in such a sure and confident way, for prophecies boldly uttered never fall barren on superstitious ears. Yes, this speech certainly impressed them, but her closing remark impressed them still more. It was for the ringleader, and Joan said it sorrowfully—

"It is a pity that you should plot another's death when your own is so close at hand."

That man's horse stumbled and fell on him in the first ford which we crossed that night, and he was drowned before we could help him. We had no more conspiracies.

This night was harassed with ambuscades, but we got through without having any men killed. One more night would carry us over the hostile frontier if we had good luck, and we saw the night close down with a good deal of solicitude. Always before, we had been more or less reluctant to start out into the gloom and the silence to be frozen in the fords and persecuted by the enemy, but this time we were impatient to get under way and have it over, although there was promise of more and harder fighting than any of the previous nights had furnished. Moreover, in front of us about three leagues there was a deep stream with a frail wooden bridge over it, and as a cold rain mixed with snow had been falling steadily all day we were anxious to find out whether we were in a trap or not. If the swollen stream had washed away the bridge, we might properly consider ourselves trapped and cut off from escape.

As soon as it was dark we filed out from the depths of the

forest where we had been hidden and began the march. From the time that we had begun to encounter ambushes Joan had ridden at the head of the column, and she took this post now. By the time we had gone a league the rain and snow had turned to sleet, and under the impulse of the storm-wind it lashed my face like whips, and I envied Joan and the knights, who could close their visors and shut up their heads in their helmets as in a box. Now, out of the pitchy darkness and close at hand, came the sharp command—

"Halt!"

We obeyed. I made out a dim mass in front of us which might be a body of horsemen, but one could not be sure. A man rode up and said to Joan in a tone of reproof—

"Well, you have taken your time, truly. And what have you found out? Is she still behind us, or in front?"

Joan answered in a level voice—

"She is still behind."

This news softened the stranger's tone. He said—

"If you know that to be true, you have not lost your time, Captain. But are you sure? How do you know?"

"Because I have seen her."

"Seen her! Seen the Virgin herself?"

"Yes, I have been in her camp."

"Is it possible! Captain Raymond, I ask you to pardon me for speaking in that tone just now. You have performed a daring and admirable service. Where was she camped?"

"In the forest, not more than a league from here."

"Good! I was afraid we might be still behind her, but now that we know she is behind us, everything is safe. She is our game. We will hang her. You shall hang her yourself. No one has so well earned the privilege of abolishing this pestilent limb of Satan."

"I do not know how to thank you sufficiently. If we catch her, I—"

"If! I will take care of that; give yourself no uneasiness. All I want is just a look at her, to see what the imp is like that has been able to make all this noise, then you and the halter may have her. How many men has she?"

"I counted but eighteen, but she may have had two or three pickets out.".

"Is that all? It won't be a mouthful for my force. Is it true that she is only a girl?"

"Yes; she is not more than seventeen."

"It passes belief! Is she robust, or slender?"

"Slender."

The officer pondered a moment or two, then he said:

"Was she preparing to break camp?"

"Not when I had my last glimpse of her."

"What was she doing?"

"She was talking quietly with an officer."

"Quietly? Not giving orders?"

"No, talking as quietly as we are now."

"That is good. She is feeling a false security. She would have been restless and fussy else—it is the way of her sex when danger is about. As she was making no preparation to break camp,—"

"She certainly was not when I saw her last."

"—and was chatting quietly and at her ease, it means that this weather is not to her taste. Night-marching in sleet and wind is not for chits of seventeen. No; she will stay where she is. She has my thanks. We will camp, ourselves; here is as good a place as any. Let us get about it."

"If you command it—certainly. But she has two knights with her. They might force her to march, particularly if the weather should improve."

I was scared, and impatient to be getting out of this peril, and it distressed and worried me to have Joan apparently set herself to work to make delay and increase the danger—still, I thought she probably knew better than I what to do. The officer said—

"Well, in that case we are here to block the way."

"Yes, if they come this way. But if they should send out spies, and find out enough to make them want to try for the bridge through the woods? Is it best to allow the bridge to stand?"

It made me shiver to hear her.

The officer considered a while, then said:

"It might be well enough to send a force to destroy the bridge. I was intending to occupy it with the whole command, but that is not necessary now."

Joan said, tranquilly—

"With your permission, I will go and destroy it myself."

Ah, now I saw her idea, and was glad she had had the cleverness to invent it and the ability to keep her head cool and think of it in that tight place. The officer replied—

"You have it, Captain, and my thanks. With you to do it, it will be well done; I could send another in your place, but not a better."

They saluted, and we moved forward. I breathed freer. A dozen times I had imagined I heard the hoof-beats of the real Captain Raymond's troop arriving behind us, and had been sitting on pins and needles all the while that that conversation was dragging along. I breathed freer, but was still not comfortable, for Joan had given only the simple command, "Forward!" Consequently we moved in a walk. Moved in a dead walk past a dim and lengthening column of enemies at our side. The suspense was exhausting, yet it lasted but a short while, for when the enemy's bugles sang the "Dismount!" Joan gave the word to trot, and that was a great relief to me. She was always at herself, you see. Before the command to dismount had been given, somebody might have wanted the countersign somewhere along that line if we came flying by at speed, but now we seemed to be on our way to our allotted camping position, so we were allowed to pass unchallenged. The further we went the more formidable was the strength revealed by the hostile force. Perhaps it was only a hundred or two, but to me it seemed a thousand. When we passed the last of these people I was thankful, and the deeper we ploughed into the darkness beyond them the better I felt. I came nearer and nearer to feeling good, for an hour; then we found the bridge still standing, and I felt entirely good. We crossed it and destroyed it, and then I felt—but I cannot describe what I felt. One has to feel it himself in order to know what it is like.

We had expected to hear the rush of a pursuing force behind us, for we thought that the real Captain Raymond would arrive and suggest that perhaps the troop that had been mistaken for his belonged to the Virgin of Vaucouleurs; but he must have been delayed seriously, for when we resumed our march beyond the river there were no sounds behind us except those which the storm was furnishing.

I said that Joan had harvested a good many compliments intended for Captain Raymond, and that he would find nothing of a crop left but a dry stubble of reprimands when he got back, and a commander just in the humor to superintend the gathering of it in.

Joan said:

"It will be as you say, no doubt; for the commander took a troop for granted, in the night and unchallenged, and would have camped without sending a force to destroy the bridge if he had been left unadvised, and none are so ready to find fault with others as those who do things worthy of blame themselves."

The Sieur Bertrand was amused at Joan's naïve way of referring to her advice as if it had been a valuable present to a hostile leader who was saved by it from making a censurable blunder of omission, and then he went on to admire how ingeniously she had deceived that man and yet had not told him anything that was not the truth. This troubled Joan, and she said—

"I thought he was deceiving himself. I forbore to tell him lies, for that would have been wrong; but if my truths deceived him, perhaps that made them lies, and I am to blame. I would God I knew if I have done wrong."

She was assured that she had done right, and that in the perils and necessities of war deceptions that help one's own cause and hurt the enemy's were always permissible; but she

JOAN REPRIMANDS THE CONSPIRATORS

was not quite satisfied with that, and thought that even when a great cause was in danger one ought to have the privilege of trying honorable ways first. Jean said—

"Joan, you told us yourself that you were going to Uncle Laxart's to nurse his wife, but you didn't say you were going further, yet you did go on to Vaucouleurs. There !"

"I see, now," said Joan, sorrowfully, "I told no lie, yet I deceived. I had tried all other ways first, but I could not get away, and I *had* to get away. My mission required it. I did wrong, I think, and am to blame."

She was silent a moment, turning the matter over in her mind, then she added, with quiet decision, "But the thing itself was right, and I would do it again."

It seemed an over-nice distinction, but nobody said anything. If we had known her as well as she knew herself, and as her later history revealed her to us, we should have perceived that she had a clear meaning there, and that her position was not identical with ours, as we were supposing, but occupied a higher plane. She would sacrifice herself— and her *best* self; that is, her truthfulness — to save her cause ; but only that : she would not buy her *life* at that cost ; whereas our war-ethics permitted the purchase of our lives, or any mere military advantage, small or great, by deception. Her saying seemed a commonplace at that time, the essence of its meaning escaping us ; but one sees, now, that it contained a principle which lifted it above that and made it great and fine.

Presently the wind died down, the sleet stopped falling, and the cold was less severe. The road was become a bog, and the horses labored through it at a walk—they could do no better. As the heavy time wore on, exhaustion overcame us, and we slept in our saddles. Not even the dangers that threatened us could keep us awake.

This tenth night seemed longer than any of the others, and of course it was the hardest, because we had been accumulating fatigue from the beginning, and had more of it on hand now than at any previous time. But we were not molested again.

When the dull dawn came at last we saw a river before us and we knew it was the Loire; we entered the town of Gien, and knew we were in a friendly land, with the hostiles all behind us. That was a glad morning for us.

We were a worn and bedraggled and shabby-looking troop; and still, as always, Joan was the freshest of us all, in both body and spirits. We had averaged above thirteen leagues a night, by tortuous and wretched roads. It was a remarkable march, and shows what men can do when they have a leader with a determined purpose and a resolution that never flags.

CHAPTER V

WE rested and otherwise refreshed ourselves two or three hours at Gien, but by that time the news was abroad that the young girl commissioned of God to deliver France was come; wherefore, such a press of people flocked to our quarters to get sight of her that it seemed best to seek a quieter place; so we pushed on and halted at a small village called Fierbois.

We were now within six leagues of the King, who was at the Castle of Chinon. Joan dictated a letter to him at once, and I wrote it. In it she said she had come a hundred and fifty leagues to bring him good news, and begged the privilege of delivering it in person. She added that although she had never seen him she would know him in any disguise and would point him out.

The two knights rode away at once with the letter. The troop slept all the afternoon, and after supper we felt pretty fresh and fine, especially our little group of young Domremians. We had the comfortable tap-room of the village inn to ourselves, and for the first time in ten unspeakably long days were exempt from bodings and terrors and hardships and fatiguing labors. The Paladin was suddenly become his ancient self again, and was swaggering up and down, a very monument of self-complacency. Noël Rainguesson said—

"I think it is wonderful, the way he has brought us through."

"Who?" asked Joan.

"Why, the Paladin."

The Paladin seemed not to hear.

"What had he to do with it?" asked Pierre d'Arc.

"Everything. It was nothing but Joan's confidence in his discretion that enabled her to keep up her heart. She could depend on us and on herself for valor, but discretion is the winning thing in war, after all; discretion is the rarest and loftiest of qualities, and he has got more of it than any other man in France—more of it, perhaps, than any other sixty men in France."

"Now you are getting ready to make a fool of yourself, Noël Rainguesson," said the Paladin, "and you want to coil some of that long tongue of yours around your neck and stick the end of it in your ear, then you'll be the less likely to get into trouble."

"I didn't know he had more discretion than other people," said Pierre, "for discretion argues brains, and he hasn't any more brains than the rest of us, in my opinion."

"No, you are wrong there. Discretion hasn't anything to do with brains; brains are an obstruction to it, for it does not reason, it feels. Perfect discretion means absence of brains. Discretion is a quality of the heart—solely a quality of the heart; it acts upon us through feeling. We know this because if it were an intellectual quality it would only perceive a danger, for instance, where a danger exists; whereas—"

"Hear him twaddle—the damned idiot!" muttered the Paladin.

"—whereas, it being purely a quality of the heart, and proceeding by feeling, not reason, its reach is correspondingly wider and sublimer, enabling it to perceive and avoid dangers that haven't any existence at all; as for instance that night in the fog, when the Paladin took his horse's ears for hostile lances and got off and climbed a tree—"

"It's a lie! a lie without shadow of foundation, and I call upon you all to beware how you give credence to the malicious inventions of this ramshackle slander-mill that has been doing its best to destroy my character for years, and will grind up your own reputations for you, next. I got off to tighten my saddle-girth—I wish I may die in my tracks if it isn't so—and whoever wants to believe it can, and whoever don't, can let it alone."

"There, that is the way with him, you see ; he never can discuss a theme temperately, but always flies off the handle and becomes disagreeable. And you notice his defect of memory. He remembers getting off his horse, but forgets all the rest, even the tree. But that is natural ; he would re-member getting off the horse because he was so used to doing it. He always did it when there was an alarm and the clash of arms at the front."

"Why did he choose that time for it ?" asked Jean.

"I don't know. To tighten up his girth, *he* thinks, to climb a tree, *I* think ; I saw him climb nine trees in a single night."

"You saw nothing of the kind ! A person that can lie like that deserves no one's respect. I ask you all to answer me. Do you believe what this reptile has said ?"

All seemed embarrassed, and only Pierre replied. He said, hesitatingly—

"I—well, I hardly know what to say. It is a delicate situ-ation. It seems offensive to refuse to believe a person when he makes so direct a statement, and yet I am obliged to say, rude as it may appear, that I am not able to believe the whole of it—no, I am not able to believe that you climbed nine trees."

"There !" cried the Paladin ; "now what do you think of yourself, Noël Rainguesson ? How many do you believe I climbed, Pierre ?"

"Only eight."

The laughter that followed inflamed the Paladin's anger to white heat, and he said—

"I bide my time—I bide my time. I will reckon with you all, I promise you that !"

"Don't get him started," Noël pleaded ; "he is a perfect lion when he gets started. I saw enough to teach me that, after the third skirmish. After it was over I saw him come out of the bushes and attack a dead man single-handed."

7

"It is another lie; and I give you fair warning that you are going too far. You will see me attack a live one if you are not careful."

"Meaning me, of course. This wounds me more than any number of injurious and unkind speeches could do. Ingratitude to one's benefactor—"

"Benefactor? What do I owe you, I should like to know?"

"You owe me your life. I stood between the trees and the foe, and kept hundreds and thousands of the enemy at bay when they were thirsting for your blood. And I did not do it to display my daring, I did it because I loved you and could not live without you."

"There—you have said enough! I will not stay here to listen to these infamies. I can endure your lies, but not your love. Keep that corruption for somebody with a stronger stomach than mine. And I want to say this, before I go. That you people's small performances might appear the better and win you the more glory, I hid my own deeds through all the march. I went always to the front, where the fighting was thickest, to be remote from you, in order that you might not see and be discouraged by the things I did to the enemy. It was my purpose to keep this a secret in my own breast, but you force me to reveal it. If you ask for my witnesses, yonder they lie, on the road we have come. I found that road mud, I paved it with corpses. I found that country sterile, I fertilized it with blood. Time and again I was urged to go to the rear because the command could not proceed on account of my dead. And yet you, you miscreant, accuse me of climbing trees! Pah!"

And he strode out, with a lofty air, for the recital of his imaginary deeds had already set him up again and made him feel good.

Next day we mounted and faced toward Chinon. Orleans was at our back, now, and close by, lying in the strangling grip of the English; soon, please God, we would face about and go to their relief. From Gien the news had spread to

Orleans that the peasant Maid of Vaucouleurs was on her way, divinely commissioned to raise the siege. The news made a great excitement and raised a great hope—the first breath of hope those poor souls had breathed in five months. They sent commissioners at once to the King to beg him to consider this matter, and not throw this help lightly away. These commissioners were already at Chinon by this time.

When we were half-way to Chinon we happened upon yet one more squad of enemies. They burst suddenly out of the woods, and in considerable force, too; but we were not the apprentices we were ten or twelve days before; no, we were seasoned to this kind of adventure now; our hearts did not jump into our throats and our weapons tremble in our hands. We had learned to be always in battle array, always alert, and always ready to deal with any emergency that might turn up. We were no more dismayed by the sight of those people than our commander was. Before they could form, Joan had delivered the order, "Forward!" and we were down upon them with a rush. They stood no chance; they turned tail and scattered, we ploughing through them as if they had been men of straw. That was our last ambuscade, and it was probably laid for us by that treacherous rascal the King's own minister and favorite, De la Tremouille.

We housed ourselves in an inn, and soon the town came flocking to get a glimpse of the Maid.

"Ah, the tedious King and his tedious people! Our two good knights came presently, their patience well wearied, and reported. They and we reverently stood—as becomes persons who are in the presence of Kings and the superiors of Kings—until Joan, troubled by this mark of homage and respect, and not content with it nor yet used to it, although we had not permitted ourselves to do otherwise since the day she prophesied that wretched traitor's death and he was straightway drowned, thus confirming many previous signs that she was indeed an ambassador commissioned of God, commanded us to sit; then the Sieur de Metz said to Joan:

"The King has got the letter, but they will not let us have speech with him."

"Who is it that forbids?"

"None forbids, but there be three or four that are nearest his person—schemers and traitors every one—that put obstructions in the way, and seek all ways, by lies and pretexts, to make delay. Chiefest of these are Georges de la Tremouille and that plotting fox the Archbishop of Rheims. While they keep the King idle and in bondage to his sports and follies, they are great and their importance grows; whereas if ever he assert himself and rise and strike for crown and country like a man, their reign is done. So they but thrive they care not if the crown go to destruction and the King with it."

"You have spoken with others besides these?"

"Not of the Court, no—the Court are the meek slaves of those reptiles, and watch their mouths and their actions, acting as they act, thinking as they think, saying as they say: wherefore they are cold to us, and turn aside and go another way when we appear. But we have spoken with the commissioners from Orleans. They said with heat: 'It is a marvel that any man in such desperate case as is the King can moon around in this torpid way, and see his all go to ruin without lifting a finger to stay the disaster. What a most strange spectacle it is! Here he is, shut up in this wee corner of the realm like a rat in a trap; his royal shelter this huge gloomy tomb of a castle, with wormy rags for upholstery and crippled furniture for use, a very house of desolation; in his treasury forty francs, and not a farthing more, God be witness! no army, nor any shadow of one; and by contrast with this hungry poverty you behold this crownless pauper and his shoals of fools and favorites tricked out in the gaudiest silks and velvets you shall find in any Court in Christendom. And look you, he knows that when our city falls— as fall it surely will except succor come swiftly—*France* falls; he knows that when that day comes he will be an outlaw and a fugitive, and that behind him the English flag will float unchallenged over every acre of his great heritage; he knows these things, he knows that our faithful city is fighting all solitary and alone against disease, starvation, and the sword to stay this awful calamity, yet he will not strike one blow to save her, he will not hear our prayers, he will not even look upon our faces.' That is what the commissioners said, and they are in despair."

Joan said, gently—

"It is pity, but they must not despair. The Dauphin will hear them presently. Tell them so."

She almost always called the King the Dauphin. To her mind he was not King yet, not being crowned.

"We will tell them so, and it will content them, for they believe you come from God. The Archbishop and his confederate have for backer that veteran soldier Raoul de Gaucourt, Grand Master of the Palace, a worthy man but simply a soldier, with no head for any greater matter. He cannot make out to see how a country girl, ignorant of war, can take a sword in her small hand and win victories where the trained generals of France have looked for defeats only, for fifty years—and always found them. And so he lifts his frosty mustache and scoffs."

"When God fights it is but small matter whether the hand that bears His sword is big or little. He will perceive this in time. Is there none in that Castle of Chinon who favors us?"

"Yes, the King's mother-in-law, Yolande, Queen of Sicily, who is wise and good. She spoke with the Sieur Bertrand."

"She favors us, and she hates those others, the King's beguilers," said Bertrand. "She was full of interest, and asked a thousand questions, all of which I answered according to my ability. Then she sat thinking over these replies until I thought she was lost in a dream and would wake no more. But it was not so. At last she said, slowly, and as if she were talking to herself: 'A child of seventeen—a girl—country bred—untaught—ignorant of war, the use of arms, and the conduct of battles—modest, gentle, shrinking—yet throws away her shepherd's crook and clothes herself in steel, and fights her way through a hundred and fifty leagues of hostile territory, never losing heart or hope and never showing fear, and comes—she to whom a king must be a dread and awful presence—and will stand up before such an one and say, Be not afraid, God has sent me to save you! Ah, whence could come a courage and conviction so sublime as this *but* from very God Himself!' She was silent again awhile, thinking, and making up her mind; then she said, 'And whether she comes of God or no, there is that in her heart that raises her above men—high above all men that breathe in France to-day—for in her is that mysterious something that puts heart into soldiers, and turns mobs of cowards into armies of fighters that forget what fear is when they are in that presence—fighters who go into battle with joy in their eyes and songs on their lips, and sweep over the field like a storm—that is the spirit that can save France, and that alone, come it whence it may! It is in her, I do truly believe, for what else could have borne up that child on that great march, and made her despise its dangers and fatigues? The King must see her face to face—and shall!' She dismissed me with those good words, and I know her promise will be kept. They will delay her all they can—those animals—but she will not fail, in the end."

"Would *she* were King!" said the other knight, fervently. "For there is little hope that the King himself can be stirred out of his lethargy. He is wholly without hope, and is only thinking of throwing away everything and flying to some foreign land. The commissioners say there is a spell upon him that makes him hopeless—yes, and that it is shut up in a mystery which they cannot fathom."

"I know the mystery," said Joan, with quiet confidence; "I know it, and he knows it, but no other but God. When I see him I will tell him a secret that will drive away his trouble, then he will hold up his head again."

I was miserable with curiosity to know what it was that she would tell him, but she did not say, and I did not expect she would. She was but a child, it is true; but she was not a chatterer to tell great matters and make herself important to little people ; no, she was reserved, and kept things to herself, as the truly great always do.

The next day Queen Yolande got one victory over the King's keepers, for in spite of their protestations and obstructions she procured an audience for our two knights, and they made the most they could out of their opportunity. They told the King what a spotless and beautiful character Joan was, and how great and noble a spirt animated her, and they implored him to trust in her, believe in her, and have faith that she was sent to save France. They begged him to consent to see her. He was strongly moved to do this, and promised that he would not drop the matter out of his mind, but would consult with his council about it. This began to look encouraging. Two hours later there was a great stir below, and the inn-keeper came flying up to say a commission of illustrious ecclesiastics was come from the King—from the King his very self, understand !—think of this vast honor to his humble little hostelry !—and he was so overcome with the glory of it that he could hardly find breath enough in his excited body to put the facts into words. They were come from the King to speak with the Maid of Vaucouleurs. Then he flew down-stairs, and presently appeared again, backing into the room and bowing to the ground with every step, in front of four imposing and austere bishops and their train of servants.

Joan rose, and we all stood. The bishops took seats, and for a while no word was said, for it was their prerogative to speak first, and they were so astonished to see what a child it was that was making such a noise in the world and degrading personages of their dignity to the base function of ambassadors to her in her plebeian tavern, that they could not find any words to say, at first. Then presently their spokesman told Joan they were aware that she had a message for the King, wherefore she was now commanded to put it into words, briefly and without waste of time or embroideries of speech.

As for me, I could hardly contain my joy—our message was to reach the King at last! And there was the same joy and pride and exultation in the faces of our knights, too, and in those of Joan's brothers. And I knew that they were all praying—as I was—that the awe which we felt in the presence of these great dignitaries, and which would have tied our tongues and locked our jaws, would not affect her in the like degree, but that she would be enabled to word her message well, and with little stumbling, and so make a favorable impression here, where it would be so valuable and so important.

Ah dear, how little we were expecting what happened then! We were aghast to hear her say what she said. She was standing in a reverent attitude, with her head down and her hands clasped in front of her; for she was always reverent toward the consecrated servants of God. When the spokesman had finished, she raised her head and set her calm eye on those faces, not any more disturbed by their state and grandeur than a princess would have been, and said, with all her ordinary simplicity and modesty of voice and manner:

"Ye will forgive me, reverend sirs, but I have no message save for the King's ear alone."

Those surprised men were dumb for a moment, and their faces flushed darkly; then the spokesman said:

"Hark ye, do you fling the King's command in his face and refuse to deliver this message of yours to his servants appointed to receive it?"

"God has appointed one to receive it, and another's commandment may not take precedence of that. I pray you let me have speech of his grace the Dauphin."

"Forbear this folly, and come at your message! Deliver it, and waste no more time about it."

"You err indeed, most reverend fathers in God, and it is not well. I am not come hither to talk, but to deliver Orleans, and lead the Dauphin to his good city of Rheims, and set the crown upon his head."

"Is that the message you send to the King?"

But Joan only said, in the simple fashion which was her wont: "Ye will pardon me for reminding you again—but I have no message to send to any one."

The King's messengers rose in deep anger and swept out of the place without further words, we and Joan kneeling as they passed.

Our countenances were vacant, our hearts full of a sense of disaster. Our precious opportunity was thrown away; we could not understand Joan's conduct, she who had been so wise until this fatal hour. At last the Sieur Bertrand found courage to ask her why she had let this great chance to get her message to the King go by.

"Who sent them here?" she asked.

"The King."

"Who moved the King to send them?" She waited for an answer; none came, for we began to see what was in her mind—so she answered herself: "The Dauphin's council moved him to it. Are they enemies to me and to the Dauphin's weal, or are they friends?"

"Enemies," answered the Sieur Bertrand.

"If one would have a message go sound and ungarbled, does one choose traitors and tricksters to send it by?"

I saw that we had been fools, and she wise. They saw it too, so none found anything to say. Then she went on:

"They had but small wit that contrived this trap. They thought to get my message and seem to deliver it straight, yet deftly twist it from its purpose. You know that one part of my message is but this—to move the Dauphin by argument and reasonings to give me men-at-arms and send me to the siege. If an enemy carried these in the right words, the exact words, and no word missing, yet left out the persuasions of gesture and supplicating tone and beseeching looks that inform the words and make them live, where were the value of that argument—whom could it convince? Be patient, the Dauphin will hear me presently; have no fear."

The Sieur de Metz nodded his head several times, and muttered as to himself:

"She was right and wise, and we are but dull fools, when all is said."

It was just my thought; I could have said it myself; and indeed it was the thought of all there present. A sort of awe crept over us, to think how that untaught girl, taken suddenly and unprepared, was yet able to penetrate the cunning devices of a King's trained advisers and defeat them. Marvelling over this, and astonished at it, we fell silent and spoke no more. We had come to know that she was great in courage, fortitude, endurance, patience, conviction, fidelity to all duties—in all things, indeed, that make a good and trusty soldier and perfect him for his post; now we were beginning to feel that maybe there were greatnesses in her brain that were even greater than these great qualities of the heart. It set us thinking.

What Joan did that day bore fruit the very day after. The King was obliged to respect the spirit of a young girl who could hold her own and stand her ground like that, and he asserted himself sufficiently to put his respect into an act instead of into polite and empty words. He moved Joan out of that poor inn, and housed her, with us her servants, in the Castle of Courdray, personally confiding her to the care of Madame de Bellier, wife of old Raoul de Gaucourt, Master of the Palace. Of course this royal attention had an immediate result: all the great lords and ladies of the Court began to flock there to see and listen to the wonderful girl-soldier that all the world was talking about, and who had answered the King's mandate with a bland refusal to obey. Joan charmed them every one with her sweetness and simplicity and unconscious eloquence, and all the best and capablest among them recognized that there was an indefinable something about her that testified that she was not made of common clay, that she was built on a grander plan than the mass of mankind, and moved on a loftier plane. These spread her fame. She always made friends and advocates that way; neither the high nor the low could come within the sound of her voice and the sight of her face and go out from her presence indifferent.

JOAN DISCOVERS THE DISGUISED KING

CHAPTER VI

WELL, anything to make delay. The King's council advised him against arriving at a decision in our matter too precipitately. *He* arrive at a decision too precipitately! So they sent a committee of priests—always priests—into Lorraine to inquire into Joan's character and history—a matter which would consume several weeks, of course. You see how fastidious they were. It was as if people should come to put out the fire when a man's house was burning down, and they waited till they could send into another country to find out if he had always kept the Sabbath or not, before letting him try.

So the days poked along; dreary for us young people in some ways, but not in all, for we had one great anticipation in front of us; we had never seen a king, and now some day we should have that prodigious spectacle to see and to treasure in our memories all our lives; so we were on the lookout, and always eager and watching for the chance. The others were doomed to wait longer than I, as it turned out. One day great news came—the Orleans commissioners, with Yolande and our knights, had at last turned the council's position and persuaded the King to see Joan.

Joan received the immense news gratefully but without losing her head, but with us others it was otherwise; we could not eat or sleep or do any rational thing for the excitement and the glory of it. During two days our pair of noble knights were in distress and trepidation on Joan's account, for the audience was to be at night, and they were afraid that Joan would be so paralyzed by the glare of light from the long files of torches, the solemn pomps and ceremonies, the great concourse of renowned personages, the brilliant costumes, and the other splendors of the Court, that she, a simple country maid, and all unused to such things, would be overcome by these terrors and make a piteous failure.

No doubt I could have comforted them, but I was not free to speak. Would Joan be disturbed by this cheap spectacle, this tinsel show, with its small King and his butterfly duke-lets?—she who had spoken face to face with the princes of heaven, the familiars of God, and seen their retinue of angels stretching back into the remoteness of the sky, myriads upon myriads, like a measureless fan of light, a glory like the glory of the sun streaming from each of those innumerable heads, the massed radiance filling the deeps of space with a blinding splendor? I thought not.

Queen Yolande wanted Joan to make the best possible impression upon the King and the Court, so she was strenuous to have her clothed in the richest stuffs, wrought upon the princeliest pattern, and set off with jewels; but in that she had to be disappointed, of course, Joan not being persuadable to it, but begging to be simply and sincerely dressed, as became a servant of God, and one sent upon a mission of a serious sort and grave political import. So then the gracious Queen imagined and contrived that simple and witching costume which I have described to you so many times, and which I cannot think of even now in my dull age without being moved just as rhythmical and exquisite music moves one; for *that* was music, that dress—that is what it was—music that one saw with the eyes and felt in the heart. Yes, she was a poem, she was a dream, she was a spirit when she was clothed in that.

She kept that raiment always, and wore it several times upon occasions of state, and it is preserved to this day in the Treasury of Orleans, with two of her swords, and her banner, and other things now sacred because they had belonged to her.

At the appointed time the Count of Vendôme, a great lord of the court, came richly clothed, with his train of servants and assistants, to conduct Joan to the King, and the two knights and I went with her, being entitled to this privilege by reason of our official positions near her person.

When we entered the great audience hall, there it all was, just as I have already painted it. Here were ranks of guards in shining armor and with polished halberds; two sides of the hall were like flower-gardens for variety of color and the magnificence of the costumes; light streamed upon these masses of color from two hundred and fifty flambeaux. There was a wide free space down the middle of the hall, and at the end of it was a throne royally canopied, and upon it sat a crowned and sceptred figure nobly clothed and blazing with jewels.

It is true that Joan had been hindered and put off a good while, but now that she was admitted to an audience at last, she was received with honors granted to only the greatest personages. At the entrance door stood four heralds in a row, in splendid tabards, with long slender silver trumpets at their mouths, with square silken banners depending from them embroidered with the arms of France. As Joan and the Count passed by, these trumpets gave forth in unison one long rich note, and as we moved down the hall under the pictured and gilded vaulting, this was repeated at every fifty feet of our progress—six times in all. It made our good knights proud and happy, and they held themselves erect, and stiffened their stride, and looked fine and soldierly. They were not expecting this beautiful and honorable tribute to our little country maid.

Joan walked two yards behind the Count, we three walked two yards behind Joan. Our solemn march ended when we were as yet some eight or ten steps from the throne. The Count made a deep obeisance, pronounced Joan's name, then bowed again and moved to his place among a group of officials near the throne. I was devouring the crowned personage with all my eyes, and my heart almost stood still with awe.

The eyes of all others were fixed upon Joan in a gaze of

wonder which was half worship, and which seemed to say, "How sweet—how lovely—how divine !" All lips were parted and motionless, which was a sure sign that those people, who seldom forget themselves, had forgotten themselves now, and were not conscious of anything but the one object they were gazing upon. They had the look of people who are under the enchantment of a vision.

Then they presently began to come to life again, rousing themselves out of the spell and shaking it off as one drives away little by little a clinging drowsiness or intoxication. Now they fixed their attention upon Joan with a strong new interest of another sort; they were full of curiosity to see what she would do—they having a secret and particular reason for this curiosity. So they watched. This is what they saw :

She made no obeisance, nor even any slight inclination of her head, but stood looking toward the throne in silence. That was all there was to see, at present.

I glanced up at De Metz, and was shocked at the paleness of his face. I whispered and said—

"What is it man, what is it ?"

His answering whisper was so weak I could hardly catch it—

"They have taken advantage of the hint in her letter to play a trick upon her ! She will err, and they will laugh at her. That is not the King that sits there."

Then I glanced at Joan. She was still gazing steadfastly toward the throne, and I had the curious fancy that even her shoulders and the back of her head expressed bewilderment. Now she turned her head slowly, and her eye wandered along the lines of standing courtiers till it fell upon a young man who was very quietly dressed ; then her face lighted joyously, and she ran and threw herself at his feet, and clasped his knees, exclaiming in that soft melodious voice which was her birthright and was now charged with deep and tender feeling—

"God of his grace give you long life, O dear and gentle Dauphin !"

In his astonishment and exultation De Metz cried out—

"By the shadow of God, it is an amazing thing!" Then he mashed all the bones of my hand in his grateful grip, and added, with a proud shake of his mane, "*Now*, what have these painted infidels to say!"

Meantime the young person in the plain clothes was saying to Joan—

"Ah, you mistake, my child, I am not the King. There he is," and he pointed to the throne.

The knight's face clouded, and he muttered in grief and indignation—

"Ah, it is a shame to use her so. But for this lie she had gone through safe. I will go and proclaim to all the house what—"

"Stay where you are!" whispered I and the Sieur Bertrand in a breath, and made him stop in his place.

Joan did not stir from her knees, but still lifted her happy face toward the King, and said—

"No, gracious liege, you are he, and none other."

De Metz's troubles vanished away, and he said—

"Verily, she was not guessing, she *knew*. Now, how could she know? It is a miracle. I am content, and will meddle no more, for I perceive that she is equal to her occasions, having that in her head that cannot profitably be helped by the vacancy that is in mine."

This interruption of his lost me a remark or two of the other talk; however, I caught the King's next question:

"But tell me who you are, and what would you?"

"I am called Joan the Maid, and am sent to say that the King of Heaven wills that you be crowned and consecrated in your good city of Rheims, and be thereafter Lieutenant of the Lord of Heaven, who is King of France. And He willeth also that you set me at my appointed work and give me men-at-arms." After a slight pause she added, her eye lighting at the sound of her words, "For then will I raise the siege of Orleans and break the English power!"

The young monarch's amused face sobered a little when

this martial speech fell upon that sick air like a breath blown from embattled camps and fields of war, and his trifling smile presently faded wholly away and disappeared. He was grave, now, and thoughtful. After a little he waved his hand lightly and all the people fell away and left those two by themselves in a vacant space. The knights and I moved to the opposite side of the hall and stood there. We saw Joan rise at a sign, then she and the King talked privately together.

All that host had been consumed with curiosity to see what Joan would do. Well, they had seen, and now they were full of astonishment to see that she had really performed that strange miracle according to the promise in her letter; and they were fully as much astonished to find that she was not overcome by the pomps and splendors about her, but was even more tranquil and at her ease in holding speech with a monarch than ever they themselves had been, with all their practice and experience.

As for our two knights, they were inflated beyond measure with pride in Joan, but nearly dumb, as to speech, they not being able to think out any way to account for her managing to carry herself through this imposing ordeal without ever a mistake or an awkwardness of any kind to mar the grace and credit of her great performance.

The talk between Joan and the King was long and earnest, and held in low voices. We could not hear, but we had our eyes and could note effects; and presently we and all the house noted one effect which was memorable and striking, and has been set down in memoires and histories and in testimony at the Process of Rehabilitation by some who witnessed it; for all knew it was big with meaning, though none knew what that meaning was at that time, of course. For suddenly we saw the King shake off his indolent attitude and straighten up like a man, and at the same time look immeasurably astonished. It was as if Joan had told him something almost too wonderful for belief, and yet of a most uplifting and welcome nature.

It was long before we found out the secret of this conversation, but we know it now, and all the world knows it. That part of the talk was like this—as one may read in all histories. The perplexed King asked Joan for a sign. He wanted to believe in her and her mission, and that her Voices were supernatural and endowed with knowledge hidden from mortals, but how could he do this unless these Voices could prove their claim in some absolutely unassailable way? It was then that Joan said—

"I will give you a sign, and you shall no more doubt. There is a secret trouble in your heart which you speak of to none—a doubt which wastes away your courage, and makes you dream of throwing all away and fleeing from your realm. Within this little while you have been praying, in your own breast, that God of his grace would resolve that doubt, even if the doing of it must show you that no kingly right is lodged in you."

It was that that amazed the King, for it was as she had said: his prayer was the secret of his own breast, and none but God could know about it. So he said:

"The sign is sufficient. I know, now, that these Voices are of God. They have said true in this matter; if they have said more, tell it me—I will believe."

"They have resolved that doubt, and I bring their very words, which are these: Thou art lawful heir to the King thy father, and true heir of France. God has spoken it. Now lift up thy head, and doubt no more, but give me men-at-arms and let me get about my work."

Telling him he was of lawful birth was what straightened him up and made a man of him for a moment, removing his doubts upon that head and convincing him of his royal right; and if any could have hanged his hindering and pestiferous council and set him free, he would have answered Joan's prayer and set her in the field. But no, those creatures were only checked, not checkmated; they could invent some more delays.

We had been made proud by the honors which had so distinguished Joan's entrance into that place—honors restrict-

8

ed to personages of very high rank and worth—but that pride was as nothing compared with the pride we had in the honor done her upon leaving it. For whereas those first honors were shown only to the great, these last, up to this time, had been shown only to the royal. The King himself led Joan by the hand down the great hall to the door, the glittering multitude standing and making reverence as they passed, and the silver trumpets sounding those rich notes of theirs. Then he dismissed her with gracious words, bending low over her hand and kissing it. Always—from all companies, high or low—she went forth richer in honor and esteem than when she came.

And the King did another handsome thing by Joan, for he sent us back to Courdray Castle torch-lighted and in state, under escort of his own troop—his guard of honor—the only soldiers he had; and finely equipped and bedizened they were, too, though they hadn't seen the color of their wages since they were children, as a body might say. The wonders which Joan had been performing before the King had been carried all around by this time, so the road was so packed with people who wanted to get a sight of her that we could hardly dig through; and as for talking together, we couldn't, all attempts at talk being drowned in the storm of shoutings and huzzas that broke out all along as we passed, and kept abreast of us like a wave the whole way.

CHAPTER VII

WE were doomed to suffer tedious waits and delays, and we settled ourselves down to our fate and bore it with a dreary patience, counting the slow hours and the dull days and hoping for a turn when God should please to send it. The Paladin was the only exception—that is to say, he was the only one who was happy and had no heavy times. This was partly owing to the satisfaction he got out of his clothes. He bought them when he first arrived. He bought them at second hand—a Spanish cavalier's complete suit, wide-brimmed hat with flowing plumes, lace collar and cuffs, faded velvet doublet and trunks, short cloak hung from the shoulder, funnel-topped buskins, long rapier, and all that—a graceful and picturesque costume, and the Paladin's great frame was the right place to hang it for effect. He wore it when off duty; and when he swaggered by with one hand resting on the hilt of his rapier, and twirling his new mustache with the other, everybody stopped to look and admire; and well they might, for he was a fine and stately contrast to the small French gentleman of the day squeezed into the trivial French costume of the time.

He was king bee of the little village that snuggled under the shelter of the frowning towers and bastions of Courdray Castle, and acknowledged lord of the tap-room of the inn. When he opened his mouth there, he got a hearing. Those simple artisans and peasants listened with deep and wondering interest; for he was a traveller and had seen the world—all of it that lay between Chinon and Domremy, at any rate—and that was a wide stretch more of it than they might ever hope to see; and he had been in battle, and knew how to paint its shock and struggle, its perils and surprises, with an art that was all his own. He was cock of that walk, hero of that hostelry; he drew custom as honey draws flies; so he was the pet of the inn-keeper, and of his wife and daughter, and they were his obliged and willing servants.

Most people who have the narrative gift—that great and rare endowment — have with it the defect of telling their choice things over the same way every time, and this injures them and causes them to sound stale and wearisome after several repetitions; but it was not so with the Paladin, whose art was of a finer sort; it was more stirring and interesting to hear him tell about a battle the tenth time than it was the first time, because he did not tell it twice the same way, but always made a new battle of it and a better one, with more casualties on the enemy's side each time, and more general wreck and disaster all around, and more widows and orphans and suffering in the neighborhood where it happened. He could not tell his battles apart himself, except by their names; and by the time he had told one of them ten times he had to lay it aside and start a new one in its place, because it had grown so that there wasn't room enough in France for it any more, but was lapping over the edges. But up to that point the audience would not allow him to substitute a new battle, knowing that the old ones were the best, and sure to improve as long as France could hold them; and so, instead of saying to him as they would have said to another, "Give us something fresh, we are fatigued with that old thing," they would say, with one voice and with a strong interest, "Tell about the surprise at Beaulieu again—tell it three or four times!" That is a compliment which few narrative experts have heard in their lifetime.

At first when the Paladin heard us tell about the glories of the Royal Audience he was broken-hearted because he was not taken with us to it; next, his talk was full of what he would have done if he had been there; and within two days he was telling what he *did* do when he *was* there. His mill was fairly started, now, and could be trusted to take care of its affair. Within three nights afterwards all his battles were taking a rest, for already his worshippers in the tap-room were so infatuated with the great tale of the Royal Audience that they would have nothing else, and so besotted with it were they that they would have cried if they could not have gotten it.

Noël Rainguesson hid himself and heard it, and came and told me, and after that we went together to listen, bribing the inn hostess to let us have her little private parlor, where we could stand at the wickets in the door and see and hear.

The tap-room was large, yet had a snug and cosey look, with its inviting little tables and chairs scattered irregularly over its red brick floor, and its great fire flaming and crackling in the wide chimney. It was a comfortable place to be in on such chilly and blustering March nights as these, and a goodly company had taken shelter there, and were sipping their wine in contentment and gossiping one with another in a neighborly way while they waited for the historian. The host, the hostess, and their pretty daughter were flying here and there and yonder among the tables and doing their best to keep up with the orders. The room was about forty feet square, and a space or aisle down the centre of it had been kept vacant and reserved for the Paladin's needs. At the end of it was a platform ten or twelve feet wide, with a big chair and a small table on it, and three steps leading up to it.

Among the wine-sippers were many familiar faces: the cobbler, the farrier, the blacksmith, the wheelwright, the armorer, the maltster, the weaver, the baker, the miller's man with his dusty coat, and so on; and conspicuous and important, as a matter of course, was the barber-surgeon, for he is that in all villages. As he has to pull everybody's teeth, and purge and bleed all the grown people once a month to keep their health sound, he knows everybody, and by constant contact with all sorts of folk becomes a master of etiquette and manners and a conversationalist of large facility. There were plenty of carriers, drovers, and their sort, and journeymen artisans.

When the Paladin presently came sauntering indolently in,

he was received with a cheer, and the barber bustled forward and greeted him with several low and most graceful and courtly bows, also taking his hand and touching his lips to it. Then he called in a loud voice for a stoup of wine for the Paladin, and when the host's daughter brought it up on to the platform and dropped her courtesy and departed, the barber called after her, and told her to add the wine to his score. This won him ejaculations of approval, which pleased him very much and made his little rat-eyes shine; and such applause is right and proper, for when we do a liberal and gallant thing it is but natural that we should wish to see notice taken of it.

The barber called upon the people to rise and drink the Paladin's health, and they did it with alacrity and affectionate heartiness, clashing their metal flagons together with a simultaneous crash, and heightening the effect with a resounding cheer. It was a fine thing to see how that young swashbuckler had made himself so popular in a strange land in so little a while, and without other helps to his advancement than just his tongue and the talent to use it given him by God—a talent which was but one talent in the beginning, but was now become ten through husbandry and the increment and usufruct that do naturally follow that and reward it as by a law.

The people sat down and began to hammer on the tables with their flagons and call for "the King's Audience!—the King's Audience!—the King's Audience!" The Paladin stood there in one of his best attitudes, with his plumed great hat tipped over to the left, the folds of his short cloak drooping from his shoulder, and the one hand resting upon the hilt of his rapier and the other lifting his beaker. As the noise died down he made a stately sort of a bow, which he had picked up somewhere, then fetched his beaker with a sweep to his lips and tilted his head back and drained it to the bottom. The barber jumped for it and set it upon the Paladin's table. Then the Paladin began to walk up and down his platform with a great deal of dignity and quite at his ease; and as he walked he talked, and every little while stopped and stood facing his house and so standing continued his talk.

We went three nights in succession. It was plain that there was a charm about the performance that was apart from the mere interest which attaches to lying. It was presently discoverable that this charm lay in the Paladin's sincerity. He was not lying consciously; he believed what he was saying. To him, his initial statements were facts, and whenever he enlarged a statement, the enlargement became a fact too. He put his heart into his extravagant narrative, just as a poet puts his heart into a heroic fiction, and his earnestness disarmed criticism—disarmed it as far as he himself was concerned. Nobody believed his narrative, but all believed that he believed it.

He made his enlargements without flourish, without emphasis, and so casually that often one failed to notice that a change had been made. He spoke of the governor of Vaucouleurs, the first night, simply as the governor of Vaucouleurs; he spoke of him the second night as his uncle the governor of Vaucouleurs; the third night he was his father. He did not seem to know that he was making these extraordinary changes; they dropped from his lips in a quite natural and effortless way. By his first night's account the governor merely attached him to the Maid's military escort in a general and unofficial way; the second night his uncle the governor sent him with the Maid as lieutenant of her rear guard; the third night his father the governor put the whole command, Maid and all, in his especial charge. The first night the governor spoke of him as a youth without name or ancestry, but "destined to achieve both"; the second night his uncle the governor spoke of him as the latest and worthiest lineal descendant of the chiefest and noblest of the Twelve Paladins of Charlemagne; the third night he spoke of him as the lineal descendant of the whole dozen. In three nights he promoted the Count of Vendôme from a fresh acquaintance to schoolmate, and then brother-in-law.

At the King's Audience everything grew, in the same way.

First the four silver trumpets were twelve, then thirty-five, finally ninety-six; and by that time he had thrown in so many drums and cymbals that he had to lengthen the hall from five hundred feet to nine hundred to accommodate them. Under his hand the people present multiplied in the same large way.

The first two nights he contented himself with merely describing and exaggerating the chief dramatic incident of the Audience, but the third night he added illustration to description. He throned the barber in his own high chair to represent the sham King; then he told how the Court watched the Maid with intense interest and suppressed merriment, expecting to see her fooled by the deception and get herself swept permanently out of credit by the storm of scornful laughter which would follow. He worked this scene up till he got his house in a burning fever of excitement and anticipation, then came his climax. Turning to the barber, he said:

"But mark you what she did. She gazed steadfastly upon that sham's villain face as I now gaze upon yours—this being her noble and simple attitude, just as I stand now—then turned she—thus—to me, and stretching her arm out—so—and pointing with her finger, she said, in that firm, calm tone which she was used to use in directing the conduct of a battle, 'Pluck me this false knave from the throne!' I, striding forward as I do now, took him by the collar and lifted him out and held him aloft—thus—as if he had been but a child." (The house rose, shouting, stamping, and banging with their flagons, and went fairly mad over this magnificent exhibition of strength—and there was not the shadow of a laugh anywhere, though the spectacle of the limp but proud barber hanging there in the air like a puppy held by the scruff of its neck was a thing that had nothing of solemnity about it.) "Then I set him down upon his feet—thus—being minded to get him by a better hold and heave him out of the window, but she bid me forbear, so by that error he escaped with his life.

"Then she turned her about and viewed the throng with those eyes of hers, which are the clear-shining windows whence her immortal wisdom looketh out upon the world, re-

THE EXAMINATION OF JOAN

solving its falsities and coming at the kernel of truth that is hid within them, and presently they fell upon a young man modestly clothed, and him she proclaimed for what he truly was, saying, 'I am thy servant—thou art the King!' Then all were astonished, and a great shout went up, the whole six thousand joining in it, so that the walls rocked with the volume and the tumult of it."

He made a fine and picturesque thing of the march-out from the Audience, augmenting the glories of it to the last limit of the impossibilities; then he took from his finger and held up a brass nut from a bolt-head which the head-ostler at the castle had given him that morning, and made his conclusion—thus:

"Then the King dismissed the Maid most graciously—as indeed was her desert—and turning to me, said, 'Take this signet-ring, son of the Paladins, and command me with it in your day of need; and look you,' said he, touching my temple, 'preserve this brain, France has use for it; and look well to its casket also, for I foresee that it will be hooped with a ducal coronet one day.' I took the ring, and knelt and kissed his hand, saying, 'Sire, where glory calls, there will I be found; where danger and death are thickest, that is my native air; when France and the throne need help—well, I say nothing, for I am not of the talking sort—let my deeds speak for me, it is all I ask.'

"So ended that most fortunate and memorable episode, so big with future weal for the crown and the nation, and unto God be the thanks! Rise! Fill your flagons! Now—to France and the King—drink!"

They emptied them to the bottom, then burst into cheers and huzzas, and kept it up as much as two minutes, the Paladin standing at stately ease the while and smiling benignantly from his platform.

CHAPTER VIII

WHEN Joan told the King what that deep secret was that was torturing his heart, his doubts were cleared away; he believed she was sent of God, and if he had been let alone he would have set her upon her great mission at once. But he was not let alone. Tremouille and the holy fox of Rheims knew their man. All they needed to say was this—and they said it:

"Your Highness says her Voices have revealed to you, by her mouth, a secret known only to yourself and God. How can you know that her Voices are not of Satan, and she his mouthpiece?—for does not Satan know the secrets of men and use his knowledge for the destruction of their souls? It is a dangerous business, and your Highness will do well not to proceed in it without probing the matter to the bottom."

That was enough. It shrivelled up the King's little soul like a raisin, with terrors and apprehensions, and straightway he privately appointed a commission of bishops to visit and question Joan daily until they should find out whether her supernatural helps hailed from heaven or from hell.

The King's relative, the Duke of Alençon, three years prisoner of war to the English, was in these days released from captivity through promise of a great ransom; and the name and fame of the Maid having reached him—for the same filled all mouths now, and penetrated to all parts—he came to Chinon to see with his own eyes what manner of creature she might be. The King sent for Joan and introduced her to the Duke. She said, in her simple fashion:

"You are welcome; the more of the blood of France that is joined to this cause, the better for the cause and it."

Then the two talked together, and there was just the usual result: when they parted, the Duke was her friend and advocate.

Joan attended the King's mass the next day, and afterward dined with the King and the Duke. The King was learning to prize her company and value her conversation; and that might well be, for, like other Kings, he was used to getting nothing out of people's talk but guarded phrases, colorless and non-committal; or carefully tinted to tally with the color of what he said himself; and so this kind of conversation only vexes and bores, and is wearisome; but Joan's talk was fresh and free, sincere and honest, and unmarred by timorous self-watching and constraint. She said the very thing that was in her mind, and said it in a plain, straightforward way. One can believe that to the King this must have been like fresh cold water from the mountains to parched lips used to the water of the sun-baked puddles of the plain.

After dinner Joan so charmed the Duke with her horsemanship and lance-practice in the meadows by the Castle of Chinon, whither the King also had come to look on, that he made her a present of a great black war-steed.

Every day the commission of bishops came and questioned Joan about her Voices and her mission, and then went to the King with their report. These pryings accomplished but little. She told as much as she considered advisable, and kept the rest to herself. Both threats and trickeries were wasted upon her. She did not care for the threats, and the traps caught nothing. She was perfectly frank and childlike about these things. She knew the bishops were sent by the King, that their questions were the King's questions, and that by all law and custom a King's questions *must* be answered; yet she told the King in her naïve way at his own table one day that she answered only such of those questions as suited her.

The bishops finally concluded that they couldn't tell whether Joan was sent by God or not. They were cautious, you see. There were two powerful parties at Court; therefore to make a decision either way would infallibly embroil them with one of those parties; so it seemed to them wisest to roost on the fence and shift the burden to other shoulders. And that is what they did. They made final report that Joan's case was beyond their powers, and recommended that it be put into the hands of the learned and illustrious doctors of the University of Poitiers. Then they retired from the field, leaving behind them this little item of testimony, wrung from them by Joan's wise reticence: they said she was a " gentle and simple little shepherdess, very candid, *but not given to talking.*"

It was quite true—in their case. But if they could have looked back and seen her with us in the happy pastures of Domremy, they would have perceived that she had a tongue that could go fast enough when no harm could come of her words.

So we travelled to Poitiers, to endure there three weeks of tedious delay while this poor child was being daily questioned and badgered before a great bench of—what? Military experts?—since what she had come to apply for was an army and the privilege of leading it to battle against the enemies of France. Oh no; it was a great bench of priests and monks—profoundly learned and astute casuists—renowned professors of theology! Instead of setting a military commission to find out if this valorous little soldier could win victories, they set a company of holy hair-splitters and phrase-mongers to work to find out if the soldier was sound in her piety and had no doctrinal leaks. The rats were devouring the house, but instead of examining the cat's teeth and claws, they only concerned themselves to find out if it was a holy cat. If it was a pious cat, a moral cat, all right, never mind about the other capacities, they were of no consequence.

Joan was as sweetly self-possessed and tranquil before this grim tribunal, with its robed celebrities, its solemn state and imposing ceremonials, as if she were but a spectator and not herself on trial. She sat there, solitary on her bench, untroubled, and disconcerted the science of the sages with her sublime ignorance—an ignorance which was a fortress; arts, wiles, the learning drawn from books, and all like missiles rebounded from its unconscious masonry and fell to the ground harmless; they could not dislodge the garrison which was within—Joan's serene great heart and spirit, the guards and keepers of her mission.

She answered all questions frankly, and she told all the story of her visions and of her experiences with the angels and what they said to her; and the manner of the telling was so unaffected, and so earnest and sincere, and made it all seem so life-like and real, that even that hard practical court forgot itself and sat motionless and mute, listening with a charmed and wondering interest to the end. And if you would have other testimony than mine, look in the histories and you will find where an eye-witness, giving sworn testimony in the Rehabilitation process, says that she told that tale "with a noble dignity and simplicity," and as to its effect, says in substance what I have said. Seventeen, she was—seventeen, and all alone on her bench by herself; yet was not afraid, but faced that great company of erudite doctors of law and theology, and by the help of no art learned in the schools, but using only the enchantments which were hers by nature, of youth, sincerity, a voice soft and musical, and an eloquence whose source was the heart, not the head, she laid that spell upon them. Now was not that a beautiful thing to see? If I could, I would put it before you just as I saw it; then I know what you would say.

As I have told you, she could not read. One day they harried and pestered her with arguments, reasonings, objections and other windy and wordy trivialities, gathered out of the works of this and that and the other great theological authority, until at last her patience vanished, and she turned upon them sharply and said—

"I don't know A from B; but I know this: that I am come by command of the Lord of Heaven to deliver Orleans from the English power and crown the King at Rheims, and the matters ye are puttering over are of no consequence!"

Necessarily those were trying days for her, and wearing for everybody that took part; but her share was the hardest, for she had no holidays, but must be always on hand and stay the long hours through, whereas this, that, and the other inquisitor could absent himself and rest up from his fatigues when he got worn out. And yet she showed no wear, no weariness, and but seldom let fly her temper. As a rule she put her day through calm, alert, patient, fencing with those veteran masters of scholarly sword-play and coming out always without a scratch.

One day a Dominican sprung upon her a question which made everybody cock up his ears with interest; as for me, I trembled, and said to myself she is caught this time, poor Joan, for there is no way of answering this. The sly Dominican began in this way—in a sort of indolent fashion, as if the thing he was about was a matter of no moment:

"You assert that God has willed to deliver France from this English bondage?"

"Yes, He has willed it."

"You wish for men-at-arms, so that you may go to the relief of Orleans, I believe?"

"Yes—and the sooner the better."

"God is all-powerful, and able to do whatsoever thing He wills to do, is it not so?"

"Most surely. None doubts it."

The Dominican lifted his head suddenly, and sprung that question I have spoken of, with exultation:

"Then answer me this. If He has willed to deliver France, and is able to do whatsoever He wills, where is the need for men-at-arms?"

There was a fine stir and commotion when he said that, and a sudden thrusting forward of heads and putting up of hands to ears to catch the answer; and the Dominican wagged his head with satisfaction, and looked about him collecting his applause, for it shone in every face. But Joan was not disturbed. There was no note of disquiet in her voice when she answered:

"He helps who help themselves. The sons of France will fight the battles, but *He* will give the victory!"

You could see a light of admiration sweep the house from face to face like a ray from the sun. Even the Dominican himself looked pleased, to see his master-stroke so neatly parried, and I heard a venerable bishop mutter, in the phrasing common to priest and people in that robust time, "By God, the child has said true. He willed that Goliath should be slain, and He sent a child like this to do it!"

Another day, when the inquisition had dragged along until everybody looked drowsy and tired but Joan, Brother Séguin, professor of theology in the University of Poitiers, who was a sour and sarcastic man, fell to plying Joan with all sorts of nagging questions in his bastard Limousin French—for he was from Limoges. Finally he said—

"How is it that you could understand those angels? What language did they speak?"

"French."

"In-deed! How pleasant to know that our language is so honored! Good French?"

"Yes—perfect."

"Perfect, eh? Well, certainly *you* ought to know. It was even better than your own, eh?"

"As to that, I—I believe I cannot say," said she, and was going on, but stopped. Then she added, almost as if she were saying it to herself, "Still, it was an improvement on yours!"

I knew there was a chuckle back of her eyes, for all their innocence. Everybody shouted. Brother Séguin was nettled, and asked brusquely—

"Do you believe in God?"

Joan answered with an irritating nonchalance—

"Oh, well, yes—better than you, it is likely."

Brother Séguin lost his patience, and heaped sarcasm after sarcasm upon her, and finally burst out in angry earnest, exclaiming—

"Very well, I can tell you this, you whose belief in God is

so great: God has not willed that any shall believe in you
without a sign. Where is your sign?—show it!"

This roused Joan, and she was on her feet in a moment,
and flung out her retort with spirit:

"I have not come to Poitiers to show signs and do miracles. Send me to Orleans and you shall have signs enough.
Give me men-at-arms—few or many—and let me go!"

The fire was leaping from her eyes—ah, the heroic little
figure! can't you see her? There was a great burst of accla-
mations, and she sat down blushing, for it was not in her
delicate nature to like being conspicuous.

This speech and that episode about the French language
scored two points against Brother Séguin, while he scored
nothing against Joan; yet, sour man as he was, he was a manly
man, and honest, as you can see by the histories; for at the
Rehabilitation he could have hidden those unlucky incidents
if he had chosen, but he didn't do it, but spoke them right
out in his evidence.

On one of the later days of that three weeks' session the
gowned scholars and professors made one grand assault
all along the line, fairly overwhelming Joan with objections
and arguments culled from the writings of every ancient and
illustrious authority of the Roman Church. She was well-
nigh smothered; but at last she shook herself free and struck
back, crying out:

"Listen! The Book of God is worth more than all these
ye cite, and I stand upon *it*. And I tell ye there are things
in that Book that not one among ye can read, with all your
learning!"

From the first she was the guest, by invitation, of the
dame De Rabateau, wife of a councillor of the Parliament of Poitiers; and to that house the great ladies of the
city came nightly to see Joan and talk with her; and not
these only, but the old lawyers, councillors, and scholars of
the Parliament and the University. And these grave men, accustomed to weigh every strange and questionable thing, and
cautiously consider it, and turn it about this way and that

and still doubt it, came night after night, and night after night, falling ever deeper and deeper under the influence of that mysterious something, that spell, that elusive and unwordable fascination, which was the supremest endowment of Joan of Arc, that winning and persuasive and convincing something which high and low alike recognized and felt, but which neither high nor low could explain or describe; and one by one they all surrendered, saying, "This child is sent of God."

All day long Joan, in the great court and subject to its rigid rules of procedure, was at a disadvantage; her judges had things their own way; but at night she held court herself, and matters were reversed, she presiding, with her tongue free and her same judges there before her. There could be but one result: all the objections and hindrances they could build around her with their hard labors of the day she would charm away at night. In the end, she carried her judges with her in a mass, and got her great verdict without a dissenting voice.

The court was a sight to see when the president of it read it from his throne, for all the great people of the town were there who could get admission and find room. First there were some solemn ceremonies, proper and usual at such times; then, when there was silence again, the reading followed, penetrating the deep hush so that every word was heard in even the remotest parts of the house:

"It is found, and is hereby declared, that Joan of Arc, called the Maid, is a good Christian and good Catholic; that there is nothing in her person or her words contrary to the faith; and that the King may and ought to accept the succor she offers; for to repel it would be to offend the Holy Spirit, and render him unworthy of the aid of God."

The court rose, and then the storm of plaudits burst forth unrebuked, dying down and bursting forth again and again, and I lost sight of Joan, for she was swallowed up in a great tide of people who rushed to congratulate her and pour out benedictions upon her and upon the cause of France, now solemnly and irrevocably delivered into her little hands.

CHAPTER IX

It was indeed a great day, and a stirring thing to see.

She had won! It was a mistake of Tremouille and her other ill-wishers to let her hold court those nights.

The commission of priests sent to Lorraine ostensibly to inquire into Joan's character—in fact to weary her with delays and wear out her purpose and make her give it up—arrived back and reported her character perfect. Our affairs were in full career now, you see.

The verdict made a prodigious stir. Dead France woke suddenly to life, wherever the great news travelled. Whereas before, the spiritless and cowed people hung their heads and slunk away if one mentioned war to them, now they came clamoring to be enlisted under the banner of the Maid of Vaucouleurs, and the roaring of war-songs and the thundering of the drums filled all the air. I remembered now what she had said, that time there in our village when I proved by facts and statistics that France's case was hopeless, and nothing could ever rouse the people from their lethargy:

"They will hear the drums — and they will answer, they will march!"

It has been said that misfortunes never come one at a time, but in a body. In our case it was the same with good luck. Having got a start, it came flooding in, tide after tide. Our next wave of it was of this sort. There had been grave doubts among the priests as to whether the Church ought to permit a female soldier to dress like a man. But now came a verdict on that head. Two of the greatest scholars and theologians of the time—one of whom had been Chancellor of the University of Paris—rendered it. They decided that since Joan "must do the work of a man and a soldier, it is just and legitimate that her apparel should conform to the situation."

It was a great point gained, the Church's authority to dress as a man. Oh yes, wave on wave the good luck came sweeping in. Never mind about the smaller waves, let us come to the largest one of all, the wave that swept us small fry quite off our feet and almost drowned us with joy. The day of the great verdict, couriers had been despatched to the King with it, and the next morning bright and early the clear notes of a bugle came floating to us on the crisp air, and we pricked up our ears and began to count them. One—two—three; pause; one—two; pause; one—two—three, again—and out we skipped and went flying; for that formula was used only when the King's herald-at-arms would deliver a proclamation to the people. As we hurried along, people came racing out of every street and house and alley, men, women, and children, all flushed, excited, and throwing lacking articles of clothing on as they ran; still those clear notes pealed out, and still the rush of people increased till the whole town was abroad and streaming along the principal street. At last we reached the square, which was now packed with citizens, and there, high on the pedestal of the great cross, we saw the herald in his brilliant costume, with his servitors about him. The next moment he began his delivery in the powerful voice proper to his office :

"Know all men, and take heed therefore, that the most high, the most illustrious Charles, by the grace of God King of France, hath been pleased to confer upon his well-beloved servant Joan of Arc, called the Maid, the title, emoluments, authorities, and dignity of General - in - Chief of the Armies of France—"

Here a thousand caps flew into the air, and the multitude burst into a hurricane of cheers that raged and raged till it seemed as if it would never come to an end; but at last it did ; then the herald went on and finished :

—" and hath appointed to be her lieutenant and chief of

staff a prince of his royal house, his grace the Duke of Alençon!"

That was the end, and the hurricane began again, and was split up into innumerable strips by the blowers of it and wafted through all the lanes and streets of the town.

General of the Armies of France, with a prince of the blood for subordinate! Yesterday she was nothing—to-day she was this. Yesterday she was not even a sergeant, not even a corporal, not even a private—to-day, with one step, she was at the top. Yesterday she was less than nobody to the newest recruit—to-day her command was law to La Hire, Saintrailles, the Bastard of Orleans, and all those others, veterans of old renown, illustrious masters of the trade of war. These were the thoughts I was thinking; I was trying to realize this strange and wonderful thing that had happened, you see.

My mind went travelling back, and presently lighted upon a picture—a picture which was still so new and fresh in my memory that it seemed a matter of only yesterday—and indeed its date was no further back than the first days of January. This is what it was. A peasant girl in a far-off village, her seventeenth year not yet quite completed, and herself and her village as unknown as if they had been on the other side of the globe. She had picked up a friendless wanderer somewhere and brought it home—a small gray kitten in a forlorn and starving condition—and had fed it and comforted it and got its confidence and made it believe in her, and now it was curled up in her lap asleep, and she was knitting a coarse stocking and thinking—dreaming—about what, one may never know. And now—the kitten had hardly had time to become a cat, and yet already the girl is General of the Armies of France, with a prince of the blood to give orders to, and out of her village obscurity her name has climbed up like the sun and is visible from all corners of the land! It made me dizzy to think of these things, they were so out of the common order, and seemed so impossible.

JOAN PUZZLES THE SCHOLARS

CHAPTER X

JOAN's first official act was to dictate a letter to the English commanders at Orleans, summoning them to deliver up all strongholds in their possession and depart out of France. She must have been thinking it all out before and arranging it in her mind, it flowed from her lips so smoothly, and framed itself into such vivacious and forcible language. Still, it might not have been so; she always had a quick mind and a capable tongue, and her faculties were constantly developing in these latter weeks. This letter was to be forwarded presently from Blois. Men, provisions, and money were offering in plenty now, and Joan appointed Blois as a recruiting station and depôt of supplies, and ordered up La Hire from the front to take charge.

The Great Bastard—him of the ducal house, and governor of Orleans — had been clamoring for weeks for Joan to be sent to him, and now came another messenger, old D'Aulon, a veteran officer, a trusty man and fine and honest. The King kept him, and gave him to Joan to be chief of her household, and commanded her to appoint the rest of her people herself, making their number and dignity accord with the greatness of her office ; and at the same time he gave order that they should be properly equipped with arms, clothing, and horses.

Meantime the King was having a complete suit of armor made for her at Tours. It was of the finest steel, heavily plated with silver, richly ornamented with engraved designs, and polished like a mirror.

Joan's Voices had told her that there was an ancient sword hidden somewhere behind the altar of St. Catherine's at Fierbois, and she sent De Metz to get it. The priests knew of no such sword, but a search was made, and sure enough it was found in that place, buried a little way under the ground. It had no sheath and was very rusty, but the priests polished it up and sent it to Tours, whither we were now to come. They also had a sheath of crimson velvet made for it, and the people of Tours equipped it with another one, made of cloth of gold. But Joan meant to carry this sword always in battle; so she laid the showy sheaths away and got one made of leather. It was generally believed that this sword had belonged to Charlemagne, but that was only a matter of opinion. I wanted to sharpen that old blade, but she said it was not necessary, as she should never kill anybody, and should carry it only as a symbol of authority.

At Tours she designed her Standard, and a Scotch painter named James Power made it. It was of the most delicate white *boucassin*, with fringes of silk. For device it bore the image of God the Father throned in the clouds and holding the world in His hand; two angels knelt at His feet, presenting lilies; inscription, JESUS, MARIA; on the reverse the crown of France supported by two angels.

She also caused a smaller standard or pennon to be made, whereon was represented an angel offering a lily to the Holy Virgin.

Everything was humming, there at Tours. Every now and then one heard the bray and crash of military music, every little while one heard the measured tramp of marching men— squads of recruits leaving for Blois; songs and shoutings and huzzas filled the air night and day, the town was full of strangers, the streets and inns were thronged, the bustle of preparation was everywhere, and everybody carried a glad and cheerful face. Around Joan's headquarters a crowd of people was always massed, hoping for a glimpse of the new General, and when they got it, they went wild; but they seldom got it, for she was busy planning her campaign, receiving reports, giving orders, despatching couriers, and giving what odd moments she could spare to the companies of great folk waiting in the drawing-rooms. As for us boys, we hardly saw her at all, she was so occupied.

We were in a mixed state of mind—sometimes hopeful, sometimes not; mostly not. She had not appointed her household yet—that was our trouble. We knew she was being overrun with applications for places in it, and that these applications were backed by great names and weighty influence, whereas we had nothing of the sort to recommend us. She could fill her humblest places with titled folk—folk whose relationships would be a bulwark for her and a valuable support at all times. In these circumstances would policy allow her to consider us? We were not as cheerful as the rest of the town, but were inclined to be depressed and worried. Sometimes we discussed our slim chances and gave them as good an appearance as we could. But the very mention of the subject was anguish to the Paladin; for whereas we had some little hope, he had none at all. As a rule, Noël Rainguesson was quite willing to let the dismal matter alone; but not when the Paladin was present. Once we were talking the thing over, when Noël said—

"Cheer up, Paladin; I had a dream last night, and you were the only one among us that got an appointment. It wasn't a high one, but it was an appointment, anyway—some kind of a lackey or body-servant, or something of that kind."

The Paladin roused up and looked almost cheerful; for he was a believer in dreams, and in anything and everything of a superstitious sort, in fact. He said, with a rising hopefulness—

"I wish it might come true. Do you think it will come true?"

"Certainly; I might almost say I know it will, for my dreams hardly ever fail."

"Noël, I could hug you if that dream could come true, I could indeed! To be servant to the first General of France and have all the world hear of it, and the news go back to the village and make those gawks stare that always said I wouldn't ever amount to anything—wouldn't it be great!

Do you think it *will* come true, Noël? Don't you believe it will?"

"I do. There's my hand on it."

"Noël, if it comes true I'll never forget you—shake again! I should be dressed in a noble livery, and the news would go to the village, and those animals would say, '*Him*, lackey to the General-in-Chief, with the eyes of the whole world on him, admiring—well, he has shot up into the sky, now, hasn't he!'"

He began to walk the floor and pile castles in the air so fast and so high that we could hardly keep up with him. Then all of a sudden all the joy went out of his face and misery took its place, and he said:

"Oh dear, it is all a mistake, it will never come true. I forgot about that foolish business at Toul. I have kept out of her sight as much as I could, all these weeks, hoping she would forget that and forgive it—but I know she never will. She can't, of course. And after all, I wasn't to blame. I did say she promised to marry me, but they put me up to it and persuaded me, I swear they did!" The vast creature was almost crying. Then he pulled himself together and said, remorsefully, "It was the only lie I've ever told, and—"

He was drowned out with a chorus of groans and outraged exclamations; and before he could begin again, one of D'Aulon's liveried servants appeared and said we were required at headquarters. We rose, and Noël said—

"There—what did I tell you? I have a presentiment—the spirit of prophecy is upon me. She is going to appoint him, and we are to go there and do him homage. Come along!"

But the Paladin was afraid to go, so we left him.

When we presently stood in the presence, in front of a crowd of glittering officers of the army, Joan greeted us with a winning smile, and said she appointed all of us to places in her household, for she wanted her old friends by her. It was a beautiful surprise to have ourselves honored like this when she could have had people of birth and consequence instead, but we couldn't find our tongues to say so, she was become so great and so high above us now. One at a time we stepped forward and each received his warrant from the hand of our chief, D'Aulon. All of us had honorable places : the two knights stood highest; then Joan's two brothers; I was first page and secretary, a young gentleman named Raimond was second page; Noël was her messenger; she had two heralds, and also a chaplain and almoner, whose name was Jean Pasquerel. She had previously appointed a maître d'hôtel and a number of domestics. Now she looked around and said—

"But where is the Paladin ?"

The Sieur Bertrand said—

"He thought he was not sent for, your Excellency."

"Now that is not well. Let him be called."

The Paladin entered humbly enough. He ventured no farther than just within the door. He stopped there, looking embarrassed and afraid. Then Joan spoke pleasantly, and said—

"I watched you on the road. You began badly, but improved. Of old you were a fantastic talker, but there is a man in you, and I will bring it out." It was fine to see the Paladin's face light up when she said that. "Will you follow where I lead ?"

"Into the fire !" he said ; and I said to myself, "By the ring of that, I think she has turned this braggart into a hero. It is another of her miracles, I make no doubt of it."

"I believe you," said Joan. "Here—take my banner. You will ride with me in every field, and when France is saved, you will give it me back."

He took the banner, which is now the most precious of the memorials that remain of Joan of Arc, and his voice was unsteady with emotion when he said—

"If I ever disgrace this trust, my comrades here will know how to do a friend's office upon my body, and this charge I lay upon them, as knowing they will not fail me."

CHAPTER XI

NOËL and I went back together—silent at first, and impressed. Finally Noël came up out of his thinkings and said—

"The first shall be last and the last first—there's authority for this surprise. But at the same time *wasn't* it a lofty hoist for our big bull!"

"It truly was; I am not over being stunned yet. It was the greatest place in her gift."

"Yes, it was. There are many generals, and she can create more; but there is only one Standard-Bearer."

"True. It is the most conspicuous place in the army, after her own."

"And the most coveted and honorable. Sons of two dukes tried to get it, as we know. And of all people in the world, this majestic windmill carries it off. Well, isn't it a gigantic promotion, when you come to look at it!"

"There's no doubt about it. It's a kind of copy of Joan's own in miniature."

"I don't know how to account for it—do you?"

"Yes—without any trouble at all—that is, I think I do."

Noël was surprised at that, and glanced up quickly, as if to see if I was in earnest. He said—

"I thought you couldn't be in earnest, but I see you are. If you can make me understand this puzzle, do it. Tell me what the explanation is."

"I believe I can. You have noticed that our chief knight says a good many wise things and has a thoughtful head on his shoulders. One day, riding along, we were talking about Joan's great talents, and he said, 'But, greatest of all her gifts, she has the seeing eye.' I said, like an unthinking fool, 'The seeing eye?—I shouldn't count that for much—I suppose we all have it.' 'No,' he said; 'very few have it.' Then he explained, and made his meaning clear. He said the common eye sees only the outside of things, and judges by that, but the seeing eye pierces through and reads the heart and the soul, finding there capacities which the outside didn't indicate or promise, and which the other kind of eye couldn't detect. He said the mightiest military genius must fail and come to nothing if it have not the seeing eye—that is to say, if it cannot read men and select its subordinates with an infallible judgment. It sees as by intuition that this man is good for strategy, that one for dash and dare-devil assault, the other for patient bull-dog persistence, and it appoints each to his right place and wins, while the commander without the seeing eye would give to each the other's place and lose. He was right about Joan, and I saw it. When she was a child and the tramp came one night, her father and all of us took him for a rascal, but she saw the honest man through the rags. When I dined with the governor of Vaucouleurs so long ago, I saw nothing in our two knights, though I sat with them and talked with them two hours; Joan was there five minutes, and neither spoke with them nor heard them speak, yet she marked them for men of worth and fidelity, and they have confirmed her judgment. Whom has she sent for to take charge of this thundering rabble of new recruits at Blois, made up of old disbanded Armagnac raiders, unspeakable hellions, every one? Why, she has sent for Satan himself—that is to say, La Hire—that military hurricane, that godless swashbuckler, that lurid conflagration of blasphemy, that Vesuvius of profanity, forever in eruption. Does he know how to deal with that mob of roaring devils? Better than any man that lives; for he is the head devil of this world his own self, he is the match of the whole of them combined, and probably the father of most of them. She places him in temporary command until she can get to Blois herself—and then! Why, then she will certainly take them in hand personally, or I don't know her as well as I ought to, after all these years of intimacy. That will be a sight to see—that fair spirit in her white armor, delivering her will to that muck-heap, that rag-pile, that abandoned refuse of perdition."

"La Hire!" cried Noël, "our hero of all these years—I do want to see that man!"

"I too. His name stirs me just as it did when I was a little boy."

"I want to hear him swear."

"Of course. I would rather hear him swear than another man pray. He is the frankest man there is, and the naïvest. Once when he was rebuked for pillaging on his raids, he said it was nothing. Said he, 'If God the Father were a soldier, He would rob.' I judge he is the right man to take temporary charge there at Blois. Joan has cast the seeing eye upon him, you see."

"Which brings us back to where we started. I have an honest affection for the Paladin, and not merely because he is a good fellow, but because he is my child—I made him what he is, the windiest blusterer and most catholic liar in the kingdom. I'm glad of his luck, but I hadn't the seeing eye. I shouldn't have chosen him for the most dangerous post in the army, I should have placed him in the rear to kill the wounded and violate the dead."

"Well, we shall see. Joan probably knows what is in him better than we do. And I'll give you another idea. When a person in Joan of Arc's position tells a man he is brave, he *believes* it; and *believing* it is enough; in fact to believe yourself brave is to *be* brave; it is the one only essential thing."

"Now you've hit it!" cried Noël. "She's got the creating mouth as well as the seeing eye! Ah yes, that is the thing. France was cowed and a coward; Joan of Arc has spoken, and France is marching, with her head up!"

I was summoned now, to write a letter from Joan's dictation. During the next day and night our several uniforms were made by the tailors, and our new armor provided. We were beautiful to look upon now, whether clothed for peace or war. Clothed for peace, in costly stuffs and rich colors, the Paladin was a tower dyed with the glories of the sunset; plumed and sashed and iron-clad for war, he was a still statelier thing to look at.

Orders had been issued for the march towards Blois. It was a clear, sharp, beautiful morning. As our showy great company trotted out in column, riding two and two, Joan and the Duke of Alençon in the lead, D'Aulon and the big standard-bearer next, and so on, we made a handsome spectacle, as you may well imagine; and as we ploughed through the cheering crowds, with Joan bowing her plumed head to left and right and the sun glinting from her silver mail, the spectators realized that the curtain was rolling up before their eyes upon the first act of a prodigious drama, and their rising hopes were expressed in an enthusiasm that increased with each moment, until at last one seemed to even physically feel the concussion of the huzzas as well as hear them. Far down the street we heard the softened strains of wind-blown music, and saw a cloud of lancers moving, the sun glowing with a subdued light upon the massed armor but striking bright upon the soaring lance-heads—a vaguely luminous nebula, so to speak, with a constellation twinkling above it—and that was our guard of honor. It joined us, the procession was complete, the first war-march of Joan of Arc was begun, the curtain was up.

CHAPTER XII

WE were at Blois three days. Oh, that camp, it is one of the treasures of my memory! Order? There was no more order among those brigands than there is among the wolves and the hyenas. They went roaring and drinking about, whooping, shouting, swearing, and entertaining themselves with all manner of rude and riotous horse-play; and the place was full of loud and lewd women, and they were no whit behind the men for romps and noise and fantastics.

It was in the midst of this wild mob that Noël and I had our first glimpse of La Hire. He answered to our dearest dreams. He was of great size and of martial bearing, he was cased in mail from head to heel, with a bushel of swishing plumes on his helmet, and at his side the vast sword of the time.

He was on his way to pay his respects in state to Joan, and as he passed through the camp he was restoring order, and proclaiming that the Maid was come, and he would have no such spectacle as this exposed to the head of the army. His way of creating order was his own, not borrowed. He did it with his great fists. As he moved along swearing and admonishing, he let drive this way, that way, and the other, and wherever his blow landed, a man went down.

"Damn you!" he said, "staggering and cursing around like this, and the Commander-in-Chief in the camp! Straighten up!" and he laid the man flat. What his idea of straightening up was, was his own secret.

We followed the veteran to headquarters, listening, observing, admiring—yes, devouring, you may say, the pet hero of the boys of France from our cradles up to that happy day, and their idol and ours. I called to mind how Joan had once re-buked the Paladin, there in the pastures of Domremy, for ut-tering lightly those mighty names, La Hire and the Bastard of Orleans, and how she said that if she could but be permit-ted to stand afar off and let her eyes rest once upon those great men, she would hold it a privilege. They were to her and the other girls just what they were to the boys. Well, here was one of them, at last—and what was his errand? It was hard to realize it, and yet it was true; he was coming to uncover his head before her and take her orders.

While he was quieting a considerable group of his brigands in his soothing way, near headquarters, we stepped on ahead and got a glimpse of Joan's military family, the great chiefs of the army, for they had all arrived now. There they were, six officers of wide renown, handsome men in beautiful armor, but the Lord High Admiral of France was the handsomest of them all and had the most gallant bearing.

When La Hire entered, one could see the surprise in his face at Joan's beauty and extreme youth, and one could see, too, by Joan's glad smile, that it made her happy to get sight of this hero of her childhood at last. La Hire bowed low, with his helmet in his gauntleted hand, and made a bluff but handsome little speech with hardly an oath in it, and one could see that those two took to each other on the spot.

The visit of ceremony was soon over, and the others went away; but La Hire stayed, and he and Joan sat there, and he sipped her wine, and they talked and laughed together like old friends. And presently she gave him some instructions, in his quality as master of the camp, which made his breath stand still. For, to begin with, she said that all those loose women must pack out of the place at once, she wouldn't allow one of them to remain. Next, the rough carousing must stop, drinking must be brought within proper and strictly defined limits, and discipline must take the place of disorder. And finally she climaxed the list of surprises with this—which nearly lifted him out of his armor:

"Every man who joins my standard must confess before

the priest and absolve himself from sin ; and all accepted recruits must be present at divine service twice a day."

La Hire could not say a word for a good part of a minute, then he said, in deep dejection:

"Oh, sweet child, they were littered in hell, these poor darlings of mine! Attend mass? Why, dear heart, they'll see us both damned first!"

And he went on, pouring out a most pathetic stream of arguments and blasphemy, which broke Joan all up, and made her laugh as she had not laughed since she played in the Domremy pastures. It was good to hear.

But she stuck to her point; so the soldier yielded, and said all right, if such were the orders he must obey, and would do the best that was in him; then he refreshed himself with a lurid explosion of oaths, and said that if any man in the camp refused to renounce sin and lead a pious life, he would knock his head off. That started Joan off again: she was really having a good time, you see. But she would not consent to that form of conversions. She said they must be voluntary.

La Hire said that that was all right, he wasn't going to kill the voluntary ones, but only the others.

No matter, none of them must be killed—Joan couldn't have it. She said that to give a man a chance to volunteer, on pain of death if he didn't, left him more or less trammelled, and she wanted him to be entirely free.

So the soldier sighed and said he would advertise the mass, but said he doubted if there was a man in camp that was any more likely to go to it than he was himself. Then there was another surprise for him, for Joan said—

"But dear man, *you* are going!"

"I? Impossible! Oh, this is lunacy!"

"Oh no, it isn't. You are going to the service—twice a day."

"Oh, am I dreaming? Am I drunk—or is my hearing playing me false? Why, I would rather go to—"

"Never mind where. In the morning you are going to begin, and after that it will come easy. Now *don't* look downhearted like that. Soon you won't mind it."

La Hire tried to cheer up, but he was not able to do it. He sighed like a zephyr, and presently said—

"Well, I'll do it for you, but before I would do it for another, I swear I—"

"But don't swear. Break it off."

"Break it off? It is impossible. I beg you to—to— Why—oh, my General, it is my native speech!"

He begged so hard for grace for his impediment, that Joan left him one fragment of it; she said he might swear by his bâton, the symbol of his generalship.

He promised that he would swear only by his bâton when in her presence, and would try to modify himself elsewhere, but doubted if he could manage it, now that it was so old and stubborn a habit, and such a solace and support to his declining years.

That tough old lion went away from there a good deal tamed and civilized—not to say softened and sweetened, for perhaps those expressions would hardly fit him. Noël and I believed that when he was away from Joan's influence his old aversions would come up so strong in him that he could not master them, and so wouldn't go to mass. But we got up early in the morning to see.

Well, he really went. It was hardly believable, but there he was, striding along, holding himself grimly to his duty, and looking as pious as he could, but growling and cursing like a fiend. It was another instance of the same old thing: whoever listened to the voice and looked into the eyes of Joan of Arc fell under a spell, and was not his own man any more.

Satan was converted, you see. Well, the rest followed. Joan rode up and down that camp, and wherever that fair young form appeared in its shining armor, with that sweet face to grace the vision and perfect it, the rude host seemed to think they saw the god of war in person, descended out of the clouds; and first they wondered, then they worshipped. After that, she could do with them what she would.

In three days it was a clean camp and orderly, and those barbarians were herding to divine service twice a day like

10

good children. The women were gone. La Hire was stunned by these marvels; he could not understand them. He went outside the camp when he wanted to swear. He was that sort of a man—sinful by nature and habit, but full of superstitious respect for holy places.

The enthusiasm of the reformed army for Joan, its devotion to her, and the hot desire she had aroused in it to be led against the enemy, exceeded any manifestations of this sort which La Hire had ever seen before in his long career. His admiration of it all, and his wonder over the mystery and miracle of it, were beyond his power to put into words. He had held this army cheap before, but his pride and confidence in it knew no limits now. He said—

"Two or three days ago it was afraid of a hen-roost; one could storm the gates of hell with it now."

Joan and he were inseparable, and a quaint and pleasant contrast they made. He was so big, she so little; he was so gray and so far along in his pilgrimage of life, she so youthful; his face was so bronzed and scarred, hers so fair and pink, so fresh and smooth; she was so gracious, and he so stern; she was so pure, so innocent, he such a cyclopædia of sin. In her eye was stored all charity and compassion, in his lightnings; when her glance fell upon you it seemed to bring benediction and the peace of God, but with his it was different, generally.

They rode through the camp a dozen times a day, visiting every corner of it, observing, inspecting, perfecting; and wherever they appeared the enthusiasm broke forth. They rode side by side, he a great figure of brawn and muscle, she a little master-work of roundness and grace; he a fortress of rusty iron, she a shining statuette of silver; and when the reformed raiders and bandits caught sight of them they spoke out, with affection and welcome in their voices, and said—

"There they come—Satan and the Page of Christ!"

All the three days that we were in Blois, Joan worked earnestly and tirelessly to bring La Hire to God—to rescue him from the bondage of sin—to breathe into his stormy heart the

JOAN CHOOSES HER STANDARD-BEARER

serenity and peace of religion. She urged, she begged, she implored him to pray. He stood out, the three days of our stay, begging almost piteously to be let off—to be let off from just that one thing, that impossible thing; he would do anything else — anything — command, and he would obey — he would go through the fire for her if she said the word—but spare him this, only this, for he couldn't pray, had never prayed, he was ignorant of how to frame a prayer, he had no words to put it in.

And yet—can any believe it?—she carried even that point, she won that incredible victory. She made La Hire pray. It shows, I think, that nothing was impossible to Joan of Arc. Yes, he stood there before her and put up his mailed hands and made a prayer. And it was not borrowed, but was his very own; he had none to help him frame it, he made it out of his own head—saying:

"Fair Sir God, I pray you to do by La Hire as he would do by you if you were La Hire and he were God." *

Then he put on his helmet and marched out of Joan's tent as satisfied with himself as any one might be who has arranged a perplexed and difficult business to the content and admiration of all the parties concerned in the matter.

If I had known that he had been praying, I could have understood why he was feeling so superior, but of course I could not know that.

I was coming to the tent at that moment, and saw him come out, and saw him march away in that large fashion, and indeed it was fine and beautiful to see. But when I got to the tent door I stopped and stepped back, grieved and shocked, for I heard Joan crying, as I mistakenly thought— crying as if she could not contain nor endure the anguish of her soul, crying as if she would die. But it was not so, she was laughing—laughing at La Hire's prayer.

* This prayer has been stolen many times and by many nations in the past four hundred and sixty years, but it originated with La Hire, and the fact is of official record in the National Archives of France. We have the authority of Michelet for this.—TRANSLATOR.

It was not until six-and-thirty years afterwards that I found that out, and then—oh, then I only cried when that picture of young care-free mirth rose before me out of the blur and mists of that long-vanished time ; for there had come a day between, when God's good gift of laughter had gone out from me to come again no more in this life.

CHAPTER XIII

WE marched out in great strength and splendor, and took the road toward Orleans. The initial part of Joan's great dream was realizing itself at last. It was the first time that any of us youngsters had ever seen an army, and it was a most stately and imposing spectacle to us. It was indeed an inspiring sight, that interminable column, stretching away into the fading distances, and curving itself in and out of the crookedness of the road like a mighty serpent. Joan rode at the head of it with her personal staff; then came a body of priests singing the *Veni Creator*, the banner of the Cross rising out of their midst; after these the glinting forest of spears. The several divisions were commanded by the great Armagnac generals, La Hire, the Marshal de Boussac, the Sire de Retz, Florent d'Illiers, and Poton de Saintrailles.

Each in his degree was tough, and there were three degrees —tough, tougher, toughest—and La Hire was the last by a shade, but only a shade. They were just illustrious official brigands, the whole party; and by long habits of lawlessness they had lost all acquaintanceship with obedience, if they had ever had any.

The King's strict orders to them had been, " Obey the General-in-Chief in everything; attempt nothing without her knowledge, do nothing without her command."

But what was the good of saying that? These independent birds knew no law. They seldom obeyed the King; they never obeyed him when it didn't suit them to do it. Would they obey the Maid? In the first place they wouldn't know *how* to obey her or anybody else, and in the second place it was of course not possible for them to take her military character seriously—that country girl of seventeen who had been trained for the complex and terrible business of war—how? By tending sheep.

They had no idea of obeying her except in cases where their veteran military knowledge and experience showed them that the thing she required was sound and right when gauged by the regular military standards. Were they to blame for this attitude? I should think not. Old war-worn captains are hard-headed, practical men. They do not easily believe in the ability of ignorant children to plan campaigns and command armies. No general that ever lived could have taken Joan seriously (militarily) before she raised the siege of Orleans and followed it with the great campaign of the Loire.

Did they consider Joan valueless? Far from it. They valued her as the fruitful earth values the sun—they fully believed she could produce the crop, but that it was in their line of business, not hers, to take it off. They had a deep and superstitious reverence for her as being endowed with a mysterious supernatural something that was able to do a mighty thing which they were powerless to do — blow the breath of life and valor into the dead corpses of cowed armies and turn them into heroes.

To their minds they were everything *with* her, but nothing without her. She could inspire the soldiers and fit them for battle—but fight the battle herself? Oh, nonsense—that was their function. They, the generals, would fight the battles, Joan would give the victory. That was their idea—an unconscious paraphrase of Joan's reply to the Dominican.

So they began by playing a deception upon her. She had a clear idea of how she meant to proceed. It was her purpose to march boldly upon Orleans by the north bank of the Loire. She gave that order to her generals. They said to themselves, "The idea is insane—it is blunder No. 1; it is what might have been expected of this child who is ignorant of war." They privately sent the word to the Bastard of Orleans. He also recognized the insanity of it—at least he thought he did—and privately advised the generals to get around the order in some way.

They did it by deceiving Joan. She trusted those people, she was not expecting this sort of treatment, and was not on the lookout for it. It was a lesson to her; she saw to it that the game was not played a second time.

Why was Joan's idea insane, from the generals' point of view, but not from hers? Because her plan was to raise the siege immediately, by fighting, while theirs was to besiege the besiegers and starve them out by closing their communications—a plan which would require months in the consummation.

The English had built a fence of strong fortresses called bastilles around Orleans—fortresses which closed all the gates of the city but one. To the French generals the idea of trying to fight their way past those fortresses and lead the army into Orleans was preposterous; they believed that the result would be the army's destruction. One may not doubt that their opinion was militarily sound—no, *would* have been, but for one circumstance which they overlooked. That was this: the English soldiers were in a demoralized condition of superstitious terror; they had become satisfied that the Maid was in league with Satan. By reason of this a good deal of their courage had oozed out and vanished. On the other hand the Maid's soldiers were full of courage, enthusiasm, and zeal.

Joan could have marched by the English forts. However, it was not to be. She had been cheated out of her first chance to strike a heavy blow for her country.

In camp that night she slept in her armor on the ground. It was a cold night, and she was nearly as stiff as her armor itself when we resumed the march in the morning, for iron is not good material for a blanket. However, her joy in being now so far on her way to the theatre of her mission was fire enough to warm her, and it soon did it.

Her enthusiasm and impatience rose higher and higher with every mile of progress ; but at last we reached Olivet, and down it went, and indignation took its place. For she saw the trick that had been played upon her—the river lay between us and Orleans.

She was for attacking one of the three bastilles that were on our side of the river and forcing access to the bridge which it guarded (a project which, if successful, would raise the siege instantly), but the long-ingrained fear of the English came upon her generals and they implored her not to make the attempt. The soldiers wanted to attack, but had to suffer disappointment. So we moved on and came to a halt at a point opposite Chécy, six miles above Orleans.

Dunois, Bastard of Orleans, with a body of knights and citizens, came up from the city to welcome Joan. Joan was still burning with resentment over the trick that had been put upon her, and was not in the mood for soft speeches, even to revered military idols of her childhood. She said—

"Are you the Bastard of Orleans?"

"Yes, I am he, and am right glad of your coming."

"And did you advise that I be brought by this side of the river instead of straight to Talbot and the English?"

Her high manner abashed him and he was not able to answer with anything like a confident promptness, but with many hesitations and partial excuses he managed to get out the confession that for what he and the council had regarded as imperative military reasons they had so advised.

"In God's name," said Joan, "my Lord's counsel is safer and wiser than yours. You thought to deceive me, but you have deceived yourselves, for I bring you the best help that ever knight or city had; for it is God's help, not sent for love of me, but by God's pleasure. At the prayer of St. Louis and St. Charlemagne He has had pity on Orleans, and will not suffer the enemy to have both the Duke of Orleans and his city. The provisions to save the starving people are here, the boats are below the city, the wind is contrary, they cannot come up hither. Now then tell me, in God's name, you who are so wise, what that council of yours was thinking about, to invent this foolish difficulty."

Dunois and the rest fumbled around the matter a moment, then gave in and conceded that a blunder had been made.

"Yes, a blunder has been made," said Joan, "and except God take your proper work upon Himself and change the wind and correct your blunder for you, there is none else that can devise a remedy."

Some of those people began to perceive that with all her technical ignorance she had practical good sense, and that with all her native sweetness and charm she was not the right kind of a person to play with.

Presently God did take the blunder in hand, and by His grace the wind did change. So the fleet of boats came up and went away loaded with provisions and cattle, and conveyed that welcome succor to the hungry city, managing the matter successfully under protection of a sortie from the walls against the bastille of St. Loup. Then Joan began on the Bastard again :

"You see here the army?"

"Yes."

"It is here on this side by advice of your council?"

"Yes."

"Now, in God's name, can that wise council explain why it is better to have it here than it would be to have it in the bottom of the sea?"

Dunois made some wandering attempts to explain the inexplicable and excuse the inexcusable, but Joan cut him short and said—

"Answer me this, good sir—has the army any value on this side of the river?"

The Bastard confessed that it hadn't—that is, in view of the plan of campaign which she had devised and decreed.

"And yet, knowing this, you had the hardihood to disobey my orders. Since the army's place is on the other side, will you explain to me how it is to get there?"

The whole size of the needless muddle was apparent. Evasions were of no use; therefore Dunois admitted that there was no way to correct the blunder but to send the army all the way back to Blois, and let it begin over again and come up on the other side this time, according to Joan's original plan.

Any other girl, after winning such a triumph as this over a veteran soldier of old renown, might have exulted a little and been excusable for it, but Joan showed no disposition of this sort. She dropped a word or two of grief over the precious time that must be lost, then began at once to issue commands for the march back. She sorrowed to see her army go; for she said its heart was great and its enthusiasm high, and that with it at her back she did not fear to face all the might of England.

All arrangements having been completed for the return of the main body of the army, she took the Bastard and La Hire and a thousand men and went down to Orleans, where all the town was in a fever of impatience to have sight of her face. It was eight in the evening when she and the troops rode in at the Burgundy gate, with the Paladin preceding her with her standard. She was riding a white horse, and she carried in her hand the sacred sword of Fierbois. You should have seen Orleans then. What a picture it was! Such black seas of people, such a starry firmament of torches, such roaring whirlwinds of welcome, such booming of bells and thundering of cannon! It was as if the world was come to an end. Everywhere in the glare of the torches one saw rank upon rank of upturned white faces, the mouths wide open, shouting, and the unchecked tears running down; Joan forged her slow way through the solid masses, her mailed form projecting above the pavement of heads like a silver statue. The people about her struggled along, gazing up at her through their tears with the rapt look of men and women who believe they are seeing one who is divine; and always her feet were being kissed by grateful folk, and such as failed of that privilege touched her horse and then kissed their fingers.

Nothing that Joan did escaped notice; everything she did was commented upon and applauded. You could hear the remarks going all the time.

" There—she's smiling—see !''

" Now she's taking her little plumed cap off to somebody—ah, it's fine and graceful !''

" She's patting that woman on the head with her gauntlet.''

" Oh, she was born on a horse—see her turn in her saddle, and kiss the hilt of her sword to the ladies in the window that threw the flowers down.''

" Now there's a poor woman lifting up a child—she's kissed it—oh, she's divine !''

" What a dainty little figure it is, and what a lovely face—and such color and animation !''

Joan's slender long banner streaming backward had an accident—the fringe caught fire from a torch. She leaned forward and crushed the flame in her hand.

" She's not afraid of fire nor anything !'' they shouted, and delivered a storm of admiring applause that made everything quake.

She rode to the cathedral and gave thanks to God, and the people crammed the place and added their devotions to hers; then she took up her march again and picked her slow way through the crowds and the wilderness of torches to the house of Jacques Boucher, treasurer of the Duke of Orleans, where she was to be the guest of his wife as long as she stayed in the city, and have his young daughter for comrade and room-mate. The delirium of the people went on the rest of the night, and with it the clamor of the joy-bells and the welcoming cannon.

Joan of Arc had stepped upon her stage at last, and was ready to begin.

CHAPTER XIV

SHE was ready, but must sit down and wait until there was an army to work with.

Next morning, Saturday, April 30, 1429, she set about inquiring after the messenger who carried her proclamation to the English from Blois — the one which she had dictated at Poitiers. Here is a copy of it. It is a remarkable document, for several reasons : for its matter-of-fact directness, for its high spirit and forcible diction, and for its naïve confidence in her ability to achieve the prodigious task which she had laid upon herself, or which had been laid upon her—which you please. All through it you seem to see the pomps of war and hear the rumbling of the drums. In it Joan's warrior soul is revealed, and for the moment the soft little shepherdess has disappeared from your view. This untaught country damsel, unused to dictating anything at all to anybody, much less documents of state to kings and generals, poured out this procession of vigorous sentences as fluently as if this sort of work had been her trade from childhood :

"JESUS MARIA

" King of England, and you Duke of Bedford who call yourself Regent of France ; William de la Pole, Earl of Suffolk ; and you Thomas Lord Scales, who style yourselves lieutenants of the said Bedford—do right to the King of Heaven. Render to the Maid who is sent by God the keys of all the good towns you have taken and violated in France. She is sent hither by God to restore the blood royal. She is very ready to make peace if you will do her right by giving up France and paying for what you have held. And you archers, companions of war, noble and otherwise, who are before the the good city of Orleans, begone into your own land in God's name, or expect news from the Maid who will shortly go to see you to your very great hurt.   King of England, if you do not so, I am chief of war, and wherever I shall find your people in France I will drive them out, willing or not willing ; and if they do not obey I will slay them all, but if they obey, I will have them to mercy.   I am come hither by God, the King of Heaven, body for body, to put you out of France, in spite of those who would work treason and mischief against the kingdom.   Think not you shall ever hold the kingdom from the King of Heaven, the Son of the blessed Mary ; King Charles shall hold it, for God wills it so, and has revealed it to him by the Maid.   If you believe not the news sent by God through the Maid, wherever we shall meet you we will strike boldly and make such a noise as has not been in France these thousand years.   Be sure that God can send more strength to the Maid than you can bring to any assault against her and her good men-at-arms ; and then we shall see who has the better right, the King of Heaven, or you. Duke of Bedford, the Maid prays you not to bring about your own destruction.   If you do her right, you may yet go in her company where the French shall do the finest deed that has ever been done in Christendom, and if you do not, you shall be reminded shortly of your great wrongs."

In that closing sentence she invites them to go on crusade with her to rescue the Holy Sepulchre.

No answer had been returned to this proclamation, and the messenger himself had not come back.   So now she sent her two heralds with a new letter warning the English to raise the siege and requiring them to restore that missing messenger. The heralds came back without him.   All they brought was notice from the English to Joan that they would presently catch her and burn her if she did not clear out now while she had a chance, and "go back to her proper trade of minding cows."

She held her peace, only saying it was a pity that the English would persist in inviting present disaster and eventual destruction when she was "doing all she could to get them out of the country with their lives still in their bodies."

Presently she thought of an arrangement that might be acceptable, and said to the heralds, " Go back and say to Lord Talbot this, from me : ' Come out of your bastilles with your host, and I will come with mine ; if I beat you, go in peace out of France ; if you beat me, burn me, according to your desire.' "

I did not hear this, but Dunois did, and spoke of it. The challenge was refused.

Sunday morning her Voices or some instinct gave her a warning, and she sent Dunois to Blois to take command of the army and hurry it to Orleans. It was a wise move, for he found Regnault de Chartres and some more of the King's pet rascals there trying their best to disperse the army, and crippling all the efforts of Joan's generals to head it for Orleans. They were a fine lot, those miscreants. They turned their attention to Dunois, now, but he had balked Joan once, with unpleasant results to himself, and was not minded to meddle in that way again. He soon had the army moving.

CHAPTER XV

WE of the personal staff were in fairy-land, now, during the few days that we waited for the return of the army. We went into society. To our two knights this was not a novelty, but to us young villagers it was a new and wonderful life. Any position of any sort near the person of the Maid of Vaucouleurs conferred high distinction upon the holder and caused his society to be courted; and so the D'Arc brothers, and Noël, and the Paladin, humble peasants at home, were gentlemen here, personages of weight and influence. It was fine to see how soon their country diffidences and awkwardnesses melted away under this pleasant sun of deference and disappeared, and how lightly and easily they took to their new atmosphere. The Paladin was as happy as it was possible for any one in this earth to be. His tongue went all the time, and daily he got new delight out of hearing himself talk. He began to enlarge his ancestry and spread it out all around, and ennoble it right and left, and it was not long until it consisted almost entirely of Dukes. He worked up his old battles and tricked them out with fresh splendors; also with new terrors, for he added artillery now. We had seen cannon for the first time at Blois—a few pieces—here there was plenty of it, and now and then we had the impressive spectacle of a huge English bastille hidden from sight in a mountain of smoke from its own guns, with lances of red flame darting through it; and this grand picture, along with the quaking thunders pounding away in the heart of it, inflamed the Paladin's imagination and enabled him to dress out those ambuscade-skirmishes of ours with a sublimity which made it impossible for any to recognize them at all except people who had not been there.

You may suspect that there was a special inspiration for these great efforts of the Paladin's, and there was. It was the daughter of the house, Catherine Boucher, who was eighteen, and gentle and lovely in her ways, and very beautiful. I think she might have been as beautiful as Joan herself, if she had had Joan's eyes. But that could never be. There was never but that one pair, there will never be another. Joan's eyes were deep and rich and wonderful beyond anything merely earthly. They spoke all the languages—they had no need of words. They produced all effects—and just by a glance, just a single glance: a glance that could convict a liar of his lie and make him confess it; that could bring down a proud man's pride and make him humble; that could put courage into a coward and strike dead the courage of the bravest; that could appease resentments and real hatreds; that could speak peace to storms of passion and be obeyed; that could make the doubter believe and the hopeless hope again; that could purify the impure mind; that could persuade—ah, there it is—*persuasion!* that is the word; what or who is it that it couldn't persuade? The maniac of Domremy—the fairy-banishing priest—the reverend tribunal of Toul—the doubting and superstitious Laxart—the obstinate veteran of Vaucouleurs—the characterless heir of France— the sages and scholars of the Parliament and University of Poitiers—the darling of Satan, La Hire—the masterless Bastard of Orleans, accustomed to acknowledge no way as right and rational but his own—these were the trophies of that great gift that made her the wonder and mystery that she was.

We mingled companionably with the great folk who flocked to the big house to make Joan's acquaintance, and they made much of us and we lived in the clouds, so to speak. But what we preferred even to this happiness was the quieter occasions, when the formal guests were gone and the family and a few dozen of its familiar friends were gathered together for a social good time. It was then that we did our best, we five youngsters, with such fascinations as we had, and the chief

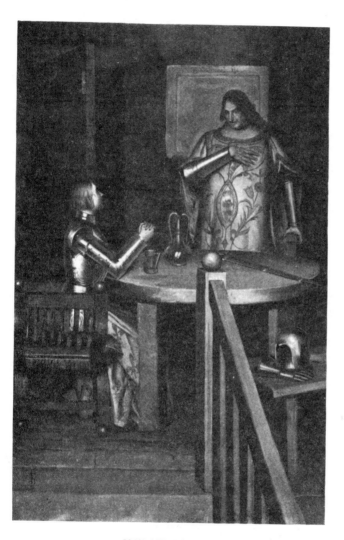

JOAN AND LA HIRE

object of them was Catherine. None of us had ever been in love before, and now we had the misfortune to all fall in love with the same person at the same time—which was the first moment we saw her. She was a merry heart, and full of life, and I still remember tenderly those few evenings that I was permitted to have my share of her dear society and of comradeship with that little company of charming people.

The Paladin made us all jealous the first night, for when he got fairly started on those battles of his he had everything to himself, and there was no use in anybody else's trying to get any attention. Those people had been living in the midst of real war for seven months; and to hear this windy giant lay out his imaginary campaigns and fairly swim in blood and spatter it all around, entertained them to the verge of the grave. Catherine was like to die, for pure enjoyment. She didn't laugh loud—we, of course, wished she would—but kept in the shelter of a fan, and shook until there was danger that she would unhitch her ribs from her spine. Then when the Paladin had got done with a battle and we began to feel thankful and hope for a change, she would speak up in a way that was so sweet and persuasive that it rankled in me, and ask him about some detail or other in the early part of his battle which she said had greatly interested her, and would he be so good as to describe that part again and with a little more particularity?—which of course precipitated the whole battle on us again, with a hundred lies added that had been overlooked before.

I do not know how to make you realize the pain I suffered. I had never been jealous before, and it seemed intolerable that this creature should have this good fortune which he was so ill entitled to, and I have to sit and see myself neglected when I was so longing for the least little attention out of the thousand that this beloved girl was lavishing upon him. I was near her, and tried two or three times to get started on some of the things that *I* had done in those battles—and I felt ashamed of myself, too, for stooping to such a business— but she cared for nothing but his battles, and could not be got to listen; and presently when one of my attempts caused her to lose some precious rag or other of his mendacities and she asked him to repeat, thus bringing on a new engagement of course and increasing the havoc and carnage tenfold, I felt so humiliated by this pitiful miscarriage of mine that I gave up and tried no more.

The others were as outraged by the Paladin's selfish conduct as I was—and by his grand luck, too, of course—perhaps, indeed, that was the main hurt. We talked our trouble over together, which was but natural, for rivals become brothers when a common affliction assails them and a common enemy bears off the victory.

Each of us could do things that would please and get notice if it were not for this person, who occupied all the time and gave others no chance. I had made a poem, taking a whole night to it—a poem in which I most happily and delicately celebrated that sweet girl's charms, without mentioning her name, but any one could see who was meant; for the bare title—"The Rose of Orleans" would reveal that, as it seemed to me. It pictured this pure and dainty white rose as growing up out of the rude soil of war and looking abroad out of its tender eyes upon the horrid machinery of death, and then—note this conceit—it blushes for the sinful nature of man, and *turns red* in a single night. Becomes a red rose, you see—a rose that was white before. The idea was my own, and quite new. Then it sent its sweet perfume out over the embattled city, and when the beleaguring forces smelt it they *laid down their arms and wept*. This was also my own idea, and new. That closed that part of the poem; then I put her into the similitude of the firmament—not the whole of it, but only part. That is to say, she was the moon, and all the constellations were following her about, their hearts in flames for love of her, but she would not halt, she would not listen, for 'twas thought she loved another. 'Twas thought she loved a poor unworthy suppliant who was upon the earth, facing danger, death, and possible multilation in the bloody field, waging relentless war against a heartless foe to save

her from an all too early grave, and her city from destruction.
And when the sad pursuing constellations came to know and
realize the bitter sorrow that was come upon them—note this
idea—their hearts broke and their tears gushed forth, filling
the vault of heaven with a fiery splendor, for those tears were
falling stars. It was a rash idea, but beautiful; beautiful and
pathetic; wonderfully pathetic, the way I had it, with the
rhyme and all to help. At the end of each verse there was
a two-line refrain pitying the poor earthly lover separated so
far, and perhaps forever, from her he loved so well, and growing always paler and weaker and thinner in his agony as he
neared the cruel grave—the most touching thing—even the
boys themselves could hardly keep back their tears, the way
Noël said those lines. There were eight four-line stanzas
in the first end of the poem—the end about the rose, the
horticultural end, as you may say, if that is not too large a
name for such a little poem—and eight in the astronomical
end—sixteen stanzas altogether, and I could have made it a
hundred and fifty if I had wanted to, I was so inspired and so
all swelled up with beautiful thoughts and fancies; but that
would have been too many to sing or recite before a company, that way, whereas sixteen was just right, and could be
done over again, if desired.

The boys were amazed that I could make such a poem as
that out of my own head, and so was I, of course, it being as
much a surprise to me as it could be to anybody, for I did
not know that it was in me. If any had asked me a single
day before if it was in me, I should have told them frankly
no, it was not.

That is the way with us; we may go on half of our life not
knowing such a thing is in us, when in reality it was there all
the time, and all we needed was something to turn up that
would call for it. Indeed, it was always so with our family.
My grandfather had a cancer, and they never knew what was
the matter with him till he died, and he didn't himself. It is
wonderful how gifts and diseases can be concealed that way.
All that was necessary in my case was for this lovely and in-

spiring girl to cross my path, and out came the poem, and no more trouble to me to word it and rhyme it and perfect it than it is to stone a dog. No, I should have said it was not in me ; but it was.

The boys couldn't say enough about it, they were so charmed and astonished. The thing that pleased them the most was the way it would do the Paladin's business for him. They forgot everything in their anxiety to get him shelved and silenced. Noël Rainguesson was clear beside himself with admiration of the poem, and wished *he* could do such a thing, but it was out of his line and he couldn't, of course. He had it by heart in half an hour, and there was never anything so pathetic and beautiful as the way he recited it. For that was just his gift—that and mimicry. He could recite anything better than anybody in the world, and he could take off La Hire to the very life—or anybody else, for that matter. Now I never could recite worth a farthing ; and when I tried with this poem the boys wouldn't let me finish ; they would have nobody but Noël. So then, as I wanted the poem to make the best possible impression on Catherine and the company, I told Noël he might do the reciting. Never was anybody so delighted. He could hardly believe that I was in earnest, but I was. I said that to have them know that I was the author of if would be enough for me. The boys were full of exultation, and Noël said if he could just get one chance at those people it would be all he would ask ; he would make them realize that there was something higher and finer than war-lies to be had here.

But how to get the opportunity—that was the difficulty. We invented several schemes that promised fairly, and at last we hit upon one that was sure. That was, to let the Paladin get a good start in a manufactured battle, and then send in a false call for him, and as soon as he was out of the room, have Noël take his place and finish the battle himself in the Paladin's own style, imitated to a shade. That would get great applause, and win the house's favor and put it in the right mood to hear the poem. The two triumphs together would finish the Standard - Bearer—modify him, anyway, to a certainty, and give the rest of us a chance for the future.

So the next night I kept out of the way until the Paladin had got his start and was sweeping down upon the enemy like a whirlwind at the head of his corps, then I stepped within the door in my official uniform and announced that a messenger from General La Hire's quarters desired speech with the Standard-Bearer. He left the room, and Noël took his place and said that the interruption was to be deplored, but that fortunately he was personally acquainted with the details of the battle himself, and if permitted would be glad to state them to the company. Then without waiting for the permission he turned himself into the Paladin—a dwarfed Paladin, of course—with manner, tones, gestures, attitudes, everything exact, and went right on with the battle, and it would be impossible to imagine a more perfectly and minutely ridiculous imitation than he furnished to those shrieking people. They went into spasms, convulsions, frenzies of laughter, and the tears flowed down their cheeks in rivulets. The more they laughed, the more inspired Noël grew with his theme and the greater the marvels he worked, till really the laughter was not properly laughing any more, but screaming. Blessedest feature of all, Catherine Boucher was dying with ecstasies, and presently there was little left of her but gasps and suffocations. Victory? It was a perfect Agincourt.

The Paladin was gone only a couple of minutes; he found out at once that a trick had been played on him, so he came back. When he approached the door he heard Noël ranting in there and recognized the state of the case; so he remained near the door but out of sight, and heard the performance through to the end. The applause Noël got when he finished was wonderful; and they kept it up and kept it up, clapping their hands like mad, and shouting to him to do it over again.

But Noël was clever. He knew the very best background for a poem of deep and refined sentiment and pathetic melancholy was one where great and satisfying merriment has prepared the spirit for the powerful contrast.

So he paused until all was quiet, then his face grew grave and assumed an impressive aspect, and at once all faces sobered in sympathy and took on a look of wondering and expectant interest. Now he began in a low but distinct voice the opening verses of The Rose. As he breathed the rhythmic measures forth, and one gracious line after another fell upon those enchanted ears in that deep hush, one could catch, on every hand, half-audible ejaculations of "How lovely—how beautiful—how exquisite."

By this time the Paladin, who had gone away for a moment with the opening of the poem, was back again, and had stepped within the door. He stood there, now, resting his great frame against the wall and gazing toward the reciter like one entranced. When Noël got to the second part, and that heartbreaking refrain began to melt and move all listeners, the Paladin began to wipe away tears with the back of first one hand and then the other. The next time the refrain was repeated he got to snuffling, and sort of half sobbing, and went to wiping his eyes with the sleeves of his doublet. He was so conspicuous that he embarrassed Noël a little, and also had an ill effect upon the audience. With the next repetition he broke quite down and began to cry like a calf, which ruined all the effect and started many in the audience to laughing. Then he went on from bad to worse, until I never saw such a spectacle; for he fetched out a towel from under his doublet and began to swab his eyes with it and let go the most infernal bellowings mixed up with sobbings and groanings and retchings and barkings and coughings and snortings and screamings and howlings—and he twisted himself about on his heels and squirmed this way and that, still pouring out that brutal clamor and flourishing his towel in the air and swabbing again and wringing it out. Hear? You couldn't hear yourself think. Noël was wholly drowned out and silenced, and those people were laughing the very lungs out of themselves. It was the most degrading sight that ever was. Now I heard the clankety-clank that plate-armor makes when the man that is in it is running, and then alongside my head there burst out the most inhuman explosion of laughter that ever rent the drum of a person's ear, and I looked, and it was La Hire; and he stood there with his gauntlets on his hips and his head tilted back and his jaws spread to that degree to let out his hurricanes and his thunders that it amounted to indecent exposure, for you could see everything that was in him. Only one thing more and worse could happen, and it happened: at the other door I saw the flurry and bustle and bowings and scrapings of officials and flunkeys which means that some great personage is coming — then Joan of Arc stepped in, and the house rose! Yes, and tried to shut its indecorous mouth and make itself grave and proper; but when it saw the Maid herself go to laughing, it thanked God for this mercy and the earthquake followed.

Such things make life a bitterness, and I do not wish to dwell upon them. The effect of the poem was spoiled.

CHAPTER XVI

This episode disagreed with me and I was not able to leave my bed the next day. The others were in the same condition. But for this, one or another of us might have had the good luck that fell to the Paladin's share that day; but it is observable that God in His compassion sends the good luck to such as are ill equipped with gifts, as compensation for their defect, but requires such as are more fortunately endowed to get by labor and talent what those others get by chance. It was Noël who said this, and it seemed to me to be well and justly thought.

The Paladin, going about the town all the day in order to be followed and admired and overhear the people say in an awed voice, "Ssh!—look, it is the Standard-Bearer of Joan of Arc!" had speech with all sorts and conditions of folk, and he learned from some boatmen that there was a stir of some kind going on in the bastilles on the other side of the river; and in the evening, seeking further, he found a deserter from the fortress called the "Augustins," who said that the English were going to send men over to strengthen the garrisons on our side during the darkness of the night, and were exulting greatly, for they meant to spring upon Dunois and the army when it was passing the bastilles and destroy it: a thing quite easy to do, since the "Witch" would not be there, and without her presence the army would do like the French armies of these many years past—drop their weapons and run when they saw an English face.

It was ten at night when the Paladin brought this news and asked leave to speak to Joan, and I was up and on duty then. It was a bitter stroke to me to see what a chance I

had lost. Joan made searching inquiries, and satisfied herself
that the word was true, then she made this annoying remark:

"You have done well, and you have my thanks. It may
be that you have prevented a disaster. Your name and service shall receive official mention."

Then he bowed low, and when he rose he was eleven feet
high. As he swelled out past me he covertly pulled down
the corner of his eye with his finger and muttered part of that
defiled refrain, "Oh tears, ah tears, oh sad sweet tears!—name
in General Orders—personal mention to the King, you see!"

I wished Joan could have seen his conduct, but she was busy
thinking what she would do. Then she had me fetch the
knight Jean de Metz, and in a minute he was off for La Hire's
quarters with orders for him and the Lord de Villars and Florent d'Iliers to report to her at five o'clock next morning with
five hundred picked men well mounted. The histories say
half-past four, but it is not true, I heard the order given.

We were on our way at five to the minute, and encountered
the head of the arriving column between six and seven, a
couple of leagues from the city. Dunois was pleased, for the
army had begun to get restive and show uneasiness now that
it was getting so near to the dreaded bastilles. But that all
disappeared now, as the word ran down the line, with a huzzah that swept along the length of it like a wave, that the
Maid was come. Dunois asked her to halt and let the column
pass in review, so that the men could be sure that the report
of her presence was not a ruse to revive their courage. So
she took position at the side of the road with her staff, and
the battalions swung by with a martial stride, huzzahing.
Joan was armed, except her head. She was wearing the cunning little velvet cap with the mass of curved white ostrich
plumes tumbling over its edges which the city of Orleans had
given her the night she arrived—the one that is in the picture
that hangs in the Hôtel de Ville at Rouen. She was looking
about fifteen. The sight of soldiers always set her blood to
leaping, and lit the fires in her eyes and brought the warm
rich color to her cheeks; it was then that you saw that she

was too beautiful to be of the earth, or at any rate that there was a subtle something somewhere about her beauty that differed it from the human types of your experience and exalted it above them.

In the train of wains laden with supplies a man lay on top of the goods. He was stretched out on his back, and his hands were tied together with ropes, and also his ankles. Joan signed to the officer in charge of that division of the train to come to her, and he rode up and saluted.

"What is he that is bound, there?" she asked.

"A prisoner, General."

"What is his offence?"

"He is a deserter."

"What is to be done with him?"

"He will be hanged, but it was not convenient on the march, and there was no hurry."

"Tell me about him."

"He is a good soldier, but he asked leave to go and see his wife who was dying, he said, but it could not be granted; so he went without leave. Meanwhile the march began, and he only overtook us yesterday evening."

"Overtook you? Did he come of his own will?"

"Yes, it was of his own will."

"*He* a deserter! Name of God! Bring him to me."

The officer rode forward and loosed the man's feet and brought him back with his hands still tied. What a figure he was—a good seven feet high, and built for business! He had a strong face; he had an unkempt shock of black hair which showed up in a striking way when the officer removed his morion for him; for weapon he had a big axe in his broad leathern belt. Standing by Joan's horse, he made Joan look littler than ever, for his head was about on a level with her own. His face was profoundly melancholy; all interest in life seemed to be dead in the man. Joan said—

"Hold up your hands."

The man's head was down. He lifted it when he heard that soft friendly voice, and there was a wistful something in

his face which made one think that there had been music in
it for him and that he would like to hear it again. When he
raised his hands Joan laid her sword to his bonds, but the
officer said with apprehension—

"Ah, madam—my General !"

"What is it ?" she said.

"He is under sentence !"

"Yes, I know. I am responsible for him "; and she cut the
bonds. They had lacerated his wrists, and they were bleeding. "Ah, pitiful !" she said; "blood—I do not like it";
and she shrank from the sight. But only for a moment.
"Give me something, somebody, to bandage his wrists with."

The officer said—

"Ah, my General! it is not fitting. Let me bring another to do it."

"Another? De par le Dieu ! You would seek far to find
one that can do it better than I, for I learned it long ago
among both men and beasts. And I can tie better than
those that did this; if I had tied him the ropes had not cut
his flesh."

The man looked on, silent, while he was being bandaged,
stealing a furtive glance at Joan's face occasionally, such as
an animal might that is receiving a kindness from an unexpected quarter and is gropingly trying to reconcile the act
with its source. All the staff had forgotten the huzzahing army
drifting by in its rolling clouds of dust, to crane their necks and
watch the bandaging as if it was the most interesting and
absorbing novelty that ever was. I have often seen people
do like that—get entirely lost in the simplest trifle, when it
is something that is out of their line. Now there in Poitiers,
once, I saw two bishops and a dozen of those grave and
famous scholars grouped together watching a man paint a
sign on a shop; they didn't breathe, they were as good as
dead ; and when it began to sprinkle they didn't know it at
first ; then they noticed it, and each man hove a deep sigh,
and glanced up with a surprised look as wondering to see the
others there, and how he came to be there himself—but that

is the way with people, as I have said. There is no way of accounting for people. You have to take them as they are.

"There," said Joan at last, pleased with her success; "another could have done it no better—not as well, I think. Tell me—what is it you did? Tell me all."

The giant said:

"It was this way, my angel. My mother died, then my three little children, one after the other, all in two years. It was the famine; others fared so—it was God's will. I saw them die; I had that grace; and I buried them. Then when my poor wife's fate was come, I begged for leave to go to her —she who was so dear to me — she who was all I had; I begged on my knees. But they would not let me. Could I let her die, friendless and alone? Could I let her die believing I would not come? Would she let *me* die and *she* not come—with her feet free to do it if she would, and no cost upon it but only her life? Ah, she would come—she would come through the fire! So I went. I saw her. She died in my arms. I buried her. Then the army was gone. I had trouble to overtake it, but my legs are long and there are many hours in a day; I overtook it last night."

Joan said, musingly, and as if she were thinking aloud—

"It sounds true. If true, it were no great harm to suspend the law this one time—any would say that. It may not be true, but if it *is* true—" She turned suddenly to the man and said, "I would see your eyes—look up!" The eyes of the two met, and Joan said to the officer, "The man is pardoned. Give you good-day; you may go." Then she said to the man, "Did you know it was death to come back to the army?"

"Yes," he said, "I knew it."

"Then why did you do it?"

The man said, quite simply—

"*Because* it was death. She was all I had. There was nothing left to love."

"Ah, yes, there was — France! The children of France have always their mother—*they* cannot be left with nothing to love. You shall live—and you shall serve France—"

" I will serve *you !*"

" —you shall fight for France—"

" I will fight for *you !*"

" You shall be France's soldier—"

" I will be *your* soldier !"

" —you shall give all your heart to France—"

" I will give all my heart to *you*—and all my soul, if I have one—and all my strength, which is great—for I was dead and am alive again ; I had nothing to live for, but now I have ! You are France for me. You are my France, and I will have no other."

Joan smiled, and was touched and pleased at the man's grave enthusiasm—solemn enthusiasm, one may call it, for the manner of it was deeper than mere gravity—and she said—

" Well, it shall be as you will. What are you called ?"

The man answered with unsmiling simplicity—

" They call me the Dwarf, but I think it is more in jest than otherwise."

It made Joan laugh, and she said—

" It has something of that look, truly ! What is the office of that vast axe ?"

The soldier replied with the same gravity — which must have been born to him, it sat upon him so naturally—

" It is to persuade persons to respect France."

Joan laughed again, and said—

" Have you given many lessons ?"

" Ah, indeed yes—many."

" The pupils behaved to suit you, afterwards ?"

" Yes ; it made them quiet—quite pleasant and quiet."

" I should think it would happen so. Would you like to be my man-at-arms ?—orderly, sentinel, or something like that ?"

" If I may !"

" Then you shall. You shall have proper armor, and shall go on teaching your art. Take one of those led horses there, and follow the staff when we move."

That is how we came by the Dwarf; and a good fellow he was. Joan picked him out on sight, but it wasn't a mistake; no one could be faithfuler than he was, and he was a devil and the son of a devil when he turned himself loose with his axe. He was so big that he made the Paladin look like an ordinary man. He liked to like people, therefore people liked him. He liked us boys from the start; and he liked the knights, and liked pretty much everybody he came across; but he thought more of a paring of Joan's finger-nail than he did of all the rest of the world put together.

Yes, that is where we got him—stretched on the wain, going to his death, poor chap, and nobody to say a good word for him. He was a good find. Why, the knights treated him almost like an equal—it is the honest truth; that is the sort of a man he was. They called him the Bastille, sometimes, and sometimes they called him Hellfire, which was on account of his warm and sumptuous style in battle, and you know they wouldn't have given him pet names if they hadn't had a good deal of affection for him.

To the Dwarf, Joan was France, the spirit of France made flesh—he never got away from that idea that he had started with; and God knows it was the true one. That was a humble eye to see so great a truth where some others failed. To me that seems quite remarkable. And yet, after all, it was, in a way, just what nations do. When they love a great and noble thing, they embody it—they want it so that they can see it with their eyes; like Liberty, for instance. They are not content with the cloudy abstract idea, they make a beautiful statue of it, and then their beloved idea is substantial and they can look at it and worship it. And so it is as I say; to the Dwarf, Joan was our country embodied, our country made visible flesh cast in a gracious form. When she stood before others, they saw Joan of Arc, but he saw France.

Sometimes he would speak of her by that name. It shows you how the idea was imbedded in his mind, and how real it was to him. The world has called our kings by it, but I

JOAN AND THE "DWARF"

know of none of them who has had so good a right as she to that sublime title.

When the march past was finished, Joan returned to the front and rode at the head of the column. When we began to file past those grim bastilles and could glimpse the men within, standing to their guns and ready to empty death into our ranks, such a faintness came over me and such a sickness that all things seemed to turn dim and swim before my eyes; and the other boys looked droopy too, I thought—including the Paladin, although I do not know this for certain, because he was ahead of me and I had to keep my eyes out toward the bastille side, because I could wince better when I saw what to wince at.

But Joan was at home—in Paradise, I might say. She sat up straight, and I could see that she was feeling different from me. The awfulest thing was the silence, there wasn't a sound but the screaking of the saddles, the measured tramp-lings, and the sneezing of the horses, afflicted by the smother-ing dust-clouds which they kicked up. I wanted to sneeze myself, but it seemed to me that I would rather go unsneezed, or suffer even a bitterer torture, if there is one, than attract attention to myself.

I was not of a rank to make suggestions, or I would have suggested that if we went faster we should get by sooner. It seemed to me that it was an ill-judged time to be taking a walk. Just as we were drifting in that suffocating stillness past a great cannon that stood just within a raised portcullis, with nothing between me and it but the moat, a most uncom-mon jackass in there split the world with his bray, and I fell out of the saddle. Sir Bertrand grabbed me as I went, which was well, for if I had gone to the ground in my armor I could not have gotten up again by myself. The English warders on the battlements laughed a coarse laugh, forgetting that every one must begin, and that there had been a time when they themselves would have fared no better when shot by a jackass.

The English never uttered a challenge nor fired a shot. It

was said afterwards that when their men saw the Maid riding at the front and saw how lovely she was, their eager courage cooled down in many cases and vanished in the rest, they feeling certain that that creature was not mortal, but the very child of Satan, and so the officers were prudent and did not try to make them fight. It was said also that some of the officers were affected by the same superstitious fears. Well, in any case, they never offered to molest us, and we poked by all the grisly fortresses in peace. During the march I caught up on my devotions, which were in arrears; so it was not all loss and no profit for me, after all.

It was on this march that the histories say Dunois told Joan that the English were expecting reinforcements under the command of Sir John Falstaff, and that she turned upon him and said—

"Bastard, Bastard, in God's name I warn you to let me know of his coming as soon as you hear of it; for if he passes without my knowledge you shall lose your head!"

It may be so; I don't deny it; but I didn't hear it. If she really said it I think she only meant she would take off his official head—degrade him from his command. It was not like her to threaten a comrade's life. She did have her doubts of her generals, and was entitled to them, for she was all for storm and assault, and they were for holding still and tiring the English out. Since they did not believe in her way and were experienced old soldiers, it would be natural for them to prefer their own and try to get around carrying hers out.

But I did hear something that the histories didn't mention and don't know about. I heard Joan say that now that the garrisons on the other side had been weakened to strengthen those on our side, the most effective point of operations had shifted to the south shore; so she meant to go over there and storm the forts which held the bridge end, and that would open up communication with our own dominions and raise the siege. The generals began to balk, privately, right away, but they only baffled and delayed her, and that for only four days.

All Orleans met the army at the gate and huzzahed it through the bannered streets to its various quarters, but nobody had to rock it to sleep; it slumped down dog-tired, for Dunois had rushed it without mercy, and for the next twenty-four hours it would be quiet, all but the snoring.

CHAPTER XVII

WHEN we got home, breakfast for us minor fry was waiting in our mess-room and the family honored us by coming in to eat it with us. The nice old treasurer, and in fact all three were flatteringly eager to hear about our adventures. Nobody asked the Paladin to begin, but he did begin, because now that his specially ordained and peculiar military rank set him above everybody on the personal staff but old D'Aulon, who didn't eat with us, he didn't care a farthing for the knights' nobility nor mine, but took precedence in the talk whenever it suited him, which was all the time, because he was born that way. He said:

"God be thanked, we found the army in admirable condition. I think I have never seen a finer body of animals."

"Animals?" said Miss Catherine.

"I will explain to you what he means," said Noël. "He—"

"I will trouble you not to trouble yourself to explain anything for me," said the Paladin, loftily. "I have reason to think—"

"That is his way," said Noël; "always when he thinks he has reason to think, he thinks he does think, but this is an error. He didn't see the army. I noticed him, and he didn't see it. He was troubled by his old complaint."

"What is his old complaint?" Catherine asked.

"Prudence," I said, seeing my chance to help.

But it was not a fortunate remark, for the Paladin said:

"It probably isn't your turn to criticise people's prudence —you who fall out of the saddle when a donkey brays."

They all laughed, and I was ashamed of myself for my hasty smartness. I said:

"It isn't quite fair for you to say I fell out on account of the donkey's braying. It was emotion, just ordinary emotion."

"Very well, if you want to call it that, I am not objecting. What would you call it, Sir Bertrand?"

"Well, it—well, whatever it was, it was excusable, I think. All of you have learned how to behave in hot hand-to-hand engagements, and you don't need to be ashamed of your record in that matter; but to walk along in front of death, with one's hands idle, and no noise, no music, and nothing going on, is a very trying situation. If I were you, De Conte, I would name the emotion; it's nothing to be ashamed of."

It was as straight and sensible a speech as ever I heard, and I was grateful for the opening it gave me; so I came out and said:

"It was fear—and thank you for the honest idea, too."

"It was the cleanest and best way out," said the old treasurer; "you've done well, my lad."

That made me comfortable, and when Miss Catharine said, "It's what I think, too," I was grateful to myself for getting into that scrape.

Sir Jean de Metz said—

"We were all in a body together when the donkey brayed, and it was dismally still at the time. I don't see how any young campaigner could escape some little touch of that emotion."

He looked about him with a pleasant expression of inquiry on his good face, and as each pair of eyes in turn met his the head they were in nodded a confession. Even the Paladin delivered his nod. That surprised everybody, and saved the Standard-Bearer's credit. It was clever of him; nobody believed he could tell the truth that way without practice, or would tell that particular sort of a truth either with or without practice. I suppose he judged it would favorably impress the family. Then the old treasurer said—

"Passing the forts in that trying way required the same sort of nerve that a person must have when ghosts are about him in the dark, I should think. What does the Standard-Bearer think?"

"Well, I don't quite know about that, sir. I've often thought I would like to see a ghost if I—"

"Would you?" exclaimed the young lady. "We've got one! Would you try that one? Will you?"

She was so eager and pretty that the Paladan said straight out that he would; and then as none of the rest had bravery enough to expose the fear that was in him, one volunteered after the other with a prompt mouth and a sick heart till all were shipped for the voyage; then the girl clapped her hands in glee, and the parents were gratified too, saying that the ghosts of their house had been a dread and a misery to them and their forebears for generations, and nobody had ever been found yet who was willing to confront them and find out what their trouble was, so that the family could heal it and content the poor spectres and beguile them to tranquillity and peace.

CHAPTER XVIII

ABOUT noon I was chatting with Madame Boucher; nothing was going on, all was quiet, when Catherine Boucher suddenly entered in great excitement, and said—

"Fly, sir, fly! The Maid was dozing in her chair in my room, when she sprang up and cried out, 'French blood is flowing!—my arms, give me my arms!' Her giant was on guard at the door, and he brought D'Aulon, who began to arm her, and I and the giant have been warning the staff. Fly! —and stay by her; and if there really is a battle, keep her out of it—don't let her risk herself—there is no need—if the men know she is near and looking on, it is all that is necessary. Keep her out of the fight—don't fail of this!"

I started on a run, saying, sarcastically—for I was always fond of sarcasm, and it was said that I had a most neat gift that way—

"Oh yes, nothing easier than that—I'll attend to it!"

At the furthest end of the house I met Joan, fully armed, hurrying toward the door, and she said—

"Ah, French blood is being spilt, and you did not tell me."

"Indeed I did not know it," I said; "there are no sounds of war; everything is quiet, your Excellency."

"You will hear war-sounds enough in a moment," she said, and was gone.

It was true. Before one could count five there broke upon the stillness the swelling rush and tramp of an approaching multitude of men and horses, with hoarse cries of command; and then out of the distance came the muffled deep *boom!* —*boom-boom!*—*boom!* of cannon, and straightway that rushing multitude was roaring by the house like a hurricane.

Our knights and all our staff came flying, armed, but with no horses ready, and we burst out after Joan in a body, the Paladin in the lead with the banner. The surging crowd was made up half of citizens and half of soldiers, and had no recognized leader. When Joan was seen a huzzah went up, and she shouted—

"A horse—a horse!"

A dozen saddles were at her disposal in a moment. She mounted, a hundred people shouting—

"Way, there—way for the MAID OF ORLEANS!" The first time that that immortal name was ever uttered—and I, praise God, was there to hear it! The mass divided itself like the waters of the Red Sea, and down this lane Joan went skimming like a bird, crying "Forward, French hearts—follow me!" and we came winging in her wake on the rest of the borrowed horses, the holy standard streaming above us, and the lane closing together in our rear.

This was a different thing from the ghastly march past the dismal bastilles. No, we felt fine, now, and all a-whirl with enthusiasm. The explanation of this sudden uprising was this. The city and the little garrison, so long hopeless and afraid, had gone wild over Joan's coming, and could no longer restrain their desire to get at the enemy; so, without orders from anybody, a few hundred soldiers and citizens had plunged out at the Burgundy gate on a sudden impulse and made a charge on one of Lord Talbot's most formidable fortresses—St. Loup—and were getting the worst of it. The news of this had swept through the city and started this new crowd that we were with.

As we poured out at the gate we met a force bringing in the wounded from the front. The sight moved Joan, and she said—

"Ah, French blood; it makes my hair rise to see it!"

We were soon on the field, soon in the midst of the turmoil. Joan was seeing her first real battle, and so were we.

It was a battle in the open field; for the garrison of St. Loup had sallied confidently out to meet the attack, being used to victories when "witches" were not around. The sally had been re-enforced by troops from the "Paris" bastille, and when we approached the French were getting whipped and were falling back. But when Joan came charging through the disorder with her banner displayed, crying "Forward, men —follow me!" there was a change; the French turned about and surged forward like a solid wave of the sea, and swept the English before them, hacking and slashing, and being hacked and slashed, in a way that was terrible to see.

In the field the Dwarf had no assignment; that is to say, he was not under orders to occupy any particular place, therefore he chose his place for himself, and went ahead of Joan and made a road for her. It was horrible to see the iron helmets fly into fragments under his dreadful axe. He called it cracking nuts, and it looked like that. He made a good road, and paved it well with flesh and iron. Joan and the rest of us followed it so briskly that we outspeeded our forces and had the English behind us as well as before. The knights commanded us to face outwards around Joan, which we did, and then there was work done that was fine to see. One was obliged to respect the Paladin, now. Being right under Joan's exalting and transforming eye, he forgot his native prudence, he forgot his diffidence in the presence of danger, he forgot what fear was, and he never laid about him in his imaginary battles in a more tremendous way than he did in this real one; and wherever he struck there was an enemy the less.

We were in that close place only a few minutes; then our forces to the rear broke through with a great shout and joined us, and then the English fought a retreating fight, but in a fine and gallant way, and we drove them to their fortress foot by foot, they facing us all the time, and their reserves on the walls raining showers of arrows, cross-bow bolts, and stone cannon-balls upon us.

The bulk of the enemy got safely within the works and left us outside with piles of French and English dead and wounded for company—a sickening sight, an awful sight to us youngsters, for our little ambush fights in February had been in the night, and the blood and the mutilations and the dead faces were mercifully dim, whereas we saw these things now for the first time in all their naked ghastliness.

Now arrived Dunois from the city, and plunged through the battle on his foam-flecked horse and galloped up to Joan, saluting, and uttering handsome compliments as he came. He waved his hand toward the distant walls of the city, where a multitude of flags were flaunting gayly in the wind, and said the populace were up there observing her fortunate performance and rejoicing over it, and added that she and the forces would have a great reception now.

"*Now?* Hardly now, Bastard. Not yet!"

"Why not yet? Is there more to be done?"

"More, Bastard? We have but begun! We will take this fortress."

"Ah, you can't be serious! We can't take this place; let me urge you not to make the attempt; it is too desperate. Let me order the forces back."

Joan's heart was overflowing with the joys and enthusiasms of war, and it made her impatient to hear such talk. She cried out—

"Bastard, Bastard, will ye play *always* with these English? Now verily I tell you we will not budge until this place is ours. We will carry it by storm. Sound the charge!"

"Ah, my General—"

"Waste no more time, man—let the bugles sound the assault!" and we saw that strange deep light in her eye which we named the battle-light, and learned to know so well in later fields.

The martial notes pealed out, the troops answered with a yell, and down they came against that formidable work, whose outlines were lost in its own cannon smoke, and whose sides were spouting flame and thunder.

We suffered repulse after repulse, but Joan was here and there and everywhere encouraging the men, and she kept them to their work. During three hours the tide ebbed and flowed, flowed and ebbed; but at last La Hire, who was now come, made a final and resistless charge, and the bastille St. Loup was ours. We gutted it, taking all its stores and artillery, and then destroyed it.

When all our host was shouting itself hoarse with rejoicings, and there went up a cry for the General, for they wanted to praise her and glorify her and do her homage for her victory, we had trouble to find her; and when we did find her, she was off by herself, sitting among a ruck of corpses, with her face in her hands, crying—for she was a young girl, you know, and her hero-heart was a young girl's heart too, with the pity and the tenderness that are natural to it. She was thinking of the mothers of those dead friends and enemies.

Among the prisoners were a number of priests, and Joan took these under her protection and saved their lives. It was urged that they were most probably combatants in disguise, but she said—

"As to that, how can any tell? They wear the livery of God, and if even one of these wears it rightfully, surely it were better that all the guilty should escape than that we have upon our hands the blood of that innocent man. I will lodge them where I lodge, and feed them, and send them away in safety."

We marched back to the city with our crop of cannon and prisoners on view and our banners displayed. Here was the first substantial bit of war-work the imprisoned people had seen in the seven months that the siege had endured, the first chance they had had to rejoice over a French exploit. You may guess that they made good use of it. They and the bells went mad. Joan was their darling now, and the press of people struggling and shouldering each other to get a glimpse of her was so great that we could hardly push our way through the streets at all. Her new name had gone all about, and was on everybody's lips. The Holy Maid of Vaucouleurs was a forgotten title; the city had claimed her for its own, and she was the MAID OF ORLEANS now. It is a happiness to me to remember that I heard that name the first time it was ever uttered. Between that first utterance and the last time it will be uttered on this earth—ah, think how many mouldering ages will lie in that gap!

The Boucher family welcomed her back as if she had been a child of the house, and saved from death against all hope or probability. They chided her for going into the battle and exposing herself to danger during all those hours. They could not realize that she had meant to carry her warriorship so far, and asked her if it had really been her purpose to go right into the turmoil of the fight, or hadn't she got swept into it by accident and the rush of the troops? They begged her to be more careful another time. It was good advice, maybe, but it fell upon pretty unfruitful soil.

CHAPTER XIX

BEING worn out with the long fight, we all slept the rest of the afternoon away and two or three hours into the night. Then we got up refreshed, and had supper. As for me, I could have been willing to let the matter of the ghost drop; and the others were of a like mind no doubt, for they talked diligently of the battle and said nothing of that other thing. And indeed it was fine and stirring to hear the Paladin rehearse his deeds and see him pile his dead, fifteen here, eighteen there, and thirty-five yonder; but this only postponed the trouble; it could not do more. He could not go on forever; when he had carried the bastille by assault and eaten up the garrison there was nothing for it but to stop, unless Catherine Boucher would give him a new start and have it all done over again—as we hoped she would, this time—but she was otherwise minded. As soon as there was a good opening and a fair chance, she brought up her unwelcome subject, and we faced it the best we could.

We followed her and her parents to the haunted room at eleven o'clock, with candles, and also with torches to place in the sockets on the walls. It was a big house, with very thick walls, and this room was in a remote part of it which had been left unoccupied for nobody knew how many years, because of its evil repute.

This was a large room, like a salon, and had a big table in it of enduring oak and well preserved; but the chairs were worm-eaten and the tapestry on the walls was rotten and discolored by age. The dusty cobwebs under the ceiling had the look of not having had any business for a century.

Catherine said—

"Tradition says that these ghosts have never been seen—they have merely been heard. It is plain that this room was once larger than it is now, and that the wall at this end was built in some bygone time to make and fence off a narrow room there. There is no communication anywhere with that narrow room, and if it exists — and of that there is no reasonable doubt—it has no light and no air, but is an absolute dungeon. Wait where you are, and take note of what happens."

That was all. Then she and her parents left us. When their footfalls had died out in the distance down the empty stone corridors an uncanny silence and solemnity ensued which was dismaller to me than the mute march past the bastilles. We sat looking vacantly at each other, and it was easy to see that no one there was comfortable. The longer we sat so, the more deadly still that stillness got to be; and when the wind began to moan around the house presently, it made me sick and miserable, and I wished I had been brave enough to be a coward this time, for indeed it is no proper shame to be afraid of ghosts, seeing how helpless the living are in their hands. And then these ghosts were invisible, which made the matter the worse, as it seemed to me. They might be in the room with us at that moment—we could not know. I felt airy touches on my shoulders and my hair, and I shrank from them and cringed, and was not ashamed to show this fear, for I saw the others doing the like, and knew that they were feeling those faint contacts too. As this went on—oh, eternities it seemed, the time dragged so drearily—all those faces became as wax, and I seemed sitting with a congress of the dead.

At last, faint and far and weird and slow, came a "boom!—boom!—boom!"—a distant bell tolling midnight. When the last stroke died, that depressing stillness followed again, and as before I was staring at those waxen faces and feeling those airy touches on my hair and my shoulders once more.

One minute—two minutes—three minutes of this, then we heard a long deep groan, and everybody sprang up and stood,

JOAN'S ENTRY INTO ORLEANS
(From a painting by Scherrer)

with his legs quaking. It came from that little dungeon. There was a pause, then we heard muffled sobbings, mixed with pitiful ejaculations. Then there was a second voice, low and not distinct, and the one seemed trying to comfort the other; and so the two voices went on, with moanings, and soft sobbings, and, ah, the tones were so full of compassion and sorrow and despair! Indeed, it made one's heart sore to hear it.

But those sounds were so real and so human and so moving that the idea of ghosts passed straight out of our minds, and Sir Jean de Metz spoke out and said—

"Come! we will smash that wall and set those poor captives free. Here, with your axe!"

The Dwarf jumped forward, swinging his great axe with both hands, and others sprang for the torches and brought them. Bang!—whang!—slam!—smash went the ancient bricks, and there was a hole an ox could pass through. We plunged within and held up the torches.

Nothing there but vacancy! On the floor lay a rusty sword and a rotten fan.

Now you know all that I know. Take the pathetic relics, and weave about them the romance of the dungeon's long-vanished inmates as best you can.

CHAPTER XX

THE next day Joan wanted to go against the enemy again, but it was the feast of the Ascension, and the holy council of bandit generals were too pious to be willing to profane it with bloodshed. But privately they profaned it with plottings, a sort of industry just in their line. They decided to do the only thing proper to do now in the new circumstances of the case—feign an attack on the most important bastille on the Orleans side, and then, if the English weakened the far more important fortresses on the other side of the river to come to its help, cross in force and capture those works. This would give them the bridge and free communication with the Sologne, which was French territory. They decided to keep this latter part of the programme secret from Joan.

Joan intruded and took them by surprise. She asked them what they were about and what they had resolved upon. They said they had resolved to attack the most important of the English bastilles on the Orleans side next morning—and there the spokesman stopped. Joan said—

"Well, go on."

"There is nothing more. That is all."

"Am I to believe this? That is to say, am I to believe that you have lost your wits?" She turned to Dunois, and said, "Bastard, you have sense, answer me this : if this attack is made and the bastille taken, how much better off would we be than we are now?"

The Bastard hesitated, and then began some rambling talk not quite germane to the question. Joan interrupted him and said—

"That will do, good Bastard, you have answered. Since the Bastard is not able to mention any advantage to be gained by taking that bastille and stopping there, it is not likely that any of you could better the matter. You waste much time here in inventing plans that lead to nothing, and making delays that are a damage. Are you concealing something from me? Bastard, this council has a general plan, I take it; without going into details, what is it?"

"It is the same it was in the beginning, seven months ago —to get provisions in for a long siege, and then sit down and tire the English out."

"In the name of God! As if seven months was not enough, you want to provide for a year of it. Now ye shall drop these pusillanimous dreams — the English shall go in three days!"

Several exclaimed—

"Ah, General, General, be prudent!"

"Be prudent and starve? Do ye call that war? I tell you this, if you do not already know it: The new circumstances have changed the face of matters. The true point of attack has shifted; it is on the other side of the river, now. One must take the fortifications that command the bridge. The English know that if we are not fools and cowards we will try to do that. They are grateful for your piety in wasting this day. They will re-enforce the bridge forts from this side to-night, knowing what ought to happen to-morrow. You have but lost a day and made our task harder, for we *will* cross and take the bridge forts. Bastard, tell me the truth—does not this council know that there is no other course for us than the one I am speaking of?"

Dunois conceded that the council did know it to be the most desirable, but considered it impracticable; and he excused the council as well as he could by saying that inasmuch as nothing was really and rationally to be hoped for but a long continuance of the siege and wearying out of the English, they were naturally a little afraid of Joan's impetuous notions. He said—

"You see, we are sure that the waiting game is the best, whereas you would carry everything by storm."

"That I would!—and moreover that I will! You have my orders—here and now. We will move upon the forts of the south bank to-morrow at dawn."

"And carry them by storm?"

"Yes, carry them by storm!"

La Hire came clanking in, and heard the last remark. He cried out—

"By *my baton*, that is the music I love to hear! Yes, that is the right tune and the beautiful words, my General—we will carry them by storm!"

He saluted in his large way and came up and shook Joan by the hand.

Some member of the council was heard to say—

"It follows, then, that we must begin with the bastille St. John, and that will give the English time to—"

Joan turned and said—

"Give yourselves no uneasiness about the bastille St. John. The English will know enough to retire from it and fall back on the bridge bastilles when they see us coming." She added, with a touch of sarcasm, "Even a war-council would know enough to do that, itself."

Then she took her leave. La Hire made this general remark to the council:

"She is a child, and that is all ye seem to see. Keep to that superstition if you must, but you perceive that this child understands this complex game of war as well as any of you; and if you want my opinion without the trouble of asking for it, here you have it without ruffles or embroidery—by God, I think she can teach the best of you how to *play* it!"

Joan had spoken truly; the sagacious English saw that the policy of the French had undergone a revolution; that the policy of paltering and dawdling was ended; that in place of taking blows, blows were to be struck, now; therefore they made ready for the new state of things by transferring heavy re-enforcements to the bastilles of the south bank from those of the north.

The city learned the great news that once more in French history, after all these humiliating years, France was going to take the offensive; that France, so used to retreating, was going to advance; that France, so long accustomed to skulking, was going to face about and strike. The joy of the people passed all bounds. The city walls were black with them to see the army march out in the morning in that strange new position—its front, not its tail, toward an English camp. You shall imagine for yourselves what the excitement was like and how it expressed itself, when Joan rode out at the head of the host with her banner floating above her.

We crossed the river in strong force, and a tedious long job it was, for the boats were small and not numerous. Our landing on the island of St. Aignan was not disputed. We threw a bridge of a few boats across the narrow channel thence to the south shore and took up our march in good order and unmolested; for although there was a fortress there— St. John—the English vacated and destroyed it and fell back on the bridge forts below as soon as our first boats were seen to leave the Orleans shore; which was what Joan had said would happen, when she was disputing with the council.

We moved down the shore and Joan planted her standard before the bastille of the Augustins, the first of the formidable works that protected the end of the bridge. The trumpets sounded the assault, and two charges followed in handsome style; but we were too weak, as yet, for our main body was still lagging behind. Before we could gather for a third assault the garrison of St. Privé were seen coming up to re-enforce the big bastille. They came on a run, and the Augustins sallied out, and both forces came against us with a rush, and sent our small army flying in a panic, and followed us, slashing and slaying, and shouting jeers and insults at us.

Joan was doing her best to rally the men, but their wits were gone, their hearts were dominated for the moment by the old-time dread of the English. Joan's temper flamed up, and she halted and commanded the trumpets to sound the advance. Then she wheeled about and cried out—

13

"If there is but a dozen of you that are not cowards, it is enough—follow me!"

Away she went, and after her a few dozen who had heard her words and been inspired by them. The pursuing force was astonished to see her sweeping down upon them with this handful of men, and it was their turn now to experience a grisly fright—surely this *is* a witch, this is a child of Satan! That was their thought—and without stopping to analyze the matter they turned and fled in a panic.

Our flying squadrons heard the bugle and turned to look; and when they saw the Maid's banner speeding in the other direction and the enemy scrambling ahead of it in disorder, their courage returned and they came scouring after us.

La Hire heard it and hurried his force forward and caught up with us just as we were planting our banner again before the ramparts of the Augustins. We were strong enough now. We had a long and tough piece of work before us, but we carried it through before night, Joan keeping us hard at it, and she and La Hire saying we were able to take that big bastille, and *must*. The English fought like—well, they fought like the English; when that is said, there is no more to say. We made assault after assault, through the smoke and flame and the deafening cannon-blasts, and at last as the sun was sinking we carried the place with a rush, and planted our standard on its walls.

The Augustins was ours. The Tourelles must be ours too, if we would free the bridge and raise the siege. We had achieved one great undertaking, Joan was determined to accomplish the other. We must lie on our arms where we were, hold fast to what we had got, and be ready for business in the morning. So Joan was not minded to let the men be demoralized by pillage and riot and carousings; she had the Augustins burned, with all its stores in it, excepting the artillery and ammunition.

Everybody was tired out with this long day's hard work, and of course this was the case with Joan; still, she wanted to stay with the army before the Tourelles, to be ready for the assault in the morning. The chiefs argued with her, and at last persuaded her to go home and prepare for the great work by taking proper rest, and also by having a leech look to a wound which she had received in her foot. So we crossed with them and went home.

Just as usual, we found the town in a fury of joy, all the bells clanging, everybody shouting, and several people drunk. We never went out or came in without furnishing good and sufficient reasons for one of these pleasant tempests, and so the tempest was always on hand. There had been a blank absence of reasons for this sort of upheavals for the past seven months, therefore the people took to the upheavals with all the more relish on that account.

CHAPTER XXI

To get away from the usual crowd of visitors and have a rest, Joan went with Catherine straight to the apartment which the two occupied together, and there they took their supper and there the wound was dressed. But then, instead of going to bed, Joan, weary as she was, sent the Dwarf for me, in spite of Catherine's protests and persuasions. She said she had something on her mind, and must send a courier to Domremy with a letter for our old Père Fronte to read to her mother. I came, and she began to dictate. After some loving words and greetings to her mother and the family, came this :

"But the thing which moves me to write now, is to say that when you presently hear that I am wounded, you shall give yourself no concern about it, and refuse faith to any that shall try to make you believe it is serious."

She was going on, when Catharine spoke up and said:

"Ah, but it will fright her so to read these words. Strike them out, Joan, strike them out, and wait only one day—two days at most—then write and say your foot *was* wounded but is well again—for it will surely be well then, or very near it. Don't distress her, Joan ; do as I say."

A laugh like the laugh of the old days, the impulsive free laugh of an untroubled spirit, a laugh like a chime of bells, was Joan's answer ; then she said—

"My foot? Why should I write about such a scratch as that ? I was not thinking of it, dear heart."

"Child, have you another wound and a worse, and have not spoken of it? What have you been dreaming about, that you—"

She had jumped up, full of vague fears, to have the leech called back at once, but Joan laid her hand upon her arm and made her sit down again, saying—

"There, now, be tranquil, there is no other wound, as yet; I am writing about one which I shall get when we storm that bastille to-morrow."

Catherine had the look of one who is trying to understand a puzzling proposition but cannot quite do it. She said, in a distraught fashion—

"A wound which you are *going* to get? But — but why grieve your mother when it—when it may not happen?"

"*May* not? Why, it *will*."

The puzzle was a puzzle still. Catherine said in that same abstracted way as before—

"*Will*. It is a strong word. I cannot seem to—my mind is not able to take hold of this. Oh, Joan, such a presentiment is a dreadful thing—it takes one's peace and courage all away. Cast it from you!—drive it out! It will make your whole night miserable, and to no good; for we will hope—"

"But it isn't a presentiment—it is a fact. And it will not make me miserable. It is uncertainties that do that, but this is not an uncertainty."

"Joan, do you *know* it is going to happen?"

"Yes, I know it. My Voices told me."

"Ah," said Catherine, resignedly, "if *they* told you— But are you sure it was they?—quite sure?"

"Yes, quite. It will happen—there is no doubt."

"It is dreadful! Since when have you known it?"

"Since—I think it is several weeks." Joan turned to me. "Louis, you will remember. How long is it?"

"Your Excellency spoke of it first to the King, in Chinon," I answered; "that was as much as seven weeks ago. You spoke of it again the 20th of April, and also the 22d, two weeks ago, as I see by my record here."

These marvels disturbed Catherine profoundly, but I had long ceased to be surprised at them. One can get used to anything in this world. Catherine said—

"And it is to happen to-morrow?—always to-morrow? Is it the same date always? There has been no mistake, and no confusion?"

"No," Joan said, "the 7th of May is the date—there is no other."

"Then you shall not go a step out of this house till that awful day is gone by! You will not dream of it, Joan, *will* you?—promise that you will stay with us."

But Joan was not persuaded. She said—

"It would not help the matter, dear good friend. The wound is to come, and come to-morrow. If I do not seek it, it will seek me. My duty calls me to that place to-morrow; I should have to go if my death were waiting for me there; shall I stay away for only a wound? Oh no, we must try to do better than that."

"Then you are determined to go?"

"Of a certainty, yes. There is only one thing that I can do for France—hearten her soldiers for battle and victory." She thought a moment, then added, "However, one should not be unreasonable, and I would do much to please you, who are so good to me. Do you love France?"

I wondered what she might be contriving now, but I saw no clew. Catherine said, reproachfully—

"Ah, what have I done to deserve this question?"

"Then you do love France. I had not doubted it, dear. Do not be hurt, but answer me—have you ever told a lie?"

"In my life I have not wilfully told a lie—fibs, but no lies."

"That is sufficient. You love France and do not tell lies; therefore I will trust you. I will go or I will stay, as you shall decide."

"Oh, I thank you from my heart, Joan! How good and dear it is of you to do this for me! Oh, you shall stay, and not go!"

In her delight she flung her arms about Joan's neck and squandered endearments upon her the least of which would have made me rich, but as it was, they only made me realize how poor I was—how miserably poor in what I would most have prized in this world. Joan said—

"Then you will send word to my headquarters that I am not going?"

"Oh, gladly. Leave that to me."

"It is good of you. And how will you word it?—for it must have proper official form. Shall I word it for you?"

"Oh, do — for you know about these solemn procedures and stately proprieties, and I have had no experience."

"Then word it like this : ' The chief of staff is commanded to make known to the King's forces in garrison and in the field, that the General-in-Chief of the Armies of France will not face the English on the morrow, she being afraid she may get hurt. Signed, JOAN OF ARC, by the hand of CATHERINE BOUCHER, who loves France.' "

There was a pause—a silence of the sort that tortures one into stealing a glance to see how the situation looks, and I did that. There was a loving smile on Joan's face, but the color was mounting in crimson waves into Catherine's, and her lips were quivering and the tears gathering ; then she said—

"Oh, I am so ashamed of myself !—and you are so noble and brave and wise, and I am so paltry—so paltry and such a fool !" and she broke down and began to cry, and I did so want to take her in my arms and comfort her, but Joan did it, and of course I said nothing. Joan did it well, and most sweetly and tenderly, but I could have done it as well, though I knew it would be foolish and out of place to suggest such a thing, and might make an awkwardness too, and be embarrassing to us all, so I did not offer, and I hope I did right and for the best, though I could not know, and was many times tortured with doubts afterwards as having perhaps let a chance pass which might have changed all my life and made it happier and more beautiful than, alas, it turned out to be. For this reason I grieve yet, when I think of that scene, and do not like to call it up out of the deeps of my memory because of the pangs it brings.

Well, well, a good and wholesome thing is a little harmless fun in this world; it tones a body up and keeps him human and prevents him from souring. To set that little trap for Catherine was as good and effective a way as any to show her what a grotesque thing she was asking of Joan. It *was* a funny idea, now, wasn't it, when you look at it all around? Even Catherine dried up her tears and laughed when she thought of the English getting hold of the French Commander-in-Chief's reason for staying out of a battle. She granted that they could have a good time over a thing like that.

We got to work on the letter again, and of course did not have to strike out the passage about the wound. Joan was in fine spirits; but when she got to sending messages to this, that, and the other old playmate and friend, it brought our village and the Fairy Tree and the flowery plain and the browsing sheep and all the peaceful beauty of our old humble home-place back, and the familiar names began to tremble on her lips; and when she got to Haumette and Little Mengette it was no use, her voice broke and she couldn't go on. She waited a moment, then said—

"Give them my love—my warm love—my deep love—oh, out of my heart of hearts! I shall never see our home any more."

Now came Pasquerel, Joan's confessor, and introduced a gallant knight, the Sire de Rais, who had been sent with a message. He said he was instructed to say that the council had decided that enough had been done for the present; that it would be safest and best to be content with what God had already done; that the city was now well victualled and able to stand a long siege; that the wise course must necessarily be to withdraw the troops from the other side of the river and resume the defensive—therefore they had decided accordingly.

"The incurable cowards!" exclaimed Joan. "So it was to get me away from my men that they pretended so much solicitude about my fatigue. Take this message back, not to the council—I have no speeches for those disguised ladies' maids—but to the Bastard and La Hire, who are men. Tell them the army is to remain where it is, and I hold them responsible if this command miscarries. And say the offensive will be resumed in the morning. You may go, good sir."

Then she said to her priest—

" Rise early, and be by me all the day. There will be much work on my hands, and I shall be hurt between my neck and my shoulder."

CHAPTER XXII

WE were up at dawn, and after mass we started. In the hall we met the master of the house, who was grieved, good man, to see Joan going breakfastless to such a day's work, and begged her to wait and eat, but she couldn't afford the time—that is to say, she couldn't afford the patience, she being in such a blaze of anxiety to get at that last remaining bastille which stood between her and the completion of the first great step in the rescue and redemption of France. Boucher put in another plea:

"But think—we poor beleaguered citizens who have hardly known the flavor of fish for these many months, have spoil of that sort again, and we owe it to you. There's a noble shad for breakfast; wait—be persuaded."

Joan said—

"Oh, there's going to be fish in plenty; when this day's work is done the whole river-front will be yours to do as you please with."

"Ah, your Excellency will do well, *that* I know; but we don't require quite that much, even of you; you shall have a month for it in place of a day. Now be beguiled—wait and eat. There's a saying that he that would cross a river twice in the same day in a boat, will do well to eat fish for luck, lest he have an accident."

"That doesn't fit my case, for to-day I cross but once in a boat."

"Oh, don't say that. Aren't you coming back to us?"

"Yes, but not in a boat."

"How, then?"

"By the bridge."

"Listen to that—by the bridge! Now stop this jesting, dear General, and do as I would have you. It's a noble fish."

"Be good, then, and save me some for supper; and I will bring one of those Englishmen with me and he shall have his share."

"Ah, well, have your way if you must. But he that fasts must attempt but little and stop early. When shall you be back?"

"When I've raised the siege of Orleans. FORWARD!"

We were off. The streets were full of citizens and of groups and squads of soldiers, but the spectacle was melancholy. There was not a smile anywhere, but only universal gloom. It was as if some vast calamity had smitten all hope and cheer dead. We were not used to this, and were astonished. But when they saw the Maid, there was an immediate stir, and the eager question flew from mouth to mouth—

"Where is she going? Whither is she bound?"

Joan heard it, and called out—

"Whither would ye suppose? I am going to take the Tourelles."

It would not be possible for any to describe how those few words turned that mourning into joy—into exaltation—into frenzy; and how a storm of huzzahs burst out and swept down the streets in every direction and woke those corpse-like multitudes to vivid life and action and turmoil in a moment. The soldiers broke from the crowd and came flocking to our standard, and many of the citizens ran and got pikes and halberds and joined us. As we moved on, our numbers increased steadily, and the hurrahing continued — yes, we moved through a solid cloud of noise, as you may say, and all the windows on both sides contributed to it, for they were filled with excited people.

You see, the council had closed the Burgundy gate and placed a strong force there, under that stout soldier Raoul de Gaucourt, Bailly of Orleans, with orders to prevent Joan from getting out and resuming the attack on the Tourelles, and this shameful thing had plunged the city into sorrow and despair. But that feeling was gone now. They believed the Maid was a match for the council, and they were right.

When we reached the gate, Joan told Gaucourt to open it and let her pass.

He said it would be impossible to do this, for his orders were from the council and were strict. Joan said—

"There is no authority above mine but the King's. If you have an order from the King, produce it."

"I cannot claim to have an order from him, General."

"Then make way, or take the consequences!"

He began to argue the case, for he was like the rest of the tribe, always ready to fight with words, not acts; but in the midst of his gabble Joan interrupted with the terse order—

"Charge!"

We came with a rush, and brief work we made of that small job. It was good to see the Bailly's surprise. He was not used to this unsentimental promptness. He said afterwards that he was cut off in the midst of what he was saying—in the midst of an argument by which he could have proved that he could not let Joan pass—an argument which Joan could not have answered.

"Still, it appears she did answer it," said the person he was talking to.

We swung through the gate in great style, with a vast accession of noise, the most of which was laughter, and soon our van was over the river and moving down against the Tourelles.

First we must take a supporting work called a boulevard, and which was otherwise nameless, before we could assault the great bastille. Its rear communicated with the bastille by a drawbridge, under which ran a swift and deep strip of the Loire. The boulevard was strong, and Dunois doubted our ability to take it, but Joan had no such doubt. She pounded it with artillery all the forenoon, then about noon she ordered an assault and led it herself. We poured into the fosse through the smoke and a tempest of missiles, and Joan,

shouting encouragements to her men, started to climb a
scaling-ladder, when that misfortune happened which we
knew was to happen—the iron bolt from an arbalest struck
between her neck and her shoulder, and tore its way down
through her armor. When she felt the sharp pain and saw
her blood gushing over her breast, she was frightened, poor
girl, and as she sank to the ground she began to cry, bitterly.

The English sent up a glad shout and came surging down
in strong force to take her, and then for a few minutes the
might of both adversaries was concentrated upon that spot.
Over her and about her, English and French fought with des-
peration—for she stood for France, indeed she *was* France to
both sides—whichever won her won France, *and could keep
it forever.* Right there in that small spot, and in ten minutes
by the clock, the fate of France, for all time, was to be decided, and *was* decided.

If the English had captured Joan then, Charles VII. would
have flown the country, the Treaty of Troyes would have
held good, and France, already English property, would have
become, without further dispute, an English province, to so
remain until the Judgment Day. A nationality and a kingdom were at stake there, and no more time to decide it in
than it takes to hard-boil an egg. It was the most momentous
ten minutes that the clock has ever ticked in France, or ever
will. Whenever you read in histories about hours or days or
weeks in which the fate of one or another nation hung in the
balance, do not you fail to remember, nor your French hearts
to beat the quicker for the remembrance, the ten minutes that
France, called otherwise Joan of Arc, lay bleeding in the fosse
that day, with two nations struggling over her for her possession.

And you will not forget the Dwarf. For he stood over her,
and did the work of any six of the others. He swung his
axe with both hands ; whenever it came down, he said those
two words, " For France !" and a splintered helmet flew
like egg-shells, and the skull that carried it had learned its
manners and would offend the French no more. He piled

a bulwark of iron-clad dead in front of him and fought from behind it; and at last when the victory was ours we closed about him, shielding him, and he ran up a ladder with Joan as easily as another man would carry a child, and bore her out of the battle, a great crowd following and anxious, for she was drenched with blood to her feet, half of it her own and the other half English, for bodies had fallen across her as she lay and had poured their red life-streams over her. One couldn't see the white armor now, with that awful dressing over it.

The iron bolt was still in the wound—some say it projected out behind the shoulder. It may be—I did not wish to see, and did not try to. It was pulled out, and the pain made Joan cry again, poor thing. Some say she pulled it out herself because others refused, saying they could not bear to hurt her. As to this I do not know; I only know it was pulled out, and that the wound was treated with oil and properly dressed.

Joan lay on the grass, weak and suffering, hour after hour, but still insisting that the fight go on. Which it did, but not to much purpose, for it was only under her eye that men were heroes and not afraid. They were like the Paladin; I think he was afraid of his shadow—I mean in the afternoon, when it was very big and long; but when he was under Joan's eye and the inspiration of her great spirit, what was he afraid of? Nothing in this world—and that is just the truth.

Toward night Dunois gave it up. Joan heard the bugles.

"What!" she cried. "Sounding the retreat!"

Her wound was forgotten in a moment. She countermanded the order, and sent another, to the officer in command of a battery, to stand ready to fire five shots in quick succession. This was a signal to a force on the Orleans side of the river under La Hire, who was not, as some of the histories say, with *us*. It was to be given whenever Joan should feel sure the boulevard was about to fall into her hands—then that force must make a counter-attack on the Tourelles by way of the bridge.

Joan mounted her horse, now, with her staff about her, and when our people saw us coming they raised a great shout, and were at once eager for another assault on the boulevard. Joan rode straight to the fosse where she had received her wound, and standing there in the rain of bolts and arrows, she ordered the Paladin to let her long standard blow free, and to note when its fringes should touch the fortress. Presently he said—

"It touches."

"Now, then," said Joan to the waiting battalions, "the place is yours—enter in! Bugles, sound the assault! Now, then—all together—*go!*"

And go it was. You never saw anything like it. We swarmed up the ladders and over the battlements like a wave —and the place was our property. Why, one might live a thousand years and never see so gorgeous a thing as that again. There, hand to hand, we fought like wild beasts, for there was no give-up to those English—there was no way to convince one of those people but to kill him, and even then he doubted. At least so it was thought, in those days, and maintained by many.

We were busy and never heard the five cannon-shots fired, but they *were* fired a moment after Joan had ordered the assault; and so, while we were hammering and being hammered in the smaller fortress, the reserve on the Orleans side poured across the bridge and attacked the Tourelles from that side. A fire-boat was brought down and moored under the drawbridge which connected the Tourelles with our boulevard; wherefore, when at last we drove our English ahead of us and they tried to cross that drawbridge and join their friends in the Tourelles, the burning timbers gave way under them and emptied them in a mass into the river in their heavy armor—and a pitiful sight it was to see brave men die such a death as that.

"Ah, God, pity them!" said Joan, and wept to see that sorrowful spectacle. She said those gentle words and wept those compassionate tears although one of those perishing men had grossly insulted her with a coarse name three days before, when she had sent him a message asking him to surrender. That was their leader, Sir William Glasdale, a most valorous knight. He was clothed all in steel; so he plunged under the water like a lance, and of course came up no more.

We soon patched a sort of bridge together and threw ourselves against the last stronghold of the English power that barred Orleans from friends and supplies. Before the sun was quite down, Joan's forever memorable day's work was finished, her banner floated from the fortress of the Tourelles, her promise was fulfilled, she had raised the siege of Orleans!

The seven months' beleaguerment was ended, the thing which the first generals of France had called impossible was accomplished; in spite of all that the King's ministers and war-councils could do to prevent it, this little country maid of seventeen had carried her immortal task through, and had done it in four days!

Good news travels fast, sometimes, as well as bad. By the time we were ready to start homewards by the bridge the whole city of Orleans was one red flame of bonfires, and the heavens blushed with satisfaction to see it; and the booming and bellowing of cannon and the banging of bells surpassed by great odds anything that even Orleans had attempted before in the way of noise.

When we arrived—well, there is no describing that. Why, those acres of people that we ploughed through shed tears enough to raise the river; there was not a face in the glare of those fires that hadn't tears streaming down it; and if Joan's feet had not been protected by iron they would have kissed them off of her. "Welcome! welcome to the Maid of Orleans!" That was the cry; I heard it a hundred thousand times. Welcome to *our* Maid!" some of them worded it.

No other girl in all history has ever reached such a summit of glory as Joan of Arc reached that day. And do you think it turned her head, and that she sat up to enjoy that delicious music of homage and applause? No; another girl would have done that, but not this one. That was the greatest heart and the simplest that ever beat. She went straight to bed and to sleep, like any tired child; and when the people found she was wounded and would rest, they shut off all passage and traffic in that region and stood guard themselves the whole night through, to see that her slumbers were not disturbed. They said, " She has given us peace, she shall have peace herself."

All knew that that region would be empty of English next day, and all said that neither the present citizens nor their posterity would ever cease to hold that day sacred to the memory of Joan of Arc. That word has been true for more than sixty years; it will continue so always. Orleans will never forget the 8th of May, nor ever fail to celebrate it. It is Joan of Arc's day—and holy. *

* It is still celebrated every year with civic and military pomps and solemnities.—TRANSLATOR.

CHAPTER XXIII

In the earliest dawn of the morning, Talbot and his English forces evacuated their bastilles and marched away, not stopping to burn, destroy, or carry off anything, but leaving their fortresses just as they were, provisioned, armed, and equipped for a long siege. It was difficult for the people to believe that this great thing had really happened; that they were actually free once more, and might go and come through any gate they pleased, with none to molest or forbid; that the terrible Talbot, that scourge of the French, that man whose mere name had been able to annul the effectiveness of French armies, was gone, vanquished, retreating—driven away by a girl.

The city emptied itself. Out of every gate the crowds poured. They swarmed about the English bastilles like an invasion of ants, but noisier than those creatures, and carried off the artillery and stores, then turned all those dozen fortresses into monster bonfires, imitation volcanoes whose lofty columns of thick smoke seemed supporting the arch of the sky.

The delight of the children took another form. To some of the younger ones seven months was a sort of lifetime. They had forgotten what grass was like, and the velvety green meadows seemed paradise to their surprised and happy eyes after the long habit of seeing nothing but dirty lanes and streets. It was a wonder to them—those spacious reaches of open country to run and dance and tumble and frolic in, after their dull and joyless captivity; so they scampered far and wide over the fair regions on both sides of the river, and came back at eventide weary, but laden with flowers and flushed with new health drawn from the fresh country air and the vigorous exercise.

After the burnings, the grown folk followed Joan from church to church and put in the day in thanksgivings for the city's deliverance, and at night they fêted her and her generals and illuminated the town, and high and low gave themselves up to festivities and rejoicings. By the time the populace were fairly in bed, toward dawn, we were in the saddle and away toward Tours to report to the King.

That was a march which would have turned any one's head but Joan's. We moved between emotional ranks of grateful country people all the way. They crowded about Joan to touch her feet, her horse, her armor, and they even knelt in the road and kissed her horse's hoof-prints.

The land was full of her praises. The most illustrious chiefs of the Church wrote to the King extolling the Maid, comparing her to the saints and heroes of the Bible, and warning him not to let "unbelief, ingratitude, or other injustice" hinder or impair the divine help sent through her. One might think there was a touch of prophecy in that, and we will let it go at that; but to my mind it had its inspiration in those great men's accurate knowledge of the King's trivial and treacherous character.

The King had come to Tours to meet Joan. At the present day this poor thing is called Charles the Victorious, an account of victories which other people won for him, but in our time we had a private name for him which described him better, and was sanctified to him by personal deserving—Charles the Base. When we entered the presence he sat throned, with his tinselled snobs and dandies around him. He looked like a forked carrot, so tightly did his clothing fit him from his waist down; he wore shoes with a rope-like pliant toe a foot long that had to be hitched up to the knee to keep it out of the way; he had on a crimson velvet cape that came no lower than his elbows; on his head he had a tall felt thing like a thimble, with a feather in its jewelled band that stuck up like a pen from an inkhorn, and from under that thimble his bush of stiff hair stuck down to his shoulders, curving outwards at the bottom, so that the cap and the hair together made the head like a shuttlecock. All the materials of his dress were rich, and all the colors brilliant. In his lap he cuddled a miniature greyhound that snarled, lifting its lip and showing its white teeth whenever any slight movement disturbed it. The King's dandies were dressed in about the same fashion as himself, and when I remembered that Joan had called the war - council of Orleans "disguised ladies' maids," it reminded me of people who squander all their money on a trifle and then haven't anything to invest when they come across a better chance; that name ought to have been saved for these creatures.

Joan fell on her knees before the majesty of France, and the other frivolous animal in his lap—a sight which it pained me to see. What had that man done for his country or for anybody in it, that she or any other person should kneel to him? But she—she had just done the only great deed that had been done for France in fifty years, and had consecrated it with the libation of her blood. The positions should have been reversed.

However, to be fair, one must grant that Charles acquitted himself very well for the most part, on that occasion—very much better than he was in the habit of doing. He passed his pup to a courtier, and took off his cap to Joan as if she had been a queen. Then he stepped from his throne and raised her, and showed quite a spirited and manly joy and gratitude in welcoming her and thanking her for her extraordinary achievement in his service. My prejudices are of a later date than that. If he had continued as he was at that moment, I should not have acquired them.

He acted handsomely. He said—

"You shall not kneel to me, my matchless General; you have wrought royally, and royal courtesies are your due." Noticing that she was pale, he said, "But you must not stand; you have lost blood for France, and your wound is yet green —come." He led her to a seat and sat down by her. "Now, then, speak out frankly, as to one who owes you much and freely confesses it before all this courtly assemblage. What shall be your reward? Name it."

I was ashamed of him. And yet that was not fair, for how could he be expected to know this marvellous child in these few weeks, when we who thought we had known her all her life were daily seeing the clouds uncover some new altitudes of her character whose existence was not suspected by us before? But we are all that way: when *we* know a thing we have only scorn for other people who don't happen to know it. And I was ashamed of these courtiers, too, for the way they licked their chops, so to speak, as envying Joan her great chance, they not knowing her any better than the King did. A blush began to rise in Joan's cheeks at the thought that she was working for her country for pay, and she dropped her head and tried to hide her face, as girls always do when they find themselves blushing; no one knows why they do, but they do, and the more they blush the more they fail to get reconciled to it, and the more they can't bear to have people look at them when they are doing it. The King made it a great deal worse by calling attention to it, which is the unkindest thing a person can do when a girl is blushing; sometimes, when there is a big crowd of strangers, it is even likely to make her cry if she is as young as Joan was. God knows the reason for this, it is hidden from men. As for me, I would as soon blush as sneeze; in fact, I would rather. However, these meditations are not of consequence: I will go on with what I was saying. The King rallied her for blushing, and this brought up the rest of the blood and turned her face to fire. Then he was sorry, seeing what he had done, and tried to make her comfortable by saying the blush was exceedingly becoming to her and not to mind it—which caused even the dog to notice it now, so of course the red in Joan's face turned to purple, and the tears overflowed and ran down—I could have told anybody that that would happen. The King was distressed, and saw that the best thing to do would be to get away from this subject, so he began to say the finest kind of things about Joan's capture of the Tourelles, and presently when she was more composed he mentioned the reward again and pressed her to name it. Everybody listened with anxious interest to hear what her claim was going to be, but when her answer came their faces showed that the thing she asked for was not what they had been expecting.

"Oh, dear and gracious Dauphin, I have but one desire—only one. If—"

"Do not be afraid, my child—name it."

"That you will not delay a day. My army is strong and valiant, and eager to finish its work — march with me to Rheims and receive your crown."

You could see the indolent King shrink, in his butterfly clothes.

"To Rheims — oh, impossible, my General! *We* march through the heart of England's power?"

Could those be French faces there? Not one of them lighted in response to the girl's brave proposition, but all promptly showed satisfaction in the King's objection. Leave this silken idleness for the rude contact of war? None of these butterflies desired that. They passed their jewelled comfit-boxes one to another and whispered their content in the head butterfly's practical prudence. Joan pleaded with the King, saying—

"Ah, I pray you do not throw away this perfect opportunity Everything is favorable—everything. It is as if the circumstances were specially made for it. The spirits of our army are exalted with victory, those of the English forces depressed by defeat. Delay will change this. Seeing us hesitate to follow up our advantage, our men will wonder, doubt, lose confidence, and the English will wonder, gather courage, and be bold again. Now is the time—prithee let us march!"

The King shook his head, and La Tremouille, being asked for an opinion, eagerly furnished it :

"Sire, all prudence is against it. Think of the English strongholds along the Loire ; think of those that lie between us and Rheims!"

He was going on, but Joan cut him short, and said, turning to him—

"If we wait, they will all be strengthened, re-enforced. Will that advantage us?"

"Why—no."

"Then what is your suggestion?—what is it that you would propose to do?"

"My judgment is to wait."

"Wait for what?"

The minister was obliged to hesitate, for he knew of no explanation that would sound well. Moreover, he was not used to being catechised in this fashion, with the eyes of a crowd of people on him, so he was irritated, and said—

"Matters of state are not proper matters for public discussion."

Joan said, placidly—

"I have to beg your pardon. My trespass came of ignorance. I did not know that matters connected with your department of the government were matters of state."

The minister lifted his brows in amused surprise, and said, with a touch of sarcasm—

"I am the King's chief minister, and yet you had the impression that matters connected with my department are not matters of state? Pray how is that?"

Joan replied, indifferently—

"Because there is no state."

"No state!"

"No, sir, there is no state, and no use for a minister. France is shrunk to a couple of acres of ground; a sheriff's constable could take care of it; its affairs are not matters of state. The term is too large."

The King did not blush, but burst into a hearty, careless laugh, and the court laughed too, but prudently turned its head and did it silently. La Tremouille was angry, and opened his mouth to speak, but the King put up his hand, and said—

"There—I take her under the royal protection. She has spoken the truth, the ungilded truth—how seldom I hear it! With all this tinsel on me and all this tinsel about me, I am but a sheriff after all—a poor shabby two-acre sheriff—and you are but a constable," and he laughed his cordial laugh again. "Joan, my frank, honest General, *will* you name your reward? I would ennoble you. You shall quarter the crown and the lilies of France for blazon, and with them your victorious sword to defend them—speak the word."

It made an eager buzz of surprise and envy in the assemblage, but Joan shook her head and said—

"Ah, I cannot, dear and noble Dauphin. To be allowed to work for France, to spend one's self for France, is itself so supreme a reward that nothing can add to it—nothing. Give me the one reward I ask, the dearest of all rewards, the highest in your gift—march with me to Rheims and receive your crown. I will beg it on my knees."

But the King put his hand on her arm, and there was a really brave awakening in his voice and a manly fire in his eye when he said—

"No; sit. You have conquered me—it shall be as you—"

But a warning sign from his minister halted him, and he added, to the relief of the Court—

"Well, well, we will think of it, we will think it over and see. Does that content you, impulsive little soldier?"

The first part of the speech sent a glow of delight to Joan's face, but the end of it quenched it and she looked sad, and the tears gathered in her eyes. After a moment she spoke out with what seemed a sort of terrified impulse, and said—

"Oh, use me; I beseech you, use me—there is but little time!"

"But little time?"

"Only a year—I shall last only a year."

"Why, child, there are fifty good years in that compact little body yet."

"Oh, you err, indeed you do. In one little year the end will come. Ah, the time is so short, so short; the moments are flying, and so much to be done. Oh, use me, and quickly—it is life or death for France."

Even those insects were sobered by her impassioned words. The King looked very grave—grave, and strongly impressed.

EMBELLISHMENT SHOWING THE DOORWAY OF THE HOUSE IN
WHICH JOAN WAS BORN

His eyes lit suddenly with an eloquent fire, and he rose and drew his sword and raised it aloft; then he brought it slowly down upon Joan's shoulder and said:

"Ah, thou art so simple, so true, so great, so noble—and by this accolade I join thee to the nobility of France, thy fitting place! And for thy sake I do hereby ennoble all thy family and all thy kin; and all their descendants born in wedlock, not only in the male but also in the female line. And more! —more! To distinguish thy house and honor it above all others, we add a privilege never accorded to any before in the history of these dominions: the females of thy line shall have and hold the right to ennoble their husbands when these shall be of inferior degree." [Astonishment and envy flared up in every countenance when the words were uttered which conferred this extraordinary grace. The King paused and looked around upon these signs with quite evident satisfaction.] "Rise, Joan of Arc, now and henceforth surnamed *Du Lis*, in grateful acknowledgment of the good blow which you have struck for the lilies of France; and they, and the royal crown, and your own victorious sword, fit and fair company for each other, shall be grouped in your escutcheon and be and remain the symbol of your high nobility forever."

As my lady Du Lis rose, the gilded children of privilege pressed forward to welcome her to their sacred ranks and call her by her new name; but she was troubled, and said these honors were not meet for one of her lowly birth and station, and by their kind grace she would remain simple Joan of Arc, nothing more—and so be called.

Nothing more! As if there *could* be anything more, anything higher, anything greater! My lady Du Lis — why, it was tinsel, petty, perishable. But—JOAN OF ARC! The mere sound of it sets one's pulses leaping.

CHAPTER XXIV

It was vexatious to see what a to-do the whole town, and next the whole country, made over the news. Joan of Arc ennobled by the King! People went dizzy with wonder and delight over it. You cannot imagine how she was gaped at, stared at, envied. Why, one would have supposed that some great and fortunate thing had happened to her. But *we* did not think any great things of it. To our minds no mere human hand could add a glory to Joan of Arc. To us she was the sun soaring in the heavens, and her new nobility a candle atop of it; to us it was swallowed up and lost in her own light. And she was as indifferent to it and as unconscious of it as the other sun would have been.

But it was different with her brothers. They were proud and happy in their new dignity, which was quite natural. And Joan was glad it had been conferred, when she saw how pleased they were. It was a clever thought in the King to outflank her scruples by marching on them under shelter of her love for her family and her kin.

Jean and Pierre sported their coat-of-arms right away; and their society was courted by everybody, the nobles and commons alike. The Standard-bearer said, with some touch of bitterness, that he could see that they just felt good to be alive, they were so soaked with the comfort of their glory; and didn't like to sleep at all, because when they were asleep they didn't know they were noble, and so sleep was a clean loss of time. And then he said—

"They can't take precedence of me in military functions and state ceremonies, but when it comes to civil ones and society affairs I judge they'll cuddle coolly in behind you and

the knights, and Noël and I will have to walk behind *them*—
hey?"

"Yes," I said, "I think you are right."

"I was just afraid of it—just afraid of it," said the Standard-bearer, with a sigh. "Afraid of it? I'm talking like a
fool: of course I *knew* it. Yes, I was talking like a fool."

Noël Rainguesson said, musingly—

"Yes, I noticed something natural about the tone of it."

We others laughed.

"Oh, you did, did you? You think you are very clever,
don't you? I'll take and wring your neck for you one of these
days, Noël Rainguesson."

The Sieur de Metz said—

"Paladin, your fears haven't reached the top notch. They
are away below the grand possibilities. Didn't it occur to
you that in civil and society functions they will take precedence of *all* the rest of the personal staff—every individ-
ual of us?"

"Oh, come!"

"You'll find it's so. Look at their escutcheon. Its chiefest feature is the lilies of France. It's royal, man, royal—do
you understand the size of that? The lilies are there by authority of the King—do you understand the size of *that?*
Though not in detail and in entirety, they do nevertheless
substantially *quarter the arms of France* in their coat. Imagine it! consider it! measure the magnitude of it! *We* walk
in front of those boys? Bless you, we've done that for the
last time. In my opinion there isn't a lay lord in this whole
region that can walk in front of them, except the Duke d'Alençon, prince of the blood."

You could have knocked the Paladin down with a feather.
He seemed to actually turn pale. He worked his lips a moment without getting anything out; then it came:

"*I* didn't know that, nor the half of it; how *could* I? I've
been an idiot. I see it now—I've been an idiot. I met them
this morning, and sung out *hello* to them just as I would to
anybody. *I* didn't mean to be ill-mannered, but I didn't know

the half of this that you've been telling. I've been an ass. Yes, that is all there is to it—I've been an ass."

Noël Rainguesson said, in a kind of weary way:

"Yes, that is likely enough; but I don't see why you should seem surprised at it."

"You don't, don't you? Well, *why* don't you?"

"Because I don't see any novelty about it. With some people it is a condition which is present all the time. Now you take a condition which is present all the time, and the results of that condition will be uniform; this uniformity of result will in time become monotonous; monotonousness, by the law of its being, is fatiguing. If you had manifested *fatigue* upon noticing that you had been an ass, that would have been logical, that would have been rational; whereas it seems to me that to manifest surprise was to be *again* an ass, because the condition of intellect that can enable a person to be surprised and stirred by inert monotonousness is a—"

"Now that is enough, Noël Rainguesson; stop where you are, before you get yourself into trouble. And don't bother me any more for some days or a week an it please you, for I cannot abide your clack."

"Come, I like that! *I* didn't want to talk. I tried to get out of talking. If you didn't want to hear my clack, what did you keep intruding your conversation on me for?"

"I? I never dreamed of such a thing."

"Well, you did it, anyway. And I have a right to feel hurt, and I do feel hurt, to have you treat me so. It seems to me that when a person goads, and crowds, and in a manner forces another person to talk, it is neither very fair nor very good-mannered to call what he says *clack*."

"Oh, snuffle—do! and break your heart, you poor thing. Somebody fetch this sick doll a sugar-rag. Look you, Sir Jean de Metz, do you feel absolutely certain about that thing?"

"What thing?"

"Why that Jean and Pierre are going to take precedence of all the lay noblesse hereabouts except the Duke d'Alençon?"

"I think there is not a doubt of it."

The Standard-bearer was deep in thoughts and dreams a few moments, then the silk-and-velvet expanse of his vast breast rose and fell with a sigh, and he said—

"Dear, dear, what a lift it is! It just shows what luck can do. Well, I don't care. I shouldn't care to be a painted accident—I shouldn't value it. I am prouder to have climbed up to where I am just by sheer natural merit than I would be to ride the very sun in the zenith and have to reflect that I was nothing but a poor little accident, and got shot up there out of somebody else's catapult. To me, merit is everything —in fact the only thing. All else is dross."

Just then the bugles blew the assembly, and that cut our talk short.

CHAPTER XXV

THE days began to waste away—and nothing decided, nothing done. The army was full of zeal, but it was also hungry. It got no pay, the treasury was getting empty, it was becoming impossible to feed it; under pressure of privation it began to fall apart and disperse—which pleased the trifling court exceedingly. Joan's distress was pitiful to see. She was obliged to stand helpless while her victorious army dissolved away until hardly the skeleton of it was left.

At last one day she went to the Castle of Loches, where the King was idling. She found him consulting with three of his councillors, Robert le Maçon, a former Chancellor of France, Christophe d'Harcourt, and Gerard Machet. The Bastard of Orleans was present also, and it is through him that we know what happened. Joan threw herself at the King's feet and embraced his knees, saying:

"Noble Dauphin, prithee hold no more of these long and numerous councils, but come, and come quickly, to Rheims and receive your crown."

Christophe d'Harcourt asked—

"Is it your Voices that command you to say that to the King?"

"Yes, and urgently."

"Then will you not tell us in the King's presence in what way the Voices communicate with you?"

It was another sly attempt to trap Joan into indiscreet admissions and dangerous pretensions. But nothing came of it. Joan's answer was simple and straightforward, and the smooth Bishop was not able to find any fault with it. She said that when she met with people who doubted the truth of her mission she went aside and prayed, complaining of the distrust of these, and then the comforting Voices were heard at her ear saying, soft and low, "Go forward, Daughter of God, and I will help thee." Then she added, "When I hear that, the joy in my heart, oh, it is insupportable !"

The Bastard said that when she said these words her face lit up as with a flame, and she was like one in an ecstasy.

Joan pleaded, persuaded, reasoned ; gaining ground little by little, but opposed step by step by the council. She begged, she implored, leave to march. When they could answer nothing further, they granted that perhaps it had been a mistake to let the army waste away, but how could we help it now ? how could we march without an army ?

"Raise one !" said Joan.

"But it will take six weeks."

"No matter—begin ! let us begin !"

"It is too late. Without doubt the Duke of Bedford has been gathering troops to push to the succor of his strongholds on the Loire."

"Yes, while we have been disbanding ours—and pity 'tis. But we must throw away no more time; we must bestir ourselves."

The King objected that he could not venture toward Rheims with those strong places on the Loire in his path. But Joan said :

"We will break them up. Then you can march."

With *that* plan the king was willing to venture assent. He could sit around out of danger while the road was being cleared.

Joan came back in great spirits. Straightway everything was stirring. Proclamations were issued calling for men, a recruiting camp was established at Selles in Berry, and the commons and the nobles began to flock to it with enthusiasm.

A deal of the month of May had been wasted ; and yet by the 6th of June Joan had swept together a new army and was ready to march. She had eight thousand men. Think of that. Think of gathering together such a body as that in that little region. And these were veteran soldiers, too. In fact most of the men in France were soldiers, when you came to that; for the wars had lasted generations now. Yes, most Frenchmen were soldiers; and admirable runners, too, both by practice and inheritance; they had done next to nothing but run for near a century. But that was not their fault. They had had no fair and proper leadership—at least leaders with a fair and proper chance. Away back, King and Court got the habit of being treacherous to the leaders; then the leaders easily got the habit of disobeying the King and going their own way, each for himself and nobody for the lot. Nobody could win victories that way. Hence, running became the habit of the French troops, and no wonder. Yet all that those troops needed in order to be good fighters was a leader who would attend strictly to business—a leader with *all* authority in his hands in place of a tenth of it along with nine other generals equipped with an equal tenth apiece. They had a leader rightly clothed with authority now, and with a head and heart bent on war of the most intensely business-like and earnest sort—and there would be *results*. No doubt of that. They had Joan of Arc; and under that leadership their legs would lose the art and mystery of running.

Yes, Joan was in great spirits. She was here and there and everywhere, all over the camp, by day and by night, pushing things. And wherever she came charging down the lines, reviewing the troops, it was good to hear them break out and cheer. And nobody could help cheering, she was such a vision of young bloom and beauty and grace, and such an incarnation of pluck and life and go! She was growing more and more ideally beautiful every day, as was plain to be seen—and these were days of development; for she was well past seventeen, now—in fact she was getting close upon seventeen and a half—indeed, just a little woman, as you may say.

The two young Counts de Laval arrived one day—fine young fellows allied to the greatest and most illustrious houses of France; and they could not rest till they had seen

Joan of Arc. So the King sent for them and presented them to her, and you may believe she filled the bill of their expectations. When they heard that rich voice of hers they must have thought it was a flute; and when they saw her deep eyes and her face, and the soul that looked out of that face, you could see that the sight of her stirred them like a poem, like lofty eloquence, like martial music. One of them wrote home to his people, and in his letter he said, "It seemed something divine to see her and hear her." Ah, yes, and it was a true word. Truer word was never spoken.

He saw her when she was ready to begin her march and open the campaign, and this is what he said about it:

" She was clothed all in white armor save her head, and in her hand she carried a little battle-axe; and when she was ready to mount her great black horse he reared and plunged and would not let her. Then she said, 'Lead him to the cross.' This cross was in front of the church close by. So they led him there. Then she mounted, and he never budged, any more than if he had been tied. Then she turned toward the door of the church and said, in her soft womanly voice, 'You, priests and people of the Church, make processions and pray to God for us!' Then she spurred away, under her standard, with her little axe in her hand, crying 'Forward—march!' One of her brothers, who came eight days ago, departed with her; and he also was clad all in white armor."

I was there, and I saw it too; saw it all, just as he pictures it. And I see it yet—the little battle-axe, the dainty plumed cap, the white armor—all in the soft June afternoon; I see it just as if it were yesterday. And I rode with the staff—the personal staff—the staff of Joan of Arc.

That young Count was dying to go too, but the King held him back for the present. But Joan had made him a promise. In his letter he said:

" She told me that when the King starts for Rheims I shall go with him. But God grant I may not have to wait till then, but may have a part in the battles!"

She made him that promise when she was taking leave of

my lady the Duchess d'Alençon. The Duchess was exacting a promise, so it seemed a proper time for others to do the like. The Duchess was troubled for her husband, for she foresaw desperate fighting; and she held Joan to her breast, and stroked her hair lovingly, and said:

"You must watch over him, dear, and take care of him, and send him back to me safe. I require it of you; I will not let you go till you promise."

Joan said:

"I give you the promise with all my heart; and it is not just words, it *is* a promise: you shall have him back without a hurt. Do you believe? And are you satisfied with me now?"

The Duchess could not speak, but she kissed Joan on the forehead; and so they parted.

We left on the 6th and stopped over at Romorantin; then on the 9th Joan entered Orleans in state, under triumphal arches, with the welcoming cannon thundering and seas of welcoming flags fluttering in the breeze. The Grand Staff rode with her, clothed in shining splendors of costume and decorations: the Duke d'Alençon; the Bastard of Orleans; the Sire de Boussac, Marshal of France; the Lord de Graville, Master of the Crossbowmen; the Sire de Culan, Admiral of France; Ambroise de Loré; Étienne de Vignoles, called La Hire; Gautier de Brusac, and other illustrious captains.

It was grand times: the usual shoutings, and packed multitudes, the usual crush to get sight of Joan; but at last we crowded through to our old lodgings, and I saw old Boucher and the wife and that dear Catherine gather Joan to their hearts and smother her with kisses—and my heart ached so! for I could have kissed Catherine better than anybody, and more and longer; yet was not thought of for that office, and I so famished for it. Ah, she was so beautiful, and oh, so sweet! I had loved her the first day I ever saw her, and from that day forth she was sacred to me. I have carried her image in my heart for sixty-three years—all lonely there, yes, solitary, for it never has had company—and I am grown so old, so old: but it, oh, it is as fresh and young and merry and mischievous and lovely and sweet and pure and witching and divine as it was when it crept in there, bringing benediction and peace to its habitation so long ago, so long ago—for it has not aged a day!

CHAPTER XXVI

THIS time, as before, the King's last command to the generals was this: "*See to it that you do nothing without the sanction of the Maid.*" And this time the command was obeyed; and would continue to be obeyed all through the coming great days of the Loire campaign.

That was a change! That was new! It broke the traditions. It shows you what sort of a reputation as a commander-in-chief the child had made for herself in ten days in the field. It was a conquering of men's doubts and suspicions and a capturing and solidifying of men's belief and confidence such as the grayest veteran on the Grand Staff had not been able to achieve in thirty years. Don't you remember that when at sixteen Joan conducted her own case in a grim court of law and won it, the old judge spoke of her as "this marvellous child?" It was the right name, you see.

These veterans were not going to branch out and do things without the sanction of the Maid—that is true; and it was a great gain. But at the same time there were some among them who still trembled at her new and dashing war-tactics and earnestly desired to modify them. And so, during the 10th, while Joan was slaving away at her plans and issuing order after order with tireless industry, the old-time consultations and arguings and speechifyings were going on among certain of the generals.

In the afternoon of that day they came in a body to hold one of these councils of war; and while they waited for Joan to join them they discussed the situation. Now this discussion is not set down in the histories; but I was there, and

I will speak of it, as knowing you will trust me, I not being given to beguiling you with lies.

Gautier de Brusac was spokesman for the timid ones; Joan's side was resolutely upheld by D'Alençon, the Bastard, La Hire, the Admiral of France, the Marshal de Boussac, and all the other really important chiefs.

De Brusac argued that the situation was very grave; that Jargeau, the first point of attack, was formidably strong; its imposing walls bristling with artillery, with 7000 picked English veterans behind them, and at their head the great Earl of Suffolk and his two redoubtable brothers the De la Poles. It seemed to him that the proposal of Joan of Arc to try to take such a place by storm was a most rash and over-daring idea, and she ought to be persuaded to relinquish it in favor of the soberer and safer procedure of investment by regular siege. It seemed to him that this fiery and furious new fashion of hurling masses of men against impregnable walls of stone, in defiance of the established laws and usages of war, was—

But he got no further. La Hire gave his plumed helm an impatient toss and burst out with—

"By God she knows her trade, and none can teach it her!"

And before he could get out anything more, D'Alençon was on his feet, and the Bastard of Orleans, and half a dozen others, all thundering at once, and pouring out their indignant displeasure upon any and all that might hold, secretly or publicly, distrust of the wisdom of the Commander-in-Chief. And when they had said their say, La Hire took a chance again, and said:

"There are some that never know how to change. Circumstances may change, but those people are never able to see that *they* have got to change too, to meet those circumstances. All that they know is the one beaten track that their fathers and grandfathers have followed and that they themselves have followed in their turn. If an earthquake come and rip the land to chaos, and that beaten track now lead over precipices and into morasses, those people *can't* learn that they must strike out a new road — no; they will march stupidly along and follow the old one to death and perdition. Men, there's a new state of things; and a surpassing military genius has perceived it with her clear eye. And a new road is required, and that same clear eye has noted where it must go, and has marked it out for us. The man does not live, never has lived, never *will* live, that can improve upon it! The old state of things was defeat, defeat, defeat — and by consequence we had troops with no dash, no heart, no hope. Would you assault stone walls with such? No—there was but one way, with that kind: sit down before a place and wait, wait—starve it out, if you could. The new case is the very opposite; it is this: men all on fire with pluck and dash and vim and fury and energy—a restrained conflagration! What would you do with it? Hold it down and let it smoulder and perish and go out? What would Joan of Arc do with it? Turn it *loose*, by the Lord God of heaven and earth, and let it swallow up the foe in the whirlwind of its fires! Nothing shows the splendor and wisdom of her military genius like her instant comprehension of the size of the change which has come about, and her instant perception of the right and only right way to take advantage of it. With her is no sitting down and starving out; no dilly-dallying and fooling around; no lazying, loafing, and going to sleep; no, it is storm! storm! storm! and still storm! storm! storm! and forever storm! storm! storm! hunt the enemy to his hole, then turn her French hurricanes loose and carry him by storm! And that is *my* sort! Jargeau? What of Jargeau, with its battlements and towers, its devastating artillery, its seven thousand picked veterans? Joan of Arc is to the fore, and by the splendor of God its fate is sealed!"

Oh, he carried them. There was not another word said about persuading Joan to change her tactics. They sat talking comfortably enough after that.

By-and-by Joan entered, and they rose and saluted with their swords, and she asked what their pleasure might be. La Hire said:

THE CAPTURE OF THE TOURELLES

"It is settled, my General. The matter concerned Jargeau. There were some who thought we could not take the place."

Joan laughed her pleasant laugh; her merry, care-free laugh; the laugh that rippled so buoyantly from her lips and made old people feel young again to hear it; and she said to the company—

"Have no fears—indeed there is no need nor any occasion for them. We will strike the English boldly by assault, and you will see." Then a far-away look came into her eyes, and I think that a picture of her home drifted across the vision of her mind; for she said very gently, and as one who muses, "But that I know God guides us and will give us success, I had liefer keep sheep than endure these perils."

We had a homelike farewell supper that evening—just the personal staff and the family. Joan had to miss it; for the city had given a banquet in her honor, and she had gone there in state with the Grand Staff, through a riot of joy-bells and a sparkling Milky Way of illuminations.

After supper some lively young folk whom we knew came in, and we presently forgot that we were soldiers, and only remembered that we were boys and girls and full of animal spirits and long-pent fun; and so there was dancing, and games, and romps, and screams of laughter—just as extravagant and innocent and noisy a good time as ever I had in my life. Dear, dear, how long ago it was!—and I was young then. And outside, all the while, was the measured tramp of marching battalions, belated odds and ends of the French power gathering for the morrow's tragedy on the grim stage of war. Yes, in those days we had those contrasts side by side. And as I passed along to bed there was another one: the big Dwarf, in brave new armor, sat sentry at Joan's door—the stern Spirit of War made flesh, as it were—and on his ample shoulder was curled a kitten asleep.

CHAPTER XXVII

WE made a gallant show next day when we filed out through the frowning gates of Orleans, with banners flying and Joan and the Grand Staff in the van of the long column. Those two young De Lavals were come, now, and were joined to the Grand Staff. Which was well; war being their proper trade, for they were grandsons of that illustrious fighter Bertrand du Guesclin, Constable of France in earlier days. Louis de Bourbon, the Marshal de Rais, and the Vidame de Chartres were added also. We had a right to feel a little uneasy, for we knew that a force of five thousand men was on its way under Sir John Fastolfe to reinforce Jargeau, but I think we were not uneasy, nevertheless. In truth that force was not yet in our neighborhood. Sir John was loitering; for some reason or other he was not hurrying. He was losing precious time —four days at Étampes, and four more at Janville.

We reached Jargeau and began business at once. Joan sent forward a heavy force which hurled itself against the outworks in handsome style, and gained a footing and fought hard to keep it; but it presently began to fall back before a sortie from the city. Seeing this, Joan raised her battle-cry and led a new assault herself under a furious artillery fire. The Paladin was struck down at her side, wounded, but she snatched her standard from his failing hand and plunged on through the ruck of flying missiles, cheering her men with encouraging cries, and then for a good time one had turmoil, and clash of steel, and collision and confusion of struggling multitudes, and the hoarse bellowing of the guns; and then the hiding of it all under a rolling firmament of smoke; a firmament through which veiled vacancies appeared for a moment now and then, giving fitful dim glimpses of the wild tragedy enacting beyond; and always at these times one caught sight of that slight figure in white mail which was the centre and soul of our hope and trust, and whenever we saw that, with its back to us and its face to the fight, we knew that all was well. At last a great shout went up—a joyous roar of shoutings, in fact—and that was sign sufficient that the faubourgs were ours.

Yes, they were ours; the enemy had been driven back within the walls. On the ground which Joan had won, we camped; for night was coming on.

Joan sent a summons to the English, promising that if they surrendered she would allow them to go in peace and take their horses with them. Nobody knew that she could take that strong place, but she knew it—knew it well; yet she offered that grace—offered it in a time when such a thing was unknown in war; in a time when it was custom and usage to massacre the garrison and the inhabitants of captured cities without pity or compunction—yes, even to the harmless women and children sometimes. There are neighbors all about you who well remember the unspeakable atrocities which Charles the Bold inflicted upon the men and women and children of Dinant when he took that place some years ago. It was a unique and kindly grace which Joan offered that garrison; but that was her way, that was her loving and merciful nature—she always did her best to save her enemy's life and his soldierly pride when she had the mastery of him.

The English asked fifteen days' armistice to consider the proposal in. And Fastolfe coming with five thousand men! Joan said no. But she offered another grace: they might take both their horses and their side-arms—but they must go within the hour.

Well, those bronzed English veterans were pretty hardheaded folk. They declined again. Then Joan gave command that her army be made ready to move to the assault at nine in the morning. Considering the deal of marching and fighting which the men had done that day, D'Alençon thought the hour rather early; but Joan said it was best so, and so must be obeyed. Then she burst out with one of those enthusiasms which were always burning in her when battle was imminent, and said:

"Work! work! and God will work with us!"

Yes, one might say that her motto was "Work! stick to it; keep on working!" for in war she never knew what indolence was. And whoever will take that motto and live by it will be likely to succeed. There's many a way to win, in this world, but none of them is worth much without good hard work back of it.

I think we should have lost our big Standard-Bearer that day, if our bigger Dwarf had not been at hand to bring him out of the mêlée when he was wounded. He was unconscious, and would have been trampled to death by our own horse, if the Dwarf had not promptly rescued him and haled him to the rear and safety. He recovered, and was himself again after two or three hours; and then he was happy and proud, and made the most of his wound, and went swaggering around in his bandages showing off like an innocent big child—which was just what he was. He was prouder of being wounded than a really modest person would be of being killed. But there was no harm in his vanity, and nobody minded it. He said he was hit by a stone from a catapult— a stone the size of a man's head. But the stone grew, of course. Before he got through with it he was claiming that the enemy had flung a building at him.

"Let him alone," said Noël Rainguesson. "Don't interrupt his processes. To-morrow it will be a cathedral."

He said that privately. And, sure enough, to-morrow it *was* a cathedral. I never saw anybody with such an abandoned imagination.

Joan was abroad at the crack of dawn, galloping here and there and yonder, examining the situation minutely, and choosing what she considered the most effective positions for her artillery; and with such accurate judgment did she place her

guns that her Lieutenant-General's admiration of it still survived in his memory when his testimony was taken at the Rehabilitation, a quarter of a century later.

In this testimony the Duke d'Alençon said that at Jargeau
that morning of the 12th of June she made her dispositions
not like a novice, but "with the sure and clear judgment of a
trained general of twenty or thirty years' experience."

The veteran captains of the armies of France said she was
great in war in all ways, but greatest of all in her genius for
posting and handling artillery.

Who taught the shepherd girl to do these marvels — she
who could not read, and had had no opportunity to study the
complex arts of war? I do not know any way to solve such
a baffling riddle as that, there being no precedent for it, nothing in history to compare it with and examine it by. For in
history there is no great general, however gifted, who arrived
at success otherwise than through able teaching and hard
study and some experience. It is a riddle which will never
be guessed. *I* think these vast powers and capacities were
born in her, and that she applied them by an intuition which
could not err.

At eight o'clock all movement ceased, and with it all sounds,
all noise. A mute expectancy reigned. The stillness was
something awful—because it meant so much. There was no
air stirring. The flags on the towers and ramparts hung
straight down like tassels. Wherever one saw a person,
that person had stopped what he was doing, and was in a
waiting attitude, a listening attitude. We were on a commanding spot, clustered around Joan. Not far from us, on
every hand, were the lanes and humble dwellings of these out-
lying suburbs. Many people were visible—all were listening,
not one was moving. A man had placed a nail; he was
about to fasten something with it to the door-post of his shop
—but he had stopped. There was his hand reaching up
holding the nail; and there was his other hand in the act of
striking with the hammer; but he had forgotten everything—
his head was turned aside, listening. Even children uncon-

sciously stopped in their play; I saw a little boy with his hoop-stick pointed slanting toward the ground in the act of steering the hoop around the corner; and so he had stopped and was listening—the hoop was rolling away, doing its own steering. I saw a young girl prettily framed in an open window, a watering-pot in her hand and window-boxes of red flowers under its spout—but the water had ceased to flow; the girl was listening. Everywhere were these impressive petrified forms; and everywhere was suspended movement and that awful stillness.

Joan of Arc raised her sword in the air. At the signal, the silence was torn to rags: cannon after cannon vomited flames and smoke and delivered its quaking thunders; and we saw answering tongues of fire dart from the towers and walls of the city, accompanied by answering deep thunders, and in a minute the walls and the towers disappeared, and in their place stood vast banks and pyramids of snowy smoke, motionless in the dead air. The startled girl dropped her watering-pot and clasped her hands together, and at that moment a stone cannon-ball crashed through her fair body.

The great artillery duel went on, each side hammering away with all its might; and it was splendid for smoke and noise, and most exalting to one's spirits. The poor little town around about us suffered cruelly. The cannon-balls tore through its slight buildings, wrecking them as if they had been built of cards; and every moment or two one would see a huge rock come curving through the upper air above the smoke clouds and go plunging down through the roofs. Fire broke out, and columns of flame and smoke rose toward the sky.

Presently the artillery concussions changed the weather. The sky became overcast, and a strong wind rose and blew away the smoke that hid the English fortresses.

Then the spectacle was fine: turreted gray walls and towers, and streaming bright flags, and jets of red fire and gushes of white smoke in long rows, all standing out with sharp vividness against the deep leaden background of the sky;

and then the whizzing missiles began to knock up the dirt all around us, and I felt no more interest in the scenery. There was one English gun that was getting our position down finer and finer all the time. Presently Joan pointed to it and said:

"Fair Duke, step out of your tracks, or that machine will kill you."

The Duke d'Alençon did as he was bid ; but Monsieur du Lude rashly took his place, and that cannon tore his head off in a moment.

Joan was watching all along for the right time to order the assault. At last, about nine o'clock, she cried out—

"Now—to the assault !" and the buglers blew the charge.

Instantly we saw the body of men that had been appointed to this service move forward toward a point where the concentrated fire of our guns had crumbled the upper half of a broad stretch of wall to ruins; we saw this force descend into the ditch and begin to plant the scaling-ladders. We were soon with them. The Lieutenant-General thought the assault premature. But Joan said :

"Ah, gentle Duke, are you afraid ? Do you not know that I have promised to send you home safe ?"

It was warm work in the ditches. The walls were crowded with men, and they poured avalanches of stones down upon us. There was one gigantic Englishman who did us more hurt than any dozen of his brethren. He always dominated the places easiest of assault, and flung down exceedingly troublesome big stones which smashed men and ladders both —then he would near burst himself with laughing over what he had done. But the Duke settled accounts with him. He went and found the famous cannoneer Jean le Lorrain, and said—

"Train your gun—kill me this demon."

He did it with the first shot. He hit the Englishman fair in the breast and knocked him backwards into the city.

The enemy's resistance was so effective and so stubborn that our people began to show signs of doubt and dismay. Seeing this, Joan raised her inspiring battle-cry and descended into the fosse herself, the Dwarf helping her and the Paladin sticking bravely at her side with the standard. She started up a scaling-ladder, but a great stone flung from above came crashing down upon her helmet and stretched her, wounded and stunned, upon the ground. But only for a moment. The Dwarf stood her upon her feet, and straightway she started up the ladder again, crying—

"To the assault, friends, to the assault—the English are ours! It is the appointed hour!"

There was a grand rush, and a fierce roar of war-cries, and we swarmed over the ramparts like ants. The garrison fled, we pursued; Jargeau was ours!

The Earl of Suffolk was hemmed in and surrounded, and the Duke d'Alençon and the Bastard of Orleans demanded that he surrender himself. But he was a proud nobleman and came of a proud race. He refused to yield his sword to subordinates, saying—

"I will die rather. I will surrender to the Maid of Orleans alone, and to no other."

And so he did; and was courteously and honorably used by her.

His two brothers retreated, fighting step by step, toward the bridge, we pressing their despairing forces and cutting them down by scores. Arrived on the bridge, the slaughter still continued. Alexander de la Pole was pushed overboard or fell over, and was drowned. Eleven hundred men had fallen; John de la Pole decided to give up the struggle. But he was nearly as proud and particular as his brother of Suffolk as to whom he would surrender to. The French officer nearest at hand was Guillaume Renault, who was pressing him closely. Sir John said to him—

"Are you a gentleman?"

"Yes."

"And a knight?"

"No."

Then Sir John knighted him himself, there on the bridge, giving him the accolade with English coolness and tranquillity in the midst of that storm of slaughter and mutilation; and then bowing with high courtesy took the sword by the blade and laid the hilt of it in the man's hand in token of surrender. Ah, yes, a proud tribe, those De la Poles.

It was a grand day, a memorable day, a most splendid victory. We had a crowd of prisoners, but Joan would not allow them to be hurt. We took them with us and marched into Orleans next day through the usual tempest of welcome and joy.

And this time there was a new tribute to our leader. From everywhere in the packed streets the new recruits squeezed their way to her side to touch the sword of Joan of Arc and draw from it somewhat of that mysterious quality which made it invincible.

CHAPTER XXVIII

THE troops must have a rest. Two days would be allowed for this.

The morning of the 14th I was writing from Joan's dictation in a small room which she sometimes used as a private office when she wanted to get away from officials and their interruptions. Catherine Boucher came in and sat down and said—

"Joan, dear, I want you to talk to me."

"Indeed I am not sorry for that, but glad. What is in your mind?"

"This. I scarcely slept, last night, for thinking of the dangers you are running. The Paladin told me how you made the Duke stand out of the way when the cannon-balls were flying all about, and so saved his life."

"Well, that was right, wasn't it?"

"Right? Yes; but you stayed there yourself. Why will you do like that? It seems such a wanton risk."

"Oh, no, it was not so. I was not in any danger."

"How can you say that, Joan, with those deadly things flying all about you?"

Joan laughed, and tried to turn the subject, but Catherine persisted. She said—

"It was horribly dangerous, and it could not be necessary to stay in such a place. And you led an assault again. Joan, it is tempting Providence. I want you to make me a promise. I want you to promise me that you will let others lead the assaults, if there *must* be assaults, and that you will take better care of yourself in those dreadful battles. Will you?"

But Joan fought away from the promise and did not give

it. Catherine sat troubled and discontented awhile, then she said—

"Joan, are you going to be a soldier always? These wars are so long—so long. They last forever and ever and ever."

There was a glad flash in Joan's eye as she cried—

"This campaign will do all the really hard work that is in front of it in the next four days. The rest of it will be gentler—oh, far less bloody. Yes, in four days France will gather another trophy like the redemption of Orleans and make her second long step toward freedom!"

Catherine started (and so did I); then she gazed long at Joan like one in a trance, murmuring "four days—four days," as if to herself and unconsciously. Finally she asked, in a low voice that had something of awe in it:

"Joan, tell me—how is it that you know that? For you do know it, I think."

"Yes," said Joan, dreamily, "I know — I know. I shall strike—and strike again. And before the fourth day is finished I shall strike yet again." She became silent. We sat wondering and still. This was for a whole minute, she looking at the floor and her lips moving but uttering nothing. Then came these words, but hardly audible: "And in a thousand years the English power in France will not rise up from that blow."

It made my flesh creep. It was uncanny. She was in a trance again—I could see it—just as she was that day in the pastures of Domremy when she prophesied about us boys in the war and afterward did not know that she had done it. She was not conscious now; but Catherine did not know that, and so she said, in a happy voice—

"Oh, I believe it, I believe it, and I am so glad! Then you will come back and bide with us all your life long, and we will love you so, and so honor you!"

A scarcely perceptible spasm flitted across Joan's face, and the dreamy voice muttered—

"Before two years are sped I shall die a cruel death!"

I sprang forward with a warning hand up. That is why
16

Catherine did not scream. She was going to do that—I saw it plainly. Then I whispered her to slip out of the place, and say nothing of what had happened. I said Joan was asleep—asleep and dreaming. Catherine whispered back, and said—

"Oh, I am so grateful that it is only a dream! It sounded like prophecy." And she was gone.

Like prophecy! I knew it *was* prophecy; and I sat down crying, as knowing we should lose her. Soon she started, shivering slightly, and came to herself, and looked around and saw me crying there, and jumped out of her chair and ran to me all in a whirl of sympathy and compassion, and put her hand on my head, and said—

"My poor boy! What is it? Look up, and tell me."

I had to tell her a lie; I grieved to do it, but there was no other way. I picked up an old letter from my table, written by Heaven knows who, about some matter Heaven knows what, and told her I had just gotten it from Père Fronte, and that in it it said the children's Fairy Tree had been chopped down by some miscreant or other, and—

I got no further. She snatched the letter from my hand and searched it up and down and all over, turning it this way and that, and sobbing great sobs, and the tears flowing down her cheeks, and ejaculating all the time, "Oh, cruel, cruel! how could any be so heartless? Ah, poor Arbre Fée de Bourlemont gone—and we children loved it so! Show me the place where it says it!"

And I, still lying, showed her the pretended fatal words on the pretended fatal page, and she gazed at them through her tears, and said she could see, herself, that they were hateful, ugly words—they "had the very look of it."

Then we heard a strong voice down the corridor announcing—

"His Majesty's messenger — with despatches for her Excellency the Commander-in-Chief of the armies of France!"

CHAPTER XXIX

I knew she had seen the vision of the Tree. But when? I could not know. Doubtless before she had lately told the King to use her, for that she had but one year left to work in. It had not occurred to me at the time, but the conviction came upon me now that at that time she had already seen the Tree. It had brought her a welcome message; that was plain, otherwise she could not have been so joyous and light-hearted as she had been these latter days. The death-warning had nothing dismal about it for her; no, it was remission of exile, it was leave to come home.

Yes, she had seen the Tree. No one had taken the prophecy to heart which she made to the King; and for a good reason, no doubt: no one *wanted* to take it to heart; all wanted to banish it away and forget it. And all had succeeded, and would go on to the end placid and comfortable. All but me alone. I must carry my awful secret without any to help me. A heavy load, a bitter burden; and would cost me a daily heart-break. She was to die; and so soon. I had never dreamed of that. How could I, and she so strong and fresh and young, and every day earning a new right to a peaceful and honored old age? For at that time I thought old age valuable. I do not know why, but I thought so. All young people think it, I believe, they being ignorant and full of superstitions. She had seen the Tree. All that miserable night those ancient verses went floating back and forth through my brain :

> "And when in exile wand'ring we
> Shall fainting yearn for glimpse of thee,
> O rise upon our sight !"

But at dawn the bugles and the drums burst through the dreamy hush of the morning, and it was turn out all! mount and ride. For there was red work to be done.

We marched to Meung without halting. There we carried the bridge by assault, and left a force to hold it, the rest of the army marching away next morning toward Beaugency, where the lion Talbot, the terror of the French, was in command. When we arrived at that place, the English retired into the castle and we sat down in the abandoned town.

Talbot was not at the moment present in person, for he had gone away to watch for and welcome Fastolfe and his re-enforcement of five thousand men.

Joan placed her batteries and bombarded the castle till night. Then some news came: Richemont, Constable of France, this long time in disgrace with the King, largely because of the evil machinations of La Tremouille and his party, was approaching with a large body of men to offer his services to Joan — and very much she needed them, now that Fastolfe was so close by. Richemont had wanted to join us before, when we first marched on Orleans; but the foolish King, slave of those paltry advisers of his, warned him to keep his distance and refused all reconciliation with him.

I go into these details because they are important. Important because they lead up to the exhibition of a new gift in Joan's extraordinary mental make-up—statesmanship. It is a sufficiently strange thing to find that great quality in an ignorant country girl of seventeen and a half, but she had it.

Joan was for receiving Richemont cordially, and so was La Hire and the two young Lavals and other chiefs, but the Lieutenant-General, D'Alençon, strenuously and stubbornly opposed it. He said he had absolute orders from the King to deny and defy Richemont, and that if they were overridden he would leave the army. This would have been a heavy disaster indeed. But Joan set herself the task of persuading him that the salvation of France took precedence of all minor things—even the commands of a sceptred ass; and she accomplished it. She persuaded him to disobey the King in

JOAN DICTATING LETTERS TO HER PARENTS

the interest of the nation, and to be reconciled to Count Richemont and welcome him. That was statesmanship; and of the highest and soundest sort. Whatever thing men call great, look for it in Joan of Arc, and there you will find it.

In the early morning, June 17th, the scouts reported the approach of Talbot and Fastolfe with Fastolfe's succoring force. Then the drums beat to arms; and we set forth to meet the English, leaving Richemont and his troops behind to watch the castle of Beaugency and keep its garrison at home. By-and-by we came in sight of the enemy. Fastolfe had tried to convince Talbot that it would be wisest to retreat and not risk a battle with Joan at this time, but distribute the new levies among the English strongholds of the Loire, thus securing them against capture; then be patient and wait—wait for more levies from Paris; let Joan exhaust her army with fruitless daily skirmishing; then at the right time fall upon her in resistless mass and annihilate her. He was a wise old experienced general, was Fastolfe. But that fierce Talbot would hear of no delay. He was in a rage over the punishment which the Maid had inflicted upon him at Orleans and since, and he swore by God and Saint George that he would have it out with her if he had to fight her all alone. So Fastolfe yielded, though he said they were now risking the loss of everything which the English had gained by so many years' work and so many hard knocks.

The enemy had taken up a strong position, and were waiting, in order of battle, with their archers to the front and a stockade before them.

Night was coming on. A messenger came from the English with a rude defiance and an offer of battle. But Joan's dignity was not ruffled, her bearing was not discomposed. She said to the herald—

"Go back and say it is too late to meet to-night; but to-morrow, please God and our Lady, we will come to close quarters."

The night fell dark and rainy. It was that sort of light steady rain which falls so softly and brings to one's spirit such serenity and peace. About ten o'clock D'Alençon, the Bastard of Orleans, La Hire, Pothon of Saintrailles, and two or three other generals came to our headquarters tent, and sat down to discuss matters with Joan. Some thought it was a pity that Joan had declined battle, some thought not. Then Pothon asked her why she had declined it. She said—

"There was more than one reason. These English are ours—they cannot get away from us. Wherefore there is no need to take risks, as at other times. The day was far spent. It is good to have much time and the fair light of day when one's force is in a weakened state—nine hundred of us yonder keeping the bridge of Meung under the Marshal de Rais, fifteen hundred with the Constable of France keeping the bridge and watching the castle of Beaugency."

Dunois said—

"I grieve for this depletion, Excellency, but it cannot be helped. And the case will be the same the morrow, as to that."

Joan was walking up and down, just then. She laughed her affectionate, comrady laugh, and stopping before that old war-tiger she put her small hand above his head and touched one of his plumes, saying—

"Now tell me, wise man, which feather is it that I touch?"

"In sooth, Excellency, that I cannot."

"Name of God, Bastard, Bastard! you cannot tell me this small thing, yet are bold to name a large one—telling us what is in the stomach of the unborn morrow: that we shall not have those men. Now it is my thought that they will be with us."

That made a stir. All wanted to know why she thought that. But La Hire took the word and said—

"Let be. If she thinks it, that is enough. It will happen."

Then Pothon of Saintrailles said—

"There were other reasons for declining battle, according to the saying of your Excellency?"

"Yes. One was that we being weak and the day far gone,

the battle might not be decisive. When it is fought it *must* be decisive. And shall be."

"God grant it, and amen. There were still other reasons?"

"One other—yes." She hesitated a moment, then said: "This was not the day. To-morrow is the day. It is so written."

They were going to assail her with eager questionings, but she put up her hand and prevented them. Then she said—

"It will be the most noble and beneficent victory that God has vouchsafed to France at any time. I pray you question me not as to whence or how I know this thing, but be content that it is so."

There was pleasure in every face, and conviction and high confidence. A murmur of conversation broke out, but was interrupted by a messenger from the outposts who brought news—namely, that for an hour there had been stir and movement in the English camp of a sort unusual at such a time and with a resting army, he said. Spies had been sent under cover of the rain and darkness to inquire into it. They had just come back and reported that large bodies of men had been dimly made out who were slipping stealthily away in the direction of Meung.

The generals were very much surprised, as any might tell from their faces.

"It is a retreat," said Joan.

"It has that look," said D'Alençon.

"It certainly has," observed the Bastard and La Hire.

"It was not to be expected," said Louis de Bourbon, "but one can divine the purpose of it."

"Yes," responded Joan. "Talbot has reflected. His rash brain has cooled. He thinks to take the bridge of Meung and escape to the other side of the river. He knows that this leaves his garrison of Beaugency at the mercy of fortune, to escape our hands if it can; but there is no other course if he would avoid this battle, and that he also knows. But he shall not get the bridge. We will see to that."

"Yes," said D'Alençon, "we must follow him, and take care of that matter. What of Beaugency?"

"Leave Beaugency to me, gentle Duke; I will have it in two hours, and at no cost of blood."

"It is true, Excellency. You will but need to deliver this news there and receive the surrender."

"Yes. And I will be with you at Meung with the dawn, fetching the Constable and his fifteen hundred; and when Talbot knows that Beaugency has fallen it will have an effect upon him."

"By the mass, yes!" cried La Hire. "He will join his Meung garrison to his army and break for Paris. Then we shall have our bridge force with us again, along with our Beaugency-watchers, and be stronger for our great day's work by four-and-twenty hundred able soldiers, as was here promised within the hour. Verily this Englishman is doing our errands for us and saving us much blood and trouble. Orders, Excellency—give us our orders!"

"They are simple. Let the men rest three hours longer. At one o'clock the advance-guard will march, under your command, with Pothon of Saintrailles as second; the second division will follow at two under the Lieutenant-General. Keep well in the rear of the enemy, and see to it that you avoid an engagement. I will ride under guard to Beaugency and make so quick work there that I and the Constable of France will join you before dawn with his men."

She kept her word. Her guard mounted and we rode off through the puttering rain, taking with us a captured English officer to confirm Joan's news. We soon covered the journey and summoned the castle. Richard Guétin, Talbot's lieutenant, being convinced that he and his five hundred men were left helpless, conceded that it would be useless to try to hold out. He could not expect easy terms, yet Joan granted them nevertheless. His garrison could keep their horses and arms, and carry away property to the value of a silver mark per man. They could go whither they pleased, but must not take arms against France again under ten days.

Before dawn we were with our army again, and with us the Constable and nearly all his men, for we left only a small garrison in Beaugency castle. We heard the dull booming of cannon to the front, and knew that Talbot was beginning his attack on the bridge. But some time before it was yet light the sound ceased and we heard it no more.

Guétin had sent a messenger through our lines under a safe-conduct given by Joan, to tell Talbot of the surrender. Of course this poursuivant had arrived ahead of us. Talbot had held it wisdom to turn, now, and retreat upon Paris. When daylight came he had disappeared; and with him Lord Scales and the garrison of Meung.

What a harvest of English strongholds we had reaped in those three days!—strongholds which had defied France with quite cool confidence and plenty of it until we came.

CHAPTER XXX

WHEN the morning broke at last on that forever memorable 18th of June, there was no enemy discoverable anywhere, as I have said. But that did not trouble me. I knew we should find him, and that we should strike him; strike him the promised blow—the one from which the English power in France would not rise up in a thousand years, as Joan had said in her trance.

The enemy had plunged into the wide plains of La Beauce —a roadless waste covered with bushes, with here and there bodies of forest trees — a region where an army would be hidden from view in a very little while. We found the trail in the soft wet earth and followed it. It indicated an orderly march; no confusion, no panic.

But we had to be cautious. In such a piece of country we could walk into an ambush without any trouble. Therefore Joan sent bodies of cavalry ahead under La Hire, Poton, and other captains, to feel the way. Some of the other officers began to show uneasiness; this sort of hide - and - go - seek business troubled them and made their confidence a little shaky. Joan divined their state of mind and cried out impetuously—

"Name of God, what would you? We must smite these English, and we will. They shall not escape us. Though they were hung to the clouds we would get them!"

By-and-by we were nearing Patay; it was about a league away. Now at this time our reconnoisance, feeling its way in the bush, frightened a deer, and it went bounding away and was out of sight in a moment. Then hardly a minute later a dull great shout went up in the distance toward Patay. It was the English soldiery. They had been shut up in garrison so long on mouldy food that they could not keep their delight to themselves when this fine fresh meat came springing into their midst. Poor creature, it had wrought damage to a nation which loved it well. For the French knew where the English were, now, whereas the English had no suspicion of where the French were.

La Hire halted where he was, and sent back the tidings. Joan was radiant with joy. The Duke d'Alençon said to her—

"Very well, we have found them; shall we fight them?"

"Have you good spurs, Prince?"

"Why? Will they make us run away?"

"Nenni, en nom de Dieu! These English are ours—they are lost. They will fly. Who overtakes them will need good spurs. Forward—close up!"

By the time we had come up with La Hire the English had discovered our presence. Talbot's force was marching in three bodies. First his advance-guard; then his artillery; then his battle corps a good way in the rear. He was now out of the bush and in a fair open country. He at once posted his artillery, his advance-guard, and five hundred picked archers along some hedges where the French would be obliged to pass, and hoped to hold this position till his battle corps could come up. Sir John Fastolfe urged the battle corps into a gallop. Joan saw her opportunity and ordered La Hire to advance — which La Hire promptly did, launching his wild riders like a storm-wind, his customary fashion.

The Duke and the Bastard wanted to follow, but Joan said—

"Not yet—wait."

So they waited—impatiently, and fidgeting in their saddles. But she was steady—gazing straight before her, measuring, weighing, calculating—by shades, minutes, fractions of minutes, seconds—with all her great soul present, in eye, and set of head, and noble pose of body—but patient, steady, master of herself—master of herself and of the situation.

And yonder, receding, receding, plumes lifting and falling, lifting and falling, streamed the thundering charge of La Hire's godless crew, La Hire's great figure dominating it and his sword stretched aloft like a flag-staff.

"O, Satan and his Hellions, see them go!" Somebody muttered it in deep admiration.

And now he was closing up—closing up on Fastolfe's rushing corps.

And now he struck it—struck it hard, and broke its order. It lifted the Duke and the Bastard in their saddles to see it; and they turned, trembling with excitement, to Joan, saying—

"*Now!*"

But she put up her hand, still gazing, weighing, calculating, and said again—

"Wait—not yet."

Fastolfe's hard-driven battle corps raged on like an avalanche toward the waiting advance-guard. Suddenly these conceived the idea that it was flying in panic before Joan; and so in that instant it broke and swarmed away in a mad panic itself, with Talbot storming and cursing after it.

Now was the golden time. Joan drove her spurs home and waved the advance with her sword. "Follow me!" she cried, and bent her head to her horse's neck and sped away like the wind!

We swept down into the confusion of that flying rout, and for three long hours we cut and hacked and stabbed. At last the bugles sang "Halt!"

The Battle of Patay was won.

Joan of Arc dismounted, and stood surveying that awful field, lost in thought. Presently she said—

"The praise is to God. He has smitten with a heavy hand this day." After a little she lifted her face, and looking afar off, said, with the manner of one who is thinking aloud, "In a thousand years — a thousand years — the English power in France will not rise up from this blow." She stood again a time, thinking, then she turned toward her grouped generals,

and there was a glory in her face and a noble light in her
eye; and she said—

"O, friends, friends, do you know?—do you comprehend?
France is on the way to be free!"

"And had never been, but for Joan of Arc!" said La
Hire, passing before her and bowing low, the others following
and doing likewise; he muttering as he went, "I will say it
though I be damned for it." Then battalion after battalion
of our victorious army swung by, wildly cheering. And they
shouted "Live forever, Maid of Orleans, live forever!" while
Joan, smiling, stood at the salute with her sword.

This was not the last time I saw the Maid of Orleans on
the red field of Patay. Toward the end of the day I came
upon her where the dead and dying lay stretched all about in
heaps and winrows; our men had mortally wounded an English prisoner who was too poor to pay a ransom, and from a
distance she had seen that cruel thing done; and had galloped to the place and sent for a priest, and now she was
holding the head of her dying enemy in her lap, and easing
him to his death with comforting soft words, just as his sister
might have done; and the womanly tears running down her
face all the time.*

* Lord Ronald Gower (*Joan of Arc*, p. 82) says: "Michelet discovered
this story in the deposition of Joan of Arc's page, Louis de Conte, who
was probably an eye-witness of the scene." This is true. It was a part of
the testimony of the author of these "Personal Recollections of Joan of
Arc," given by him in the Rehabilitation proceedings of 1456.—TRANS-
LATOR.

CHAPTER XXXI

JOAN had said true: France was on the way to be free.

The war called the Hundred Years' War was very sick to-day. Sick on its English side—for the very first time since its birth, ninety-one years gone by.

Shall we judge battles by the numbers killed and the ruin wrought? Or shall we not rather judge them by the results which flowed from them? Any one will say that a battle is only truly great or small according to its results. Yes, any one will grant that, for it is the truth.

Judged by results, Patay's place is with the few supremely great and imposing battles that have been fought since the peoples of the world first resorted to arms for the settlement of their quarrels. So judged, it is even possible that Patay has no peer among that few just mentioned, but stands alone, as the supremest of historic conflicts. For when it began France lay gasping out the remnant of an exhausted life, her case wholly hopeless in the view of all political physicians; when it ended, three hours later, she was convalescent. Convalescent, and nothing requisite but time and ordinary nursing to bring her back to perfect health. The dullest physician of them all could see this, and there was none to deny it.

Many death-sick nations have reached convalescence through a series of battles, a procession of battles, a weary tale of wasting conflicts stretching over years; but only one has reached it in a single day and by a single battle. That nation is France, and that battle Patay.

Remember it and be proud of it; for you are French, and it is the stateliest fact in the long annals of your country. There it stands, with its head in the clouds! And when you grow up you will go on pilgrimage to the field of Patay, and stand uncovered in the presence of—what? A monument with *its* head in the clouds? Yes. For all nations in all times have built monuments on their battle-fields to keep green the memory of the perishable deed that was wrought there and of the perishable name of him who wrought it; and will France neglect Patay and Joan of Arc? Not for long. And will she build a monument scaled to their rank as compared with the world's other fields and heroes? Perhaps— if there be room for it under the arch of the sky.

But let us look back a little, and consider certain strange and impressive facts. The Hundred Years' War began in 1337. It raged on and on, year after year and year after year; and at last England stretched France prone with that fearful blow at Crécy. But she rose and struggled on, year after year, and at last again she went down under another devastating blow—Poitiers. She gathered her crippled strength once more, and the war raged on, and on, and still on, year after year, decade after decade. Children were born, grew up, married, died—the war raged on; *their* children in turn grew up, married, died—the war raged on; *their* children, growing, saw France struck down again; this time under the incredible disaster of Agincourt—and still the war raged on, year after year, and in time *these* children married in their turn.

France was a wreck, a ruin, a desolation. The half of it belonged to England, with none to dispute or deny the truth; the other half belonged to nobody—in three months would be flying the English flag: the French King was making ready to throw away his crown and flee beyond the seas.

Now came the ignorant country maid out of her remote village and confronted this hoary war, this all-consuming conflagration that had swept the land for three generations. Then began the briefest and most amazing campaign that is recorded in history. In seven weeks it was finished. In seven weeks she hopelessly crippled that gigantic war that was ninety-one years old. At Orleans she struck it a staggering blow; on the field of Patay she broke its back.

Think of it. Yes, one can do that; but *understand* it? Ah, that is another matter; none will ever be able to comprehend that stupefying marvel.

Seven weeks — with here and there a little bloodshed. Perhaps the most of it, in any single fight, at Patay, where the English began six thousand strong and left two thousand dead upon the field. It is said and believed that in three battles alone—Crécy, Poitiers, and Agincourt—near a hundred thousand Frenchmen fell, without counting the thousand other fights of that long war. The dead of that war make a mournful long list — an interminable list. Of men slain in the field the count goes by tens of thousands; of innocent women and children slain by bitter hardship and hunger it goes by that appalling term, millions.

It was an ogre, that war; an ogre that went about for near a hundred years, crunching men and dripping blood from his jaws. And with her little hand that child of seventeen struck him down; and yonder he lies stretched on the field of Patay, and will not get up any more while this old world lasts.

CHAPTER XXXII

THE great news of Patay was carried over the whole of France in twenty hours, people said. I do not know as to that; but one thing is sure, anyway: the moment a man got it he flew shouting and glorifying God and told his neighbor; and that neighbor flew with it to the next homestead; and so on and so on without resting the word travelled; and when a man got it in the night, at what hour soever, he jumped out of his bed and bore the blessed message along. And the joy that went with it was like the light that flows across the land when an eclipse is receding from the face of the sun; and indeed you may say that France had lain in an eclipse this long time; yes, buried in a black gloom which these beneficent tidings were sweeping away, now, before the on-rush of their white splendor.

The news beat the flying enemy to Yeuville, and the town rose against its English masters and shut the gates against their brethren. It flew to Mont Pipeau, to Saint Simon, and to this, that, and the other English fortress; and straightway the garrison applied the torch and took to the fields and the woods. A detachment of our army occupied Meung and pillaged it.

When we reached Orleans that town was as much as fifty times insaner with joy than we had ever seen it before—which is saying much. Night had just fallen, and the illuminations were on so wonderful a scale that we seemed to plough through seas of fire; and as to the noise—the hoarse cheering of the multitude, the thundering of cannon, the clash of bells—indeed there was never anything like it. And everywhere rose a new cry that burst upon us like a storm

when the column entered the gates, and nevermore ceased:
"Welcome to Joan of Arc—way for the SAVIOR OF FRANCE!"
And there was another cry: "Crécy is avenged! Poitiers is
avenged! Agincourt is avenged!—Patay shall live forever!"

Mad? Why, you never could imagine it in the world.
The prisoners were in the centre of the column. When that
came along and the people caught sight of their masterful old
enemy Talbot, that had made them dance so long to his grim
war-music, you may imagine what the uproar was like if you
can, for I cannot describe it. They were so glad to see him
that presently they wanted to have him out and hang him; so
Joan had him brought up to the front to ride in her protection. They made a striking pair.

CHAPTER XXXIII

YES, Orleans was in a delirium of felicity. She invited the King, and made sumptuous preparations to receive him, but—he didn't come. He was simply a serf at that time, and La Tremouille was his master. Master and serf were visiting together at the master's castle of Sully-sur-Loire.

At Beaugency Joan had engaged to bring about a reconciliation between the Constable Richemont and the King. She took Richemont to Sully-sur-Loire and made her promise good.

The great deeds of Joan of Arc are five:

1. The Raising of the Siege.
2. The Victory of Patay.
3. The Reconciliation at Sully-sur-Loire.
4. The Coronation of the King.
5. The Bloodless March.

We shall come to the Bloodless March presently; (and the Coronation). It was the victorious long march which Joan made through the enemy's country from Gien to Rheims, and thence to the gates of Paris, capturing every English town and fortress that barred the road, from the beginning of the journey to the end of it; and this by the mere force of her name, and without shedding a drop of blood — perhaps the most extraordinary campaign in this regard in history—this is the most glorious of her military exploits.

The Reconciliation was one of Joan's most important achievements. No one else could have accomplished it; and in fact no one else of high consequence had any disposition to try. In brains, in scientific warfare, and in statesmanship the Constable Richemont was the ablest man in France. His loyalty was sincere; his probity was above suspicion—(and it made him sufficiently conspicuous in that trivial and conscienceless Court).

In restoring Richemont to France, Joan made thoroughly secure the successful completion of the great work which she had begun. She had never seen Richemont until he came to her with his little army. Was it not wonderful that at a glance she should know him for the one man who could finish and perfect her work and establish it in perpetuity? How was it that that child was able to do this? It was because she had the "seeing eye," as one of our knights had once said. Yes, she had that great gift—almost the highest and rarest that has been granted to man. Nothing of an extraordinary sort was still to be done, yet the remaining work could not safely be left to the King's idiots; for it would require wise statesmanship and long and patient though desultory hammering of the enemy. Now and then, for a quarter of a century yet, there would be a little fighting to do, and a handy man could carry that on with small disturbance to the rest of the country; and little by little, and with progressive certainty, the English would disappear from France.

And that happened. Under the influence of Richemont the King became at a later time a man—a man, a king, a brave and capable and determined soldier. Within six years after Patay he was leading storming parties himself; fighting in fortress ditches up to his waist in water, and climbing scaling-ladders under a furious fire with a pluck that would have satisfied even Joan of Arc. In time he and Richemont cleared away all the English; even from regions where the people had been under their mastership for three hundred years. In such regions wise and careful work was necessary, for the English rule had been fair and kindly; and men who have been ruled in that way are not always anxious for a change.

Which of Joan's five chief deeds shall we call chiefest? It is my thought that each *in its turn* was that. This is saying that, taken as a whole, they *equalized* each other, and neither was then greater than its mate.

THE SIEGE OF ORLEANS
(From the painting by J. E. Lenepveu in the Panthéon at Paris)

Do you perceive? Each was a stage in an ascent. To leave out one of them would defeat the journey; to achieve one of them at the wrong time and in the wrong place would have the same effect.

Consider the Coronation. As a masterpiece of diplomacy, where can you find its superior in our history? Did the King suspect its vast importance? No. Did his ministers? No. Did the astute Bedford, representative of the English crown? No. An advantage of incalculable importance was here under the eyes of the King and of Bedford; the King could get it by a bold stroke, Bedford could get it without an effort; but being ignorant of its value, neither of them put forth his hand. Of all the wise people in high office in France, only one knew the priceless worth of this neglected prize — the untaught child of seventeen, Joan of Arc — and she had known it from the beginning, had spoken of it from the beginning as an essential detail of her mission.

How did she know it? It is simple: she was a peasant. That tells the whole story. She was of the people and knew the people; those others moved in a loftier sphere and knew nothing much about them. We make little account of that vague, formless, inert mass, that mighty underlying force which we call "the people"—an epithet which carries contempt with it. It is a strange attitude; for at bottom we know that the throne which the people support, stands, and that when that support is removed, nothing in this world can save it.

Now, then, consider this fact, and observe its importance. Whatever the parish priest believes, his flock believes; they love him, they revere him; he is their unfailing friend, their dauntless protector, their comforter in sorrow, their helper in their day of need; he has their whole confidence; what he tells them to do, that they will do, with a blind and affectionate obedience, let it cost what it may. Add these facts thoughtfully together, and what is the sum? This: *The parish priest governs the nation.* What is the King, then, if the parish priest withdraw his support and deny his authority? Merely a shadow and no King; let him resign.

Do you get that idea? Then let us proceed. A priest is consecrated to his office by the awful hand of God, laid upon him by his appointed representative on earth. That consecration is final; nothing can undo it, nothing can remove it. Neither the Pope nor any other power can strip the priest of his office; God gave it, and it is forever sacred and secure. The dull parish knows all this. To priest and parish, whosoever is anointed of God bears an office whose authority can no longer be disputed or assailed. To the parish priest, and to his subjects the nation, an uncrowned king is a similitude of a person who has been named for holy orders but has not been consecrated; he has no office, he has not been ordained, another may be appointed in his place. In a word, an uncrowned king is a *doubtful* King; but if God appoint, him and His servant the Bishop anoint him, the doubt is annihilated; the priest and the parish are his loyal subjects straightway, and while he lives they will recognize no king but him.

To Joan of Arc the peasant girl, Charles VII. was no King until he was crowned; to her he was only the *Dauphin;* that is to say, the *heir.* If I have ever made her call him King, it was a mistake; she called him the Dauphin, and nothing else until after the Coronation. It shows you as in a mirror—for Joan was a mirror in which the lowly hosts of France were clearly reflected—that to all that vast underlying force called "the people" he was no King but only Dauphin before his crowning, and was indisputably and irrevocably King *after* it.

Now you understand what a colossal move on the political chess-board the Coronation was. Bedford realized this by-and-by, and tried to patch up his mistake by crowning *his* King; but what good could that do? None in the world.

Speaking of chess, Joan's great acts may be likened to that game. Each move was made in its proper order, and it was great and effective because it *was* made in its proper order and not out of it. Each, at the time made, seemed the greatest move; but the final result made them all recognizable as equally essential and equally important. This is the game, as played:

1. Joan moves Orleans and Patay—*check*.

2. Then moves the Reconciliation—but does not proclaim check, it being a move for position, and to take effect later.

3. Next she moves the Coronation—*check*.

4. Next, the Bloodless March—*check*.

5. Final move (after her death) the reconciled Constable. Richemont to the French King's elbow—*checkmate*.

CHAPTER XXXIV

THE Campaign of the Loire had as good as opened the road to Rheims. There was no sufficient reason now why the Coronation should not take place. The Coronation would complete the mission which Joan had received from heaven, and then she would be forever done with war, and would fly home to her mother and her sheep, and never stir from the hearthstone and happiness any more. That was her dream; and she could not rest, she was so impatient to see it fulfilled. She became so possessed with this matter that I began to lose faith in her two prophecies of her early death—and of course when I found that faith wavering I encouraged it to waver all the more.

The King was afraid to start to Rheims, because the road was mile-posted with English fortresses, so to speak. Joan held them in light esteem and not things to be afraid of in the existing modified condition of English confidence.

And she was right. As it turned out, the march to Rheims was nothing but a holiday excursion. Joan did not even take any artillery along, she was so sure it would not be necessary. We marched from Gien twelve thousand strong. This was the 29th of June. The Maid rode by the side of the King; on his other side was the Duke d'Alençon. After the Duke followed three other princes of the blood. After these followed the Bastard of Orleans, the Marshal de Boussac, and the Admiral of France. After these came La Hire, Saintrailles, Tremouille, and a long procession of knights and nobles.

We rested three days before Auxerre. The city provisioned the army, and a deputation waited upon the King, but we did not enter the place.

Saint-Florentin opened its gates to the King.

On the 4th of July we reached Saint-Fal, and yonder lay Troyes before us—a town which had a burning interest for us boys; for we remembered how seven years before, in the pastures of Domremy, the Sunflower came with his black flag and brought us the shameful news of the Treaty of Troyes—that treaty which gave France to England, and a daughter of our royal line in marriage to the Butcher of Agincourt. That poor town was not to blame, of course; yet we flushed hot with that old memory, and hoped there would be a misunderstanding here, for we dearly wanted to storm the place and burn it. It was powerfully garrisoned by English and Burgundian soldiery, and was expecting re-enforcements from Paris. Before night we camped before its gates and made rough work with a sortie which marched out against us.

Joan summoned Troyes to surrender. Its commandant, seeing that she had no artillery, scoffed at the idea, and sent her a grossly insulting reply. Five days we consulted and negotiated. No result. The King was about to turn back now, and give up. He was afraid to go on, leaving this strong place in his rear. Then La Hire put in a word, with a slap in it for some of his Majesty's advisers:

"The Maid of Orleans undertook this expedition of her own motion; and it is my mind that it is her judgment that should be followed here, and not that of any other, let him be of whatsoever breed and standing he may."

There was wisdom and righteousness in that. So the King sent for the Maid, and asked her how she thought the prospect looked. She said, without any tone of doubt or question in her voice:

"In three days' time the place is ours."

The smug Chancellor put in a word now:

"If we were sure of it we would wait here six days."

"Six days, forsooth! Name of God, man, we will enter the gates to-morrow!"

Then she mounted, and rode her lines, crying out—

"Make preparation—to your work, friends, to your work! We assault at dawn!"

She worked hard that night; slaving away with her own hands like a common soldier. She ordered fascines and fagots to be prepared and thrown into the fosse, thereby to bridge it; and in this rough labor she took a man's share.

At dawn she took her place at the head of the storming force and the bugles blew the assault. At that moment a flag of truce was flung to the breeze from the walls, and Troyes surrendered without firing a shot.

The next day the King with Joan at his side and the Paladin bearing her banner entered the town in state at the head of the army. And a goodly army it was, now, for it had been growing ever bigger and bigger from the first.

And now a curious thing happened. By the terms of the treaty made with the town the garrison of English and Burgundian soldiery were to be allowed to carry away their "goods" with them. This was well, for otherwise how would they buy the wherewithal to live? Very well; these people were all to go out by the one gate, and at the time set for them to depart we young fellows went to that gate, along with the Dwarf, to see the march-out. Presently here they came in an interminable file, the foot-soldiers in the lead. As they approached one could see that each bore a burden of a bulk and weight to sorely tax his strength; and we said among ourselves, truly these folk are well off for poor common soldiers. When they were come nearer, what do you think? Every rascal of them had a French prisoner on his back! They were carrying away their "goods," you see— their property—strictly according to the permission granted by the treaty.

Now think how clever that was, how ingenious. What could a body say? what could a body do? For certainly these people were within their right. These prisoners were property; nobody could deny that. My dears, if those had been *English* captives, conceive of the richness of that booty! For English prisoners had been scarce and precious for a hundred years; whereas it was a different matter with French prisoners. They had been over-abundant for a century. The possessor of a French prisoner did not hold him long for ransom as a rule, but presently killed him to save the cost of his keep. This shows you how small was the value of such a possession in those times. When we took Troyes a calf was worth thirty francs, a sheep sixteen, a French prisoner eight. It was an enormous price for those other animals—a price which naturally seems incredible to you. It was the war, you see. It worked two ways: it made meat dear and prisoners cheap.

Well, here were these poor Frenchmen being carried off. What could we do? Very little of a permanent sort, but we did what we could. We sent a messenger flying to Joan, and we and the French guards halted the procession for a parley —to gain time, you see. A big Burgundian lost his temper and swore a great oath that none should stop *him;* he would go, and would take his prisoner with him. But we blocked him off, and he saw that he was mistaken about going—he couldn't do it. He exploded into the maddest cursings and revilings, then, and unlashing his prisoner from his back, stood him up, all bound and helpless; then drew his knife, and said to us with a light of sarcastic triumph in his eye—

"I may not carry him away, you say—yet he is mine, none will dispute it. Since I may not convey him hence, this property of mine, there is another way. Yes, I can kill him; not even the dullest among you will question *that* right. Ah, you had not thought of that—vermin!"

That poor starved fellow begged us with his piteous eyes to save him; then spoke, and said he had a wife and little children at home. Think how it wrung our heartstrings. But what could we do? The Burgundian was within his right. We could only beg and plead for the prisoner. Which we did. And the Burgundian enjoyed it. He stayed his hand to hear more of it, and laugh at it. That stung. Then the Dwarf said—

"Prithee, young sirs, let me beguile him; for when a mat-

ter requiring persuasion is to the fore, I have indeed a gift in
that sort, as any will tell you that know me well. You smile;
and that is punishment for my vanity, and fairly earned, I
grant it you. Still, if I may toy a little, just a little—" saying which he stepped to the Burgundian and began a fair
soft speech, all of goodly and gentle tenor; and in the midst
he mentioned the Maid; and was going on to say how she
out of her good heart would prize and praise this compassionate deed which he was about to—

It.was as far as he got. The Burgundian burst into his
smooth oration with an insult levelled at Joan of Arc. We
sprang forward, but the Dwarf, his face all livid, brushed us
aside and said, in a most grave and earnest way—

"I crave your patience. Am not I her guard of honor?
This is my affair."

And saying this he suddenly shot his right hand out and
gripped the great Burgundian by the throat, and so held him
upright on his feet. "You have insulted the Maid," he said;
"and the Maid is France. The tongue that does that earns
a long furlough."

One heard the muffled cracking of bones. The Burgundian's eyes began to protrude from their sockets and stare
with a leaden dulness at vacancy. The color deepened in
his face and became an opaque purple. His hands hung
down limp, his body collapsed with a shiver, every muscle re-
laxed its tension and ceased from its function. The Dwarf
took away his hand and the column of inert mortality sank
mushily to the ground.

We struck the bonds from the prisoner and told him he
was free. His crawling humbleness changed to frantic joy
in a moment, and his ghastly fear to a childish rage. He
flew at that dead corpse and kicked it, spat in its face;
danced upon it, crammed mud into its mouth, laughing, jeer-
ing, cursing and volleying forth indecencies and bestialities
like a drunken fiend. It was a thing to be expected: soldiering makes few saints. Many of the on-lookers laughed,
others were indifferent, none was surprised. But presently

in his mad caperings the freed man capered within reach of
the waiting file, and another Burgundian promptly slipped a
knife through his neck, and down he went with a death-shriek,
his brilliant artery-blood spurting ten feet as straight and
bright as a ray of light. There was a great burst of jolly
laughter all around from friend and foe alike ; and thus closed
one of the pleasantest incidents of my checkered military life.

And now came Joan hurrying, and deeply troubled. She
considered the claim of the garrison, then said—

"You have right upon your side. It is plain. It was a
careless word to put in the treaty, and covers too much. But
ye may not take these poor men away. They are French,
and I will not have it. The King shall ransom them, every
one. Wait till I send you word from him ; and hurt no hair
of their heads ; for I tell you, I who speak, that that would
cost you very dear."

That settled it. The prisoners were safe for one while,
anyway. Then she rode back eagerly and required that
thing of the King, and would listen to no paltering and no
excuses. So the King told her to have her way, and she
rode straight back and bought the captives free in his name
and let them go.

CHAPTER XXXV

It was here that we saw again the Grand Master of the King's Household, in whose castle Joan was guest when she tarried at Chinon in those first days of her coming out of her own country. She made him Bailiff of Troyes, now, by the King's permission.

And now we marched again; Châlons surrendered to us; and there by Châlons in a talk, Joan being asked if she had no fears for the future, said yes, one—treachery. Who could believe it? who could dream it? And yet in a sense it was prophecy. Truly man is a pitiful animal.

We marched, marched, kept on marching; and at last on the 16th of July we came in sight of our goal, and saw the great cathedral towers of Rheims rise out of the distance! Huzzah after huzzah swept the army from van to rear; and as for Joan of Arc, there where she sat her horse gazing, clothed all in white armor, dreamy, beautiful, and in her face a deep, deep joy, a joy not of earth, oh, she was not flesh, she was a spirit! Her sublime mission was closing—closing in flawless triumph. To-morrow she could say, "It is finished —let me go free."

We camped, and the hurry and rush and turmoil of the grand preparations began. The Archbishop and a great deputation arrived; and after these came flock after flock, crowd after crowd, of citizens and country folk hurrahing in, with banners and music, and flowed over the camp, one rejoicing inundation after another, everybody drunk with happiness. And all night long Rheims was hard at work, hammering away, decorating the town, building triumphal arches, and clothing the ancient cathedral within and without in a glory of opulent splendors.

We moved betimes in the morning: the coronation ceremonies would begin at nine and last five hours. We were
aware that the garrison of English and Burgundian soldiers
had given up all thought of resisting the Maid, and that we
should find the gates standing hospitably open and the whole
city ready to welcome us with enthusiasm.

It was a delicious morning, brilliant with sunshine but cool
and fresh and inspiring. The army was in great form, and
fine to see, as it uncoiled from its lair fold by fold, and
stretched away on the final march of the peaceful Coronation
Campaign.

Joan, on her black horse, with the Lieutenant-General and
the personal staff grouped about her, took post for a final review and a good-bye; for she was not expecting to ever be a
soldier again, or ever serve with these or any other soldiers
any more after this day. The army knew this, and believed
it was looking for the last time upon the girlish face of its invincible little Chief, its pet, its pride, its darling, whom it had
ennobled in its private heart with nobilities of its own creation, calling her "Daughter of God," "Savior of France,"
"Victory's Sweetheart," "the Page of Christ," together with
still softer titles which were simply naïf and frank endearments such as men are used to confer upon children whom
they love. And so one saw a new thing now; a thing bred
of the emotion that was present there on both sides. Always
before, in the march-past, the battalions had gone swinging
by in a storm of cheers, heads up and eyes flashing, the
drums rolling, the bands braying pæans of victory; but now
there was nothing of that. But for one impressive sound, one
could have closed his eyes and imagined himself in a world
of the dead. That one sound was all that visited the ear in
the summer stillness—just that one sound—the muffled tread
of the marching host. As the serried masses drifted by, the
men put their right hands up to their temples, palms to the
front, in military salute, turning their eyes upon Joan's face
in mute God-bless-you and farewell, and keeping them there
while they could. They still kept their hands up in reverent

salute many steps after they had passed by. Every time Joan put her handkerchief to her eyes you could see a little quiver of emotion crinkle along the faces of the files.

The march-past after a victory is a thing to drive the heart mad with jubilation ; but this one was a thing to break it.

We rode now to the King's lodging, which was the Archbishop's country palace; and he was presently ready, and we galloped off and took position at the head of the army. By this time the country people were arriving in multitudes from every direction and massing themselves on both sides of the road to get sight of Joan—just as had been done every day since our first day's march began. Our march now lay through the grassy plain, and those peasants made a dividing double border for that plain. They stretched right down through it, a broad belt of bright colors on each side of the road; for every peasant girl and woman in it had a white jacket on her body and a crimson skirt on the rest of her. Endless borders made of poppies and lilies stretching away in front of us—that is what it looked like. And that is the kind of lane we had been marching through all these days. Not a lane between multitudinous flowers standing upright on their stems — no, these flowers were always kneeling; kneeling, these human flowers, with their hands and faces lifted toward Joan of Arc, and the grateful tears streaming down. And all along, those closest to the road hugged her feet and kissed them and laid their wet cheeks fondly against them. I never, during all those days, saw any of either sex stand while she passed, nor any man keep his head covered. Afterwards in the Great Trial these touching scenes were used as a weapon against her. She had been made an object of adoration by the people, and this was proof that she was a heretic—so claimed that unjust court.

As we drew near the city the curving long sweep of ramparts and towers was gay with fluttering flags and black with masses of people ; and all the air was vibrant with the crash of artillery and gloomed with drifting clouds of smoke. We entered the gates in state and moved in procession through the city, with all the guilds and industries in holiday costume marching in our rear with their banners; and all the route was hedged with a huzzahing crush of people, and all the windows were full and all the roofs; and from the balconies hung costly stuffs of rich colors; and the waving of handkerchiefs, seen in perspective through a long vista, was like a snow-storm.

Joan's name had been introduced into the prayers of the Church—an honor theretofore restricted to royalty. But she had a dearer honor and an honor more to be proud of, from a humbler source: the common people had had leaden medals struck which bore her effigy and her escutcheon, and these they wore as charms. One saw them everywhere.

From the Archbishop's Palace, where we halted, and where the King and Joan were to lodge, the King sent to the Abbey Church of St. Remi, which was over toward the gate by which we had entered the city, for the *Sainte Ampoule*, or flask of holy oil. This oil was not earthly oil; it was made in heaven; the flask also. The flask, with the oil in it, was brought down from heaven by a dove. It was sent down to St. Remi just as he was going to baptize King Clovis, who had become a Christian. I know this to be true. I had known it long before; for Père Fronte told me in Domremy. I cannot tell you how strange and awful it made me feel when I saw that flask and knew I was looking with my own eyes upon a thing which had actually been in heaven; a thing which had been seen by angels, perhaps; and by God Himself of a certainty, for He sent it. And I was looking upon it—I. At one time I could have touched it. But I was afraid; for I could not know but that God had touched it. It is most probable that He had.

From this flask Clovis had been anointed; and from it all the Kings of France had been anointed since. Yes, ever since the time of Clovis; and that was nine hundred years. And so, as I have said, that flask of holy oil was sent for, while we waited. A coronation without that would not have been a coronation at all, in my belief.

18

Now in order to get the flask, a most ancient ceremonial had to be gone through with; otherwise the Abbé of St. Remi, hereditary guardian in perpetuity of the oil, would not deliver it. So, in accordance with custom, the King deputed five great nobles to ride in solemn state and richly armed and accoutred, they and their steeds, to the Abbey Church as a guard of honor to the Archbishop of Rheims and his canons, who were to bear the King's demand for the oil. When the five great lords were ready to start, they knelt in a row and put up their mailed hands before their faces, palm joined to palm, and swore upon their lives to conduct the sacred vessel safely, and safely restore it again to the Church of St. Remi after the anointing of the King. The Archbishop and his subordinates, thus nobly escorted, took their way to St. Remi. The Archbishop was in grand costume, with his mitre on his head and his cross in his hand. At the door of St. Remi they halted and formed, to receive the holy phial. Soon one heard the deep tones of the organ and of chanting men; then one saw a long file of lights approaching through the dim church. And so came the Abbot, in his sacerdotal panoply, bearing the phial, with his people following after. He delivered it, with solemn ceremonies, to the Archbishop; then the march back began, and it was most impressive; for it moved, the whole way, between two multitudes of men and women who lay flat upon their faces and prayed in dumb silence and in dread while that awful thing went by that had been in heaven.

This august company arrived at the great west door of the cathedral; and as the Archbishop entered a noble anthem rose and filled the vast building. The cathedral was packed with people—people in thousands. Only a wide space down the centre had been kept free. Down this space walked the Archbishop and his canons, and after them followed those five stately figures in splendid harness, each bearing his feudal banner—and riding!

Oh, that was a magnificent thing to see. Riding down the cavernous vastness of the building through the rich lights

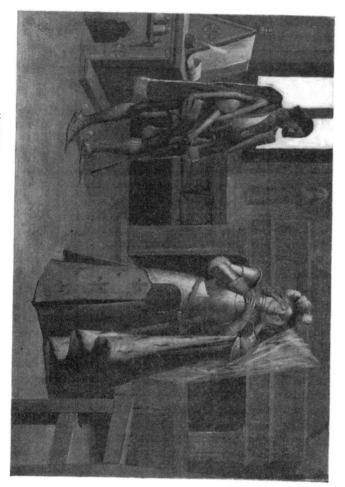

"THE DUCHESS KISSED JOAN, AND SO THEY PARTED"

streaming in long rays from the pictured windows—oh, there was never anything so grand!

They rode clear to the choir—as much as four hundred feet from the door, it was said. Then the Archbishop dismissed them, and they made deep obeisance till their plumes touched their horses' necks, then made those proud prancing and mincing and dancing creatures go backwards all the way to the door—which was pretty to see, and graceful; then they stood them on their hind-feet and spun them around and plunged away and disappeared.

For some minutes there was a deep hush, a waiting pause; a silence so profound that it was as if all those packed thousands there were steeped in dreamless slumber—why, you could even notice the faintest sounds, like the drowsy buzzing of insects; then came a mighty flood of rich strains from four hundred silver trumpets, and then, framed in the pointed archway of the great west door, appeared Joan and the King. They advanced slowly, side by side, through a tempest of welcome—explosion after explosion of cheers and cries, mingled with the deep thunders of the organ and rolling tides of triumphant song from chanting choirs. Behind Joan and the King came the Paladin with the Banner displayed; and a majestic figure he was, and most proud and lofty in his bearing, for he knew that the people were marking him and taking note of the gorgeous state dress which covered his armor.

At his side was the Sire d'Albret, proxy for the Constable of France, bearing the Sword of State.

After these, in order of rank, came a body royally attired representing the lay peers of France; it consisted of three princes of the blood, and La Tremouille and the young De Laval brothers.

These were followed by the representatives of the ecclesiastical peers—the Archbishop of Rheims, and the Bishops of Laon, Châlons, Orleans, and one other.

Behind these came the Grand Staff, all our great generals and famous names, and everybody was eager to get a sight of them. Through all the din one could hear shouts, all along, that told you where two of them were: "Live the Bastard of Orléans!" "Satan La Hire forever!"

The august procession reached its appointed place in time, and the solemnities of the Coronation began. They were long and imposing—with prayers, and anthems, and sermons, and everything that is right for such occasions; and Joan was at the King's side all these hours, with her Standard in her hand. But at last came the grand act: the King took the oath, he was anointed with the sacred oil; a splendid personage, followed by train-bearers and other attendants, approached, bearing the Crown of France upon a cushion, and kneeling offered it. The King seemed to hesitate—in fact *did* hesitate; for he put out his hand and then stopped with it there in the air over the crown, the fingers in the attitude of taking hold of it. But that was for only a moment—though a moment is a notable something when it stops the heart-beat of twenty thousand people and makes them catch their breath. Yes, only a moment; then he caught Joan's eye, and she gave him a look with all the joy of her thankful great soul in it, then he smiled, and took the Crown of France in his hand, and right finely and right royally lifted it up and set it upon his head.

Then what a crash there was! All about us cries and cheers, and the chanting of the choirs and groaning of the organ; and outside the clamoring of the bells and the booming of the cannon.

The fantastic dream, the incredible dream, the impossible dream of the peasant child stood fulfilled: the English power was broken, the Heir of France was crowned.

She was like one transfigured, so divine was the joy that shone in her face as she sank to her knees at the King's feet and looked up at him through her tears. Her lips were quivering, and her words came soft and low and broken:

"Now, oh gentle King, is the pleasure of God accomplished according to his command that you should come to Rheims and receive the crown that belongeth of right to you, and unto none other. My work which was given me to do is finished; give me your peace, and let me go back to my mother, who is poor and old, and has need of me."

The King raised her up, and there before all that host he praised her great deeds in most noble terms; and there he confirmed her nobility and titles, making her the equal of a count in rank, and also appointed a household and officers for her according to her dignity; and then he said:

"You have saved the crown. Speak—require—demand; and whatsoever grace you ask it shall be granted, though it make the kingdom poor to meet it."

Now that was fine, that was royal. Joan was on her knees again straightway, and said:

"Then, oh gentle King, if out of your compassion you will speak the word, I pray you give commandment that my village, poor and hard pressed by reason of the war, may have its taxes remitted."

"It is so commanded. Say on."

"That is all."

"All? Nothing but that?"

"It is all. I have no other desire."

"But that is nothing—less than nothing. Ask—do not be afraid."

"Indeed I cannot, gentle King. Do not press me. I will not have aught else, but only this alone."

The King seemed nonplussed, and stood still a moment, as if trying to comprehend and realize the full stature of this strange unselfishness. Then he raised his head and said:

"She has won a kingdom and crowned its King; and all she asks and all she will take is this poor grace—and even this is for others, not for herself. And it is well; her act being proportioned to the dignity of one who carries in her head and heart riches which outvalue any that any King could add, though he gave his all. She shall have her way. Now therefore it is decreed that from this day forth Domremy, natal village of Joan of Arc, Deliverer of France, called the Maid of Orleans, is freed from all taxation *forever*." Whereat the silver horns blew a jubilant blast.

There, you see, she had had a vision of this very scene the time she was in a trance in the pastures of Domremy, and we asked her to name the boon she would demand of the King if he should ever chance to tell her she might claim one. But whether she had the vision or not, this act showed that after all the dizzy grandeurs that had come upon her, she was still the same simple unselfish creature that she was that day.

Yes, Charles VII. remitted those taxes "forever." Often the gratitude of kings and nations fades and their promises are forgotten or deliberately violated; but you, who are children of France, should remember with pride that France has kept this one faithfully. Sixty-three years have gone by since that day. The taxes of the region wherein Domremy lies have been collected sixty-three times since then, and all the villages of that region have paid except that one—Domremy. The tax-gatherer never visits Domremy. Domremy has long ago forgotten what that dreaded sorrow-sowing apparition is like. Sixty-three tax-books have been filled meantime, and they lie yonder with the other public records, and any may see them that desire it. At the top of every page in the sixty-three books stands the name of a village, and below that name its weary burden of taxation is figured out and displayed; in the case of all save one. It is true, just as I tell you. In each of the sixty-three books there is a page headed "Domremi," but under that name not a figure appears. Where the figures should be, there are three words written; and the same words have been written every year for all these years; yes, it is a blank page, with always those grateful words lettered across the face of it—a touching memorial. Thus:

DOMREMI

RIEN—LA PUCELLE

"Nothing—the Maid of Orleans." How brief it is; yet how much it says! It is the nation speaking. You have the spectacle of that unsentimental thing, a Government, making reverence to that name and saying to its agent, "*Uncover, and pass on; it is France that commands.*" Yes, the promise has been kept; it will be kept always; "forever" was the King's word.*

At two o'clock in the afternoon the ceremonies of the Coronation came at last to an end; then the procession formed once more, with Joan and the King at its head, and took up its solemn march through the midst of the church, all instruments and all people making such clamor of rejoicing noises as was indeed a marvel to hear. And so ended the third of the great days of Joan's life. And how close together they stand—May 8th, June 18th, July 17th!

---

* It was faithfully kept during three hundred and sixty years and more; then the over-confident octogenarian's prophecy failed. During the tumult of the French Revolution the promise was forgotten and the grace withdrawn. It has remained in disuse ever since. Joan never asked to be remembered, but France has remembered her with an inextinguishable love and reverence; Joan never asked for a statue, but France has lavished them upon her; Joan never asked for a church for Domremy, but France is building one; Joan never asked for saintship, but even that is impending. Everything which Joan of Arc did not ask for has been given her, and with a noble profusion; but the one humble little thing which she did ask for and get, has been taken away from her. There is something infinitely pathetic about this. France owes Domremy a hundred years of taxes, and could hardly find a citizen within her borders who would vote against the payment of the debt.—Note by the Translator.

CHAPTER XXXVI

WE mounted and rode, a spectacle to remember, a most noble display of rich vestments and nodding plumes, and as we moved between the banked multitudes they sank down all along abreast of us as we advanced, like grain before the reaper, and kneeling hailed with a rousing welcome the consecrated King and his companion the Deliverer of France. But by-and-by when we had paraded about the chief parts of the city and were come near to the end of our course, we being now approaching the Archbishop's palace, one saw on the right, hard by the inn that is called the Zebra, a strange thing —two men not kneeling but *standing!* Standing in the front rank of the kneelers; unconscious, transfixed, staring. Yes, and clothed in the coarse garb of the peasantry, these two. Two halberdiers sprang at them in a fury to teach them better manners; but just as they seized them Joan cried out " Forbear !" and slid from her saddle and flung her arms about one of those peasants, calling him by all manner of endearing names, and sobbing. For it was her father; and the other was her uncle Laxart.

The news flew everywhere, and shouts of welcome were raised, and in just one little moment those two despised and unknown plebeians were become famous and popular and envied, and everybody was in a fever to get sight of them and be able to say, all their lives long, that they had seen the father of Joan of Arc and the brother of her mother. How easy it was for her to do miracles like to this ! She was like the sun; on whatsoever dim and humble object her rays fell, that thing was straightway drowned in glory.

All graciously the King said :

"Bring them to me."

And she brought them; she radiant with happiness and affection, they trembling and scared, with their caps in their shaking hands; and there before all the world the King gave them his hand to kiss, while the people gazed in envy and admiration; and he said to old D'Arc—

"Give God thanks for that you are father to this child, this dispenser of immortalities. You who bear a name that will still live in the mouths of men when all the race of Kings has been forgotten, it is not meet that you bare your head before the fleeting fames and dignities of a day—cover yourself!" And truly he looked right fine and princely when he said that. Then he gave order that the Bailly of Rheims be brought; and when he was come, and stood bent low and bare, the King said to him, "These two are guests of France"; and bade him use them hospitably.

I may as well say now as later, that Papa D'Arc and Laxart were stopping in that little Zebra inn, and that there they remained. Finer quarters were offered them by the Bailly, also public distinctions and brave entertainment; but they were frightened at these projects, they being only humble and ignorant peasants: so they begged off, and had peace. They could not have enjoyed such things. Poor souls, they did not even know what to do with their hands, and it took all their attention to keep from treading on them. The Bailly did the best he could in the circumstances. He made the innkeeper place a whole floor at their disposal, and told him to provide everything they might desire, and charge all to the city. Also the Bailly gave them a horse apiece, and furnishings; which so overwhelmed them with pride and delight and astonishment that they couldn't speak a word; for in their lives they had never dreamed of wealth like this, and could not believe, at first, that the horses were real and would not dissolve to a mist and blow away. They could not unglue their minds from those grandeurs, and were always wrenching the conversation out of its groove and dragging the matter of animals into it, so that they could say " my horse " here,

and "my horse" there and yonder and all around, and taste
the words and lick their chops over them, and spread their
legs and hitch their thumbs in their armpits, and feel as the
good God feels when He looks out on His fleets of constellations ploughing the awful deeps of space and reflects with satisfaction that they are His—all His. Well, they *were* the hap-
piest old children one ever saw, and the simplest.

The city gave a grand banquet to the King and Joan in
mid-afternoon, and to the Court and the Grand Staff; and
about the middle of it Père d'Arc and Laxart were sent for,
but would not venture until it was promised that they might
sit in a gallery and be all by themselves and see all that was
to be seen and yet be unmolested. And so they sat there
and looked down upon the splendid spectacle, and were
moved till the tears ran down their cheeks to see the unbelievable honors that were paid to their small darling, and how
naïvely serene and unafraid she sat there with those consuming glories beating upon her.

But at last her serenity was broken up. Yes, it stood the
strain of the King's gracious speech; and of D'Alençon's
praiseful words, and the Bastard's; and even La Hire's thunder-blast, which took the place by storm; but at last, as I
have said, they brought a force to bear which was too strong
for her. For at the close the King put up his hand to command silence, and so waited, with his hand up, till every
sound was dead and it was as if one could almost *feel* the
stillness, so profound it was. Then out of some remote corner of that vast place there rose a plaintive voice, and in
tones most tender and sweet and rich came floating through
that enchanted hush our poor old simple song "L'Arbre Fée
de Bourlemont!" and then Joan broke down and put her face
in her hands and cried. Yes, you see, all in a moment the
pomps and grandeurs dissolved away and she was a little
child again herding her sheep with the tranquil pastures
stretched about her, and war and wounds and blood and
death and the mad frenzy and turmoil of battle a dream. Ah,
that shows you the power of music, that magician of magi-

cians; who lifts his wand and says his mysterious word and all things real pass away and the phantoms of your mind walk before you clothed in flesh.

That was the King's invention, that sweet and dear surprise. Indeed, he had fine things hidden away in his nature, though one seldom got a glimpse of them, with that scheming Tremouille and those others always standing in the light, and he so indolently content to save himself fuss and argument and let them have their way.

At the fall of night we the Domremy contingent of the personal staff were with the father and uncle at the inn, in their private parlor, brewing generous drinks and breaking ground for a homely talk about Domremy and the neighbors, when a large parcel arrived from Joan to be kept till she came; and soon she came herself and sent her guard away, saying she would take one of her father's rooms and sleep under his roof, and so be at home again. We of the staff rose and stood, as was meet, until she made us sit. Then she turned and saw that the two old men had gotten up too, and were standing in an embarrassed and unmilitary way; which made her want to laugh, but she kept it in, as not wishing to hurt them; and got them to their seats and snuggled down between them, and took a hand of each of them upon her knees and nestled her own hands in them, and said—

"Now we will have no more ceremony, but be kin and playmates as in other times; for I am done with the great wars, now, and you two will take me home with you, and I shall see—" She stopped, and for a moment her happy face sobered, as if a doubt or a presentiment had flitted through her mind; then it cleared again, and she said, with a passionate yearning, "Oh, if the day were but come and we could start!"

The old father was surprised, and said—

"Why, child, are you in earnest? Would you leave doing these wonders that make you to be praised by everybody while there is still so much glory to be won; and would you go out from this grand comradeship with princes and generals to be a drudging villager again and a nobody? It is not rational."

"No," said the uncle, Laxart, "it is amazing to hear, and indeed not understandable. It is a stranger thing to hear her say she will stop the soldiering than it was to hear her say she would begin it; and I who speak to you can say in all truth that that was the strangest word that ever I had heard till this day and hour. I would it could be explained."

"It is not difficult," said Joan. "I was not ever fond of wounds and suffering, nor fitted by my nature to inflict them; and quarrellings did always distress me, and noise and tumult were against my liking, my disposition being toward peace and quietness, and love for all things that have life; and being made like this, how could I bear to think of wars and blood, and the pain that goes with them, and the sorrow and mourning that follow after? But by his angels God laid His great commands upon me, and could I disobey? I did as I was bid. Did he command me to do many things? No; only two: to raise the siege of Orleans, and crown the King at Rheims. The task is finished, and I am free. Has ever a poor soldier fallen in my sight, whether friend or foe, and I not felt his pain in my own body, and the grief of his home-mates in my own heart? No, not one; and, oh, it is such bliss to know that my release is won, and that I shall not any more see these cruel things or suffer these tortures of the mind again! Then why should I not go to my village and be as I was before? It is heaven! and ye wonder that I desire it. Ah, ye are men — just men! My mother would understand."

They didn't quite know what to say; so they sat still awhile, looking pretty vacant. Then old D'Arc said—

"Yes, your mother — that is true. I never saw such a woman. She worries, and worries, and worries; and wakes nights, and lies so, thinking — that is, worrying; worrying about you. And when the night-storms go raging along, she moans and says, 'Ah, God pity her, she is out in this with her poor wet soldiers.' And when the lightning glares and

the thunder crashes she wrings her hands and trembles, saying, ' It is like the awful cannon and the flash, and yonder
somewhere she is riding down upon the spouting guns and I
not there to protect her.' "

" Ah, poor mother, it is pity, it is pity !"

" Yes, a most strange woman, as I have noticed a many
times. When there is news of a victory and all the village
goes mad with pride and joy, she rushes here and there in a
maniacal frenzy till she finds out the one only thing she cares
to know — that you are safe; then down she goes on her
knees in the dirt and praises God as long as there is any
breath left in her body; and all on your account, for she
never mentions the battle once. And always she says, ' *Now*
it is over—*now* France is saved—*now* she will come home '—
and always is disappointed and goes about mourning."

" Don't, father ! it breaks my heart. I will be so good to
her when I get home. I will do her work for her, and be her
comfort, and she shall not suffer any more through me."

There was some more talk of this sort, then Uncle Laxart
said—

" You have done the will of God, dear, and are quits; it is
true, and none may deny it; but what of the King? You are
his best soldier; what if he command you to stay?"

That was a crusher—and sudden ! It took Joan a moment
or two to recover from the shock of it; then she said, quite
simply and resignedly :

" The King is my Lord; I am his servant." She was silent
and thoughtful a little while, then she brightened up and
said, cheerily, " But let us drive such thoughts away—this is
no time for them. Tell me about home."

So the two old gossips talked and talked; talked about
everything and everybody in the village; and it was good to
hear. Joan out of her kindness tried to get *us* into the conversation, but that failed, of course. She was the Commander-in-Chief, we were nobodies; her name was the mightiest in
France, we were invisible atoms; she was the comrade of
princes and heroes, we of the humble and obscure; she held

rank above all Personages and all Puissances whatsoever in the whole earth, by right of bearing her commission direct from God. To put it in one word, she was JOAN OF ARC—and when that is said, all is said. To us she was divine. Between her and us lay the bridgeless abyss which that word implies. We could not be familiar with her. No, you can see yourselves that that would have been impossible.

And yet she was so human, too, and so good and kind and dear and loving and cheery and charming and unspoiled and unaffected! Those are all the words I think of now, but they are not enough; no, they are too few and colorless and meagre to tell it all, or tell the half. Those simple old men didn't realize her; they couldn't; they had never known any people but human beings, and so they had no other standard to measure her by. To them, after their first little shyness had worn off, she was just a girl—that was all. It was amazing. It made one shiver, sometimes, to see how calm and easy and comfortable they were in her presence, and hear them talk to her exactly as they would have talked to any other girl in France.

Why, that simple old Laxart sat up there and droned out the most tedious and empty tale one ever heard, and neither he nor Papa D'Arc ever gave a thought to the badness of the etiquette of it, or ever suspected that that foolish tale was anything but dignified and valuable history. There was not an atom of value in it; and whilst they thought it distressing and pathetic, it was in fact not pathetic at all, but actually ridiculous. At least it seemed so to me, and it seems so yet. Indeed I know it was, because it made Joan laugh; and the more sorrowful it got the more it made her laugh; and the Paladin said that he could have laughed himself if she had not been there, and Noël Rainguesson said the same. It was about old Laxart going to a funeral there at Domremy two or three weeks back. He had spots all over his face and hands, and he got Joan to rub some healing ointment on them, and while she was doing it, and comforting him, and trying to say pitying things to him, he told her how it happened. And first

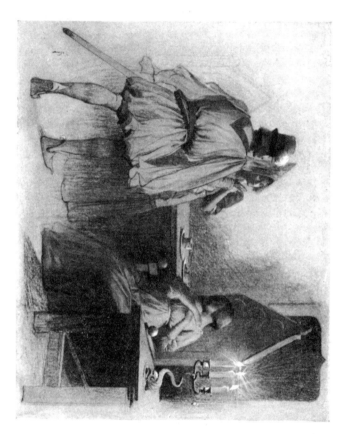

he asked her if she remembered that black bull calf that she
left behind when she came away, and she said indeed she did,
and he was a dear, and she loved him so, and was he well?—
and just drowned him in questions about that creature. And
he said it was a young bull now, and very frisky; and he
was to bear a principal hand at a funeral; and she said, "The
bull?" and he said "No, myself"; but said the bull *did* take
a hand, but not because of his being invited, for he wasn't;
but anyway he was away over beyond the Fairy Tree, and fell
asleep on the grass with his Sunday funeral clothes on, and a
long black rag on his hat and hanging down his back; and
when he woke he saw by the sun how late it was, and not a
moment to lose; and jumped up terribly worried, and saw
the young bull grazing there, and thought maybe he could
ride part way on him and gain time; so he tied a rope around
the bull's body to hold on by, and put a halter on him to
steer with, and jumped on and started; but it was all new to
the bull, and he was discontented with it, and scurried around
and bellowed and reared and pranced, and Uncle Laxart was
satisfied, and wanted to get off and go by the next bull or
some other way that was quieter, but he didn't dare try; and
it was getting very warm for him, too, and disturbing and
wearisome, and not proper for Sunday; but by-and-by the
bull lost all his temper, and went tearing down the slope with
his tail in the air and bellowing in the most awful way; and
just in the edge of the village he knocked down some bee-
hives, and the bees turned out and joined the excursion, and
soared along in a black cloud that nearly hid those other two
from sight, and prodded them both, and jabbed them and
speared them and spiked them, and made them bellow and
shriek, and shriek and bellow; and here they came roaring
through the village like a hurricane, and took the funeral procession right in the centre, and sent that section of it sprawl-
ing, and galloped over it, and the rest scattered apart and fled
screeching in every direction, every person with a layer of
bees on him, and not a rag of that funeral left but the corpse;
and finally the bull broke for the river and jumped in, and

when they fished Uncle Laxart out he was nearly drowned, and his face looked like a pudding with raisins in it. And then he turned around, this old simpleton, and looked a long time in a dazed way at Joan where she had her face in a cushion, dying, apparently, and says—

"What do you reckon she is laughing at?"

And old D'Arc stood looking at her the same way, sort of absently scratching his head; but had to give it up, and said *he* didn't know—"must have been something that happened when we weren't noticing."

Yes, both of those old people thought that that tale was pathetic; whereas to my mind it was purely ridiculous, and not in any way valuable to any one. It seemed so to me then, and it seems so to me yet. And as for history, it does not resemble history, for the office of history is to furnish serious and important facts that *teach;* whereas this strange and useless event teaches nothing; nothing that I can see, except not to ride a bull to a funeral; and surely no reflecting person needs to be taught that.

CHAPTER XXXVII

Now these were nobles, you know, by decree of the King!
—these precious old infants. But they did not realize it;
they could not be called conscious of it; it was an abstraction,
a phantom; to them it had no substance; their minds could
not take hold of it. No, they did not bother about their nobility; they lived in their horses. The horses were solid;
they were visible facts, and would make a mighty stir in
Domremy. Presently something was said about the Coronation, and old D'Arc said it was going to be a grand thing to
be able to say, when they got home, that they were present in
the very town itself when it happened. Joan looked troubled,
and said—

"Ah, that reminds me. You were here and you didn't send
me word. In the town, indeed! Why, you could have sat
with the other nobles, and been welcome; and could have
looked upon the crowning itself, and carried *that* home to tell.
Ah, why did you use me so, and send me no word?"

The old father was embarrassed, now, quite visibly embarrassed, and had the air of one who does not quite know what
to say. But Joan was looking up in his face, her hands upon
his shoulders—waiting. He had to speak; so presently he
drew her to his breast, which was heaving with emotion; and
he said, getting out his words with difficulty—

"There, hide your face, child, and let your old father humble himself and make his confession. I—I—don't you see,
don't you understand?—I could not know that these grandeurs would not turn your young head—it would be only natural. I might shame you before these great per—"

"Father!"

19

"And then I was afraid, as remembering that cruel thing I said once in my sinful anger. Oh, appointed of God to be a soldier, and the greatest in the land! and in my ignorant anger I said I would drown you with my own hands if you unsexed yourself and brought shame to your name and family. Ah, how could I ever have said it, and you so good and dear and innocent! I was afraid; for I was guilty. You understand it now, my child, and you forgive?"

Do you see? Even that poor groping old land-crab, with his skull full of pulp, had pride. Isn't it wonderful? And more — he had conscience; he had a sense of right and wrong, such as it was; he was able to feel remorse. It looks impossible, it looks incredible, but it is not. I believe that some day it will be found out that peasants are people. Yes, beings in a great many respects like ourselves. And I believe that some day *they* will find this out, too—and then! Well, then I think they will rise up and demand to be regarded as part of the race, and that by consequence there will be trouble. Whenever one sees in a book or in a king's proclamation those words "the nation," they bring before us the upper classes; only those; we know no other "nation"; for us and the kings no other "nation" exists. But from the day that I saw old D'Arc the peasant acting and feeling just as I should have acted and felt myself, I have carried the conviction in my heart that our peasants are not merely animals, beasts of burden put here by the good God to produce food and comfort for the "nation," but something more and better. You look incredulous. Well, that is your training; it is the training of everybody; but as for me, I thank that incident for giving me a better light, and I have never forgotten it.

Let me see—where was I? One's mind wanders around here and there and yonder, when one is old. I think I said Joan comforted him. Certainly, that is what she would do— there was no need to say that. She coaxed him and petted him and caressed him, and laid the memory of that old hard speech of his to rest. Laid it to rest until she should be dead. Then he would remember it again—yes, yes! Lord, how those things sting, and burn, and gnaw—the things which we did against the innocent dead! And we say in our anguish, "If they could only come back!" Which is all very well to say, but as far as I can see, it doesn't profit anything. In my opinion the best way is not to do the thing in the first place. And I am not alone in this; I have heard our two knights say the same thing; and a man there in Orleans—no, I believe it was at Beaugency, or one of those places—it seems more as if it was at Beaugency than the others—this man said the same thing exactly; almost the same words; a dark man with a cast in his eye and one leg shorter than the other. His name was—was—it is singular that I can't call that man's name; I had it in my mind only a moment ago, and I know it begins with — no, I don't remember what it begins with; but never mind, let it go; I will think of it presently, and then I will tell you.

Well, pretty soon the old father wanted to know how Joan felt when she was in the thick of a battle, with the bright blades hacking and flashing all around her, and the blows rapping and slatting on her shield, and blood gushing on her from the cloven ghastly face and broken teeth of the neighbor at her elbow, and the perilous sudden back surge of massed horses upon a person when the front ranks give way before a heavy rush of the enemy, and men tumble limp and groaning out of saddles all around, and battle-flags falling from dead hands wipe across one's face and hide the tossing turmoil a moment, and in the reeling and swaying and laboring jumble one's horse's hoofs sink into soft substances and shrieks of pain respond, and presently — panic! rush! swarm! flight! and death and hell following after! And the old fellow got ever so much excited; and strode up and down, his tongue going like a mill, asking question after question and never waiting for an answer; and finally he stood Joan up in the middle of the room and stepped off and scanned her critically, and said—

"No—I don't understand it. You are so little. So little and slender. When you had your armor on, to-day, it gave one a sort of notion of it; but in these pretty silks and velvets, you are only a dainty page, not a league-striding war-colossus, moving in clouds and darkness and breathing smoke and thunder. I would God I might see you at it and go tell your mother! *That* would help her sleep, poor thing! Here —teach me the arts of the soldier, that I may explain them to her."

And she did it. She gave him a pike, and put him through the manual of arms; and made him do the steps, too. His marching was incredibly awkward and slovenly, and so was his drill with the pike; but he didn't know it, and was wonderfully pleased with himself, and mightily excited and charmed with the ringing, crisp words of command. I am obliged to say that if looking proud and happy when one is marching were sufficient, he would have been the perfect soldier.

And he wanted a lesson in sword-play, and got it. But of course that was beyond him; he was too old. It was beautiful to see Joan handle the foils, but the old man was a bad failure. He was afraid of the things, and skipped and dodged and scrambled around like a woman who has lost her mind on account of the arrival of a bat. He was of no good as an exhibition. But if La Hire had only come in, that would have been another matter. Those two fenced often; I saw them many times. True, Joan was easily his master, but it made a good show for all that, for La Hire was a grand swordsman. What a swift creature Joan was! You would see her standing erect with her ankle-bones together and her foil arched over her head, the hilt in one hand and the button in the other—the old general opposite, bent forward, left hand reposing on his back, his foil advanced, slightly wiggling and squirming, his watching eye boring straight into hers — and all of a sudden she would give a spring forward, and back again; and there she was, with the foil arched over her head as before. La Hire had been hit, but all that the spectator saw of it was a something like a thin flash of light in the air, but nothing distinct, nothing definite.

We kept the drinkables moving, for that would please the Bailly and the landlord; and old Laxart and D'Arc got to feeling quite comfortable, but without being what you could call tipsy. They got out the presents which they had been buying to carry home — humble things and cheap, but they would be fine there, and welcome. And they gave to Joan a present from Père Fronte and one from her mother—the one a little leaden image of the Holy Virgin, the other half a yard of blue silk ribbon; and she was as pleased as a child; and touched, too, as one could see plainly enough. Yes, she kissed those poor things over and over again, as if they had been something costly and wonderful; and she pinned the Virgin on her doublet, and sent for her helmet and tied the ribbon on that; first one way, then another; then a new way, then another new way; and with each effort perching the helmet on her hand and holding it off this way and that, and canting her head to one side and then the other, examining the effect, as a bird does when it has got a new bug. And she said she could almost wish she was going to the wars again; for then she would fight with the better courage, as having always with her something which her mother's touch had blessed.

Old Laxart said he hoped she would go to the wars again, but home first, for that all the people there were cruel anxious to see her—and so he went on:

"They are proud of you, dear. Yes, prouder than any village ever was of anybody before. And indeed it is right and rational; for it is the first time a village has ever had anybody like you to be proud of and call its own. And it is strange and beautiful how they try to give your name to every creature that has a set that is convenient. It is but half a year since you began to be spoken of and left us, and so it is surprising to see how many babies there are already in that region that are named for you. First it was just Joan; then it was Joan-Orleans; then Joan-Orleans-Beaugency-Patay; and now the next ones will have a lot of towns and the Coronation added, of course. Yes, and the animals the same.

They know how you love animals, and so they try to do you honor and show their love for you by naming all those creatures after you; insomuch that if a body should step out and call 'Joan of Arc—come!' there would be a landslide of cats and all such things, each supposing it was the one wanted, and all willing to take the benefit of the doubt, anyway, for the sake of the food that might be on delivery. The kitten you left behind—the last estray you fetched home—bears your name, now, and belongs to Père Fronte, and is the pet and pride of the village; and people have come miles to look at it and pet it and stare at it and wonder over it because it was Joan of Arc's cat. Everybody will tell you that; and one day when a stranger threw a stone at it, not knowing it was your cat, the village rose against him as one man and hanged him! And but for Père Fronte—"

There was an interruption. It was a messenger from the King, bearing a note for Joan, which I read to her, saying he had reflected, and had consulted his other generals, and was obliged to ask her to remain at the head of the army and withdraw her resignation. Also, would she come immediately and attend a council of war? Straightway, at a little distance, military commands and the rumble of drums broke on the still night, and we knew that her guard was approaching.

Deep disappointment clouded her face for just one moment and no more—it passed, and with it the homesick girl, and she was Joan of Arc, Commander-in-Chief again, and ready for duty.

CHAPTER XXXVIII

In my double quality of page and secretary I followed Joan to the council. She entered that presence with the bearing of a grieved goddess. What was become of the volatile child that so lately was enchanted with a ribbon and suffocated with laughter over the distresses of a foolish peasant who had stormed a funeral on the back of a bee-stung bull? One may not guess. Simply it was gone, and had left no sign. She moved straight to the council-table, and stood. Her glance swept from face to face there, and where it fell, these it lit as with a torch, those it scorched as with a brand. She knew where to strike. She indicated the generals with a nod, and said—

"My business is not with you. You have not craved a council of war." Then she turned toward the King's privy council, and continued: "No; it is with you. A council of war! It is amazing. There is but one thing to do, and only one, and lo, ye call a council of war! Councils of war have no value but to decide between two or several doubtful courses. But a council of war when there is only *one* course? Conceive of a man in a boat and his family in the water, and he goes out among his friends to ask what he would better do? A council of war, name of God! To determine what?"

She stopped, and turned till her eyes rested upon the face of La Tremouille; and so she stood, silent, measuring him, the excitement in all faces burning steadily higher and higher, and all pulses beating faster and faster; then she said, with deliberation—

"Every sane man—whose loyalty to his King is not a show and a pretence—knows that there is but one rational thing before us—*the march upon Paris!*"

Down came the fist of La Hire with an approving crash upon the table. La Tremouille turned white with anger, but he pulled himself firmly together and held his peace. The King's lazy blood was stirred and his eye kindled finely, for the spirit of war was away down in him somewhere, and a frank bold speech always found it and made it tingle gladsomely. Joan waited to see if the chief minister might wish to defend his position; but he was experienced and wise, and not a man to waste his forces where the current was against him. He would wait; the King's private ear would be at his disposal by-and-by.

That pious fox the Chancellor of France took the word now. He washed his soft hands together, smiling persuasively, and said to Joan:

"Would it be courteous, your Excellency, to move abruptly from here without waiting for an answer from the Duke of Burgundy? You may not know that we are negotiating with his Highness, and that there is likely to be a fortnight's truce between us; and on his part a pledge to deliver Paris into our hands without cost of a blow or the fatigue of a march thither."

Joan turned to him and said, gravely—

"This is not a confessional, my lord. You were not obliged to expose that shame here."

The Chancellor's face reddened, and he retorted—

"Shame? What is there shameful about it?"

Joan answered in level, passionless tones—

"One may describe it without hunting far for words. I knew of this poor comedy, my lord, although it was not intended that I should know. It is to the credit of the devisers of it that they tried to conceal it—this comedy whose text and impulse are describable in two words."

The Chancellor spoke up with a fine irony in his manner:

"Indeed? And will your Excellency be good enough to utter them?"

"Cowardice and treachery!"

The fists of all the generals came down this time, and again the King's eye sparkled with pleasure. The Chancellor sprang to his feet and appealed to his Majesty—

"Sire, I claim your protection."

But the King waved him to his seat again, saying—

"Peace. She had a right to be consulted before that thing was undertaken, since it concerned war as well as politics. It is but just that she be heard upon it now."

The Chancellor sat down trembling with indignation, and remarked to Joan—

"Out of charity I will consider that you did not know who devised this measure which you condemn in so candid language."

"Save your charity for another occasion, my lord," said Joan, as calmly as before. "Whenever anything is done to injure the interests and degrade the honor of France, all but the dead know how to name the two conspirators-in-chief."

"Sire, sire! this insinuation—"

"It is not an insinuation, my lord," said Joan, placidly, "it is a charge. I bring it against the King's chief minister and his Chancellor."

Both men were on their feet now, insisting that the King modify Joan's frankness; but he was not minded to do it. His ordinary councils were stale water—his spirit was drinking wine, now, and the taste of it was good. He said—

"Sit—and be patient. What is fair for one must in fairness be allowed the other. Consider—and be just. When have you two spared her? What dark charges and harsh names have you withheld when you spoke of her?" Then he added, with a veiled twinkle in his eye, "If these are offences I see no particular difference between them, except that she says her hard things to your faces, whereas you say yours behind her back."

He was pleased with that neat shot and the way it shrivelled those two people up, and made La Hire laugh out loud and the other generals softly quake and chuckle. Joan tranquilly resumed—

"From the first, we have been hindered by this policy of

shilly-shally; this fashion of counselling and counselling and counselling where no counselling is needed, but only fighting. We took Orleans on the 8th of May, and could have cleared the region round about in three days and saved the slaughter of Patay. We could have been in Rheims six weeks ago, and in Paris now; and would see the last Englishman pass out of France in half a year. But we struck no blow after Orleans, but went off into the country—what for? Ostensibly to hold councils; really to give Bedford time to send re-enforcements to Talbot—which he did; and Patay had to be fought. After Patay, more counselling, more waste of precious time. O my King, I would that you would be persuaded!" She began to warm up, now. "Once more we have our opportunity. If we rise and strike, all is well. Bid me march upon Paris. In twenty days it shall be yours, and in six months all France! Here is half a year's work before us; if this chance be wasted, I give you twenty years to do it in. Speak the word, O gentle King—speak but the one—"

"I cry you mercy!" interrupted the Chancellor, who saw a dangerous enthusiasm rising in the King face. "March upon Paris? Does your Excellency forget that the way bristles with English strongholds?"

"*That* for your English strongholds!" and Joan snapped her fingers scornfully. "Whence have we marched in these last days? From Gien. And whither? To Rheims. What bristled between? English strongholds. What are they now? French ones—and they never cost a blow!" Here applause broke out from the group of generals, and Joan had to pause a moment to let it subside. "Yes, English strongholds bristled before us; now French once bristle behind us. What is the argument? A child can read it. The strongholds between us and Paris are garrisoned by no new breed of English, but by the same breed as those others—with the same fears, the same questionings, the same weaknesses, the same disposition to see the heavy hand of God descending upon them. We have but to march!—on the instant—and they are

JOAN AND THE WOUNDED ENGLISH SOLDIER

ours, Paris is ours, France is ours! Give the word, O my King, command your servant to—"

"Stay!" cried the Chancellor. "It would be madness to put this affront upon his Highness the Duke of Burgundy. By the treaty which we have every hope to make with him—"

"O, the treaty which we hope to make with him! He has scorned you for years, and defied you. Is it your subtle persuasions that have softened his manners and beguiled him to listen to proposals? No; it was *blows!*—the blows which *we* gave him! That is the only teaching that that sturdy rebel can understand. What does he care for *wind?* The treaty which we hope to make with him—alack! *He* deliver Paris! There is no pauper in the land that is less able to do it. He deliver Paris! Ah, but that would make great Bedford smile! Oh, the pitiful pretext! the blind can see that this thin pourparler with its fifteen-day truce has no purpose but to give Bedford time to hurry forward his forces against us. More treachery—always treachery! We call a council of war—with nothing to counsel about; but Bedford calls no council to teach him what our one course is. He knows what he would do in our place. *He would hang his traitors and march upon Paris!* O gentle King, rouse! The way is open, Paris beckons, France implores. Speak and we—"

"Sire, it is madness, sheer madness! Your Excellency, we cannot, we must not go back from what we have done; we have proposed to treat, we *must* treat with the Duke of Burgundy."

"And we *will!*" said Joan.

"Ah? How?"

"*At the point of the lance!*"

The house rose, to a man—all that had French hearts—and let go a crash of applause—and kept it up; and in the midst of it one heard La Hire growl out: "At the point of the lance! By God, that is the music!" The King was up, too, and drew his sword, and took it by the blade and strode to Joan and delivered the hilt of it into her hand, saying—

"There, the King surrenders. Carry it to Paris."

And so the applause burst out again, and the historical council of war that has bred so many legends was over.

CHAPTER XXXIX

IT was away past midnight, and had been a tremendous day in the matter of excitement and fatigue, but that was no matter to Joan when there was business on hand. She did not think of bed. The generals followed her to her official quarters, and she delivered her orders to them as fast as she could talk, and they sent them off to their different commands as fast as delivered; wherefore the messengers galloping hither and thither raised a world of clatter and racket in the still streets; and soon were added to this the music of distant bugles and the roll of drums—notes of preparation; for the vanguard would break camp at dawn.

The generals were soon dismissed, but I wasn't; nor Joan; for it was my turn to work, now. Joan walked the floor and dictated a summons to the Duke of Burgundy to lay down his arms and make peace and exchange pardons with the King; or, if he *must* fight, go fight the Saracens. "Pardonnez - vous l'un à l'autre de bon cœur, entièrement, ainsi que doivent faire loyaux chrétiens, et, s'il vous plait de guerroyer, allez contre les Sarrasins." It was long, but it was good, and had the sterling ring to it. It is my opinion that it was as fine and simple and straightforward and eloquent a state paper as she ever uttered.

It was delivered into the hands of a courier, and he galloped away with it. Then Joan dismissed me, and told me to go to the inn and stay; and in the morning give to her father the parcel which she had left there. It contained presents for the Domremy relatives and friends and a peasant dress which she had bought for herself. She said she would say good-bye to her father and uncle in the morning if it should still be their purpose to go, instead of tarrying awhile to see the city.

I didn't say anything, of course; but I could have said that wild horses couldn't keep those men in that town half a day. *They* waste the glory of being the first to carry the great news to Domremy—*the taxes remitted forever!*—and hear the bells clang and clatter, and the people cheer and shout? Oh, not they. Patay and Orleans and the Coronation were events which in a vague way these men understood to be colossal; but they were colossal mists, films, abstractions: *this* was a gigantic reality!

When I got there, do you suppose they were abed! Quite the reverse. They and the rest were as mellow as mellow could be; and the Paladin was doing his battles over in great style, and the old peasants were endangering the building with their applause. He was doing Patay now; and was bending his big frame forward and laying out the positions and movements with a rake here and a rake there of his formidable sword on the floor, and the peasants were stooped over with their hands on their spread knees observing with excited eyes and ripping out ejaculations of wonder and admiration all along:

"Yes, here we were, waiting—waiting for the word; our horses fidgeting and snorting and dancing to get away, we lying back on the bridles till our bodies fairly slanted to the rear; the word rang out at last—'*Go!*' and we went!

"Went? There was nothing like it ever seen! Where we swept by squads of scampering English, the mere wind of our passage laid them flat in piles and rows! Then we plunged into the ruck of Fastolfe's frantic battle-corps and tore through it like a hurricane, leaving a causeway of the dead stretching far behind; no tarrying, no slacking rein, but on! on! on! far yonder in the distance lay *our* prey—Talbot and his host looming vast and dark like a storm-cloud brooding on the sea! Down we swooped upon them, glooming all the air with a quivering pall of dead leaves flung up by the whirlwind of our flight. In another moment we should have struck them as world strikes world when disorbited constellations crash into the Milky Way, but by misfortune and the inscrutable dispensation of God I was recognized! Talbot turned white, and shouting, 'Save yourselves, it is the Standard-bearer of Joan of Arc!' drove his spurs home till they met in the middle of his horse's entrails, and fled the field with his billowing multitudes at his back! I could have cursed myself for not putting on a disguise. I saw reproach in the eyes of her Excellency, and was bitterly ashamed. I had caused what seemed an irreparable disaster. Another might have gone aside to grieve, as not seeing any way to mend it; but I thank God I am not of those. Great occasions only summon as with a trumpet-call the slumbering reserves of my intellect. I saw my opportunity in an instant—in the next I was away! Through the woods I vanished—*fst!*—like an extinguished light! Away around through the curtaining forest I sped, as if on wings, none knowing what was become of me, none suspecting my design. Minute after minute passed, on and on I flew; on, and still on; and at last with a great cheer I flung my Banner to the breeze and burst out in *front* of Talbot! Oh, it was a mighty thought! That weltering chaos of distracted men whirled and surged backward like a tidal wave which has struck a continent, and the day was ours! Poor helpless creatures, they were in a trap; they were surrounded; they could not escape to the rear, for there was our army; they could not escape to the front, for there was I. Their hearts shrivelled in their bodies, their hands fell listless at their sides. They stood still, and at our leisure we slaughtered them to a man; all except Talbot and Fastolfe, whom I saved and brought away, one under each arm."

Well, there is no denying it, the Paladin was in great form that night. Such style! such noble grace of gesture, such grandeur of attitude, such energy when he got going! such steady rise, on such sure wing, such nicely graduated expenditures of voice according to weight of matter, such skilfully calculated approaches to his surprises and explosions, such

belief-compelling sincerity of tone and manner, such a climax-
ing peal from his brazen lungs, and such a lightning-vivid picture of his mailed form and flaunting banner when he burst
out before that despairing army! And oh, the gentle art of
the last half of his last sentence—delivered in the careless
and indolent tone of one who has finished his real story, and
only adds a colorless and inconsequential detail because it
has happened to occur to him in a lazy way.

It was a marvel to see those innocent peasants. Why, they
went all to pieces with enthusiasm, and roared out applauses
fit to raise the roof and wake the dead. When they had
cooled down at last and there was silence but for their heaving and panting, old Laxart said, admiringly—

"As it seems to me, you are an army in your single person."

"Yes, that is what he is," said Noël Rainguesson, convincingly. "He is a terror; and not just in *this* vicinity. His
mere name carries a shudder with it to distant lands—just his
mere name; and when he frowns, the shadow of it falls as far
as Rome, and the chickens go to roost an hour before sched-
ule time. Yes; and some say—"

"Noël Rainguesson, you are preparing yourself for trouble.
I will say just one word to you, and it will be to your advantage to—"

I saw that the usual thing had got a start. No man could
prophesy when it would end. So I delivered Joan's message
and went off to bed.

Joan made her good-byes to those old fellows in the morning, with loving embraces and many tears, and with a packed
multitude for sympathizers, and they rode proudly away on
their precious horses to carry their great news home. I had
seen better riders, I will say that; for horsemanship was a
new art to them.

The vanguard moved out at dawn and took the road, with
bands braying and banners flying; the second division followed at eight. Then came the Burgundian ambassadors,
and lost us the rest of that day and the whole of the next.

But Joan was on hand, and so they had their journey for their pains. The rest of us took the road at dawn, next morning, July 20th. And got how far? Six leagues. Tremouille was getting in his sly work with the vacillating King, you see. The King stopped at St. Marcoul and prayed three days. Precious time lost—for us; precious time gained for Bedford. He would know how to use it.

We could not go on without the King; that would be to leave him in the conspirators' camp. Joan argued, reasoned, implored; and at last we got under way again.

Joan's prediction was verified. It was not a campaign, it was only another holiday excursion. English strongholds lined our route; they surrendered without a blow; we garrisoned them with Frenchmen and passed on. Bedford was on the march against us with his new army by this time, and on the 25th of July the hostile forces faced each other and made preparation for battle; but Bedford's good judgment prevailed, and he turned and retreated toward Paris. Now was our chance. Our men were in great spirits.

Will you believe it? Our poor stick of a King allowed his worthless advisers to persuade him to start back for Gien, whence we had set out when we first marched for Rheims and the Coronation! And we actually did start back. The fifteen-day truce had just been concluded with the Duke of Burgundy, and we would go and tarry at Gien until he should deliver Paris to us without a fight.

We marched to Bray; then the King changed his mind once more, and with it his face toward Paris. Joan dictated a letter to the citizens of Rheims to encourage them to keep heart in spite of the truce, and promising to stand by them. She furnished them the news herself that the King had made this truce; and in speaking of it she was her usual frank self. She said she was not satisfied with it, and didn't know whether she would keep it or not; that if she kept it, it would be solely out of tenderness for the King's honor. All French children know those famous words. How naïve they are! "De cette trève qui a été faite, je ne suis pas contente, et je ne sais si je la tiendrai. Si je la tiens, ce sera seulement pour garder l'honneur du roi." But in any case, she said, she would not allow the blood royal to be abused, and would keep the army in good order and ready for work at the end of the truce.

Poor child, to have to fight England, Burgundy, and a French conspiracy all at the same time—it was too bad. She was a match for the others, but a conspiracy—ah, nobody is a match for that, when the victim that is to be injured is weak and willing. It grieved her, these troubled days, to be so hindered and delayed and baffled, and at times she was sad and the tears lay near the surface. Once, talking with her good old faithful friend and servant the Bastard of Orleans, she said—

"Ah, if it might but please God to let me put off this steel raiment and go back to my father and my mother, and tend my sheep again with my sister and my brothers, who would be so glad to see me!"

By the 12th of August we were camped near Dampmartin. Later we had a brush with Bedford's rear-guard, and had hopes of a big battle on the morrow, but Bedford and all his force got away in the night and went on toward Paris.

Charles sent heralds and received the submission of Beauvais. The Bishop Pierre Cauchon, that faithful friend and slave of the English, was not able to prevent it, though he did his best. He was obscure then, but his name was to travel round the globe presently, and live forever in the curses of France! Bear with me now, while I spit in fancy upon his grave.

Compiègne surrendered, and hauled down the English flag. On the 14th we camped two leagues from Senlis. Bedford turned and approached, and took up a strong position. We went against him, but all our efforts to beguile him out from his intrenchments failed, though he had promised us a duel in the open field. Night shut down. Let him look out for the morning! But in the morning he was gone again.

20

We entered Compiègne the 18th of August, turning out the English garrison and hoisting our own flag.

On the 23d Joan gave command to move upon Paris. The King and the clique were not satisfied with this, and retired sulking to Senlis, which had just surrendered. Within a few days many strong places submitted—Creil, Pont-Saint-Maxence, Choisy, Gournay-sur-Aronde, Remy, La Neufville-en-Hez, Moguay, Chantilly, Saintines. The English power was tumbling, crash after crash! And still the King sulked and disapproved, and was afraid of our movement against the capital.

On the 26th of August, 1429, Joan camped at Saint Denis; in effect, under the walls of Paris.

And still the King hung back and was afraid. If we could but have had him there to back us with his authority! Bedford had lost heart and decided to waive resistance and go and concentrate his strength in the best and loyalest province remaining to him — Normandy. Ah, if we could only have persuaded the King to come and countenance us with his presence and approval at this supreme moment!

CHAPTER XL

COURIER after courier was despatched to the King, and he promised to come, but didn't. The Duke d'Alençon went to him and got his promise again, which he broke again. Nine days were lost thus; then he came, arriving at St. Denis September 7th.

Meantime the enemy had begun to take heart: the spiritless conduct of the King could have no other result. Preparations had now been made to defend the city. Joan's chances had been diminished, but she and her generals considered them plenty good enough yet. Joan ordered the attack for eight o'clock next morning, and at that hour it began.

Joan placed her artillery and began to pound a strong work which protected the gate St. Honoré. When it was sufficiently crippled the assault was sounded at noon, and it was carried by storm. Then we moved forward to storm the gate itself, and hurled ourselves against it again and again, Joan in the lead with her standard at her side, the smoke enveloping us in choking clouds, and the missiles flying over us and through us as thick as hail.

In the midst of our last assault, which would have carried the gate sure and given us Paris and in effect France, Joan was struck down by a crossbow bolt, and our men fell back instantly and almost in a panic—for what were they without her? *She* was the army, herself.

Although disabled, she refused to retire, and begged that a new assault be made, saying it *must* win; and adding, with the battle-light rising in her eyes, "I will take Paris now or die!" She had to be carried away by force, and this was done by Gaucourt and the Duke d'Alençon.

But her spirits were at the very top notch, now. She was brimming with enthusiasm. She said she would be carried before the gate in the morning, and in half an hour Paris would be ours without any question. She could have kept her word. About this there is no doubt. But she forgot one factor—the King, shadow of that substance named La Tremouille. The King forbade the attempt!

You see, a new Embassy had just come from the Duke of Burgundy, and another sham private trade of some sort was on foot.

You would know, without my telling you, that Joan's heart was nearly broken. Because of the pain of her wound and the pain at her heart she slept little that night. Several times the watchers heard muffled sobs from the dark room where she lay at St. Denis, and many times the grieving words "It could have been taken!—it could have been taken!" which were the only ones she said.

She dragged herself out of bed a day later with a new hope. D'Alençon had thrown a bridge across the Seine near St. Denis. Might she not cross by that and assault Paris at another point? But the King got wind of it and broke the bridge down! And more—he declared the campaign ended! And more still—he had made a new truce and a long one, in which he had agreed to leave Paris unthreatened and unmolested, and go back to the Loire whence he had come!

Joan of Arc, who had never been defeated by the enemy, was defeated by her own King. She had said once that all she feared for her cause was treachery. It had struck its first blow now. She hung up her white armor in the royal basilica of St. Denis, and went and asked the King to relieve her of her functions and let her go home. As usual, she was wise. Grand combinations, far-reaching great military moves were at an end, now; for the future, when the truce should end, the war would be merely a war of random and idle skirmishes, apparently; work suitable for subalterns, and not requiring the supervision of a sublime military genius. But the King would not let her go. The truce did not embrace all

France; there were French strongholds to be watched and preserved; he would need her. Really, you see, Tremouille wanted to keep her where he could balk and hinder her.

Now came her Voices again. They said, "*Remain at St. Denis.*" There was no explanation. They did not say why. That was the voice of God; it took precedence of the command of the King; Joan resolved to stay. But that filled La Tremouille with dread. She was too tremendous a force to be left to herself; she would surely defeat all his plans. He beguiled the King to use compulsion. Joan had to submit—because she was wounded and helpless. In the Great Trial she said she was carried away against her will; and that if she had not been wounded it could not have been accomplished. Ah, she had a spirit, that slender girl! a spirit to brave all earthly powers and defy them. We shall never know why the Voices ordered her to stay. We only know this: that if she could have obeyed, the history of France would not be as it now stands written in the books. Yes, well we know that.

On the 13th of September the army, sad and spiritless, turned its face toward the Loire, and marched—without music! Yes, one noted that detail. It was a funeral march; that is what it was. A long, dreary funeral march, with never a shout or a cheer; friends looking on in tears, all the way, enemies laughing. We reached Gien at last—that place whence we had set out on our splendid march toward Rheims less than three months before, with flags flying, bands playing, the victory-flush of Patay glowing in our faces, and the massed multitudes shouting and praising and giving us God-speed. There was a dull rain falling now, the day was dark, the heavens mourned, the spectators were few, we had no welcome but the welcome of silence, and pity, and tears.

Then the King disbanded that noble army of heroes; it furled its flags, it stored its arms: the disgrace of France was complete. La Tremouille wore the victor's crown; Joan of Arc, the unconquerable, was conquered.

CHAPTER XLI

YES, it was as I have said: Joan had Paris and France in her grip, and the Hundred Years' War under her heel, and the King made her open her fist and take away her foot.

Now followed about eight months of drifting about with the King and his council, and his gay and showy and dancing and flirting and hawking and frolicking and serenading and dissipating court—drifting from town to town and from castle to castle—a life which was pleasant to us of the personal staff, but not to Joan. However, she only *saw* it, she didn't live it. The King did his sincerest best to make her happy, and showed a most kind and constant anxiety in this matter. All others had to go loaded with the chains of an exacting court etiquette, but she was free, she was privileged. So that she paid her duty to the King once a day and passed the pleasant word, nothing further was required of her. Naturally, then, she made herself a hermit, and grieved the weary days through in her own apartments, with her thoughts and devotions for company, and the planning of now forever unrealizable military combinations for entertainment. In fancy she moved bodies of men from this and that and the other point, so calculating the distances to be covered, the time required for each body, and the nature of the country to be traversed, as to have them appear in sight of each other on a given day or at a given hour and concentrate for battle. It was her only game, her only relief from her burden of sorrow and inaction. She played it hour after hour, as others play chess; and lost herself in it, and so got repose for her mind and healing for her heart.

She never complained, of course. It was not her way. She

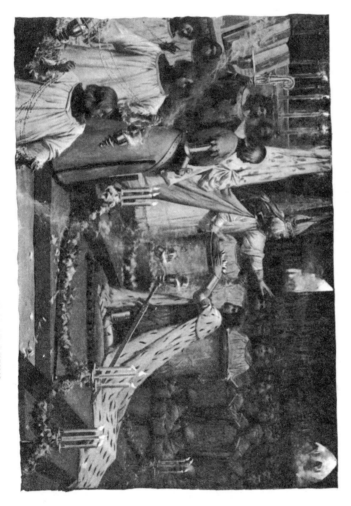

THE CORONATION OF THE FRENCH KING AT RHEIMS

was the sort that endure in silence. But—she was a caged eagle just the same, and pined for the free air and the alpine heights and the fierce joys of the storm.

France was full of rovers—disbanded soldiers ready for anything that might turn up. Several times, at intervals, when Joan's dull captivity grew too heavy to bear, she was allowed to gather a troop of cavalry and make a health-restoring dash against the enemy. These things were like a bath to her spirits.

It was like old times, there at Saint-Pierre-le-Moutier, to see her lead assault after assault, be driven back again and again, but always rally and charge anew, all in a blaze of eagerness and delight; till at last the tempest of missiles rained so intolerably thick that old D'Aulon, who was wounded, sounded the retreat (for the King had charged him on his head to let no harm come to Joan); and away everybody rushed after him— as he supposed; but when he turned and looked, there were we of the staff still hammering away; wherefore he rode back and urged her to come, saying she was mad to stay there with only a dozen men. Her eye danced merrily, and she turned upon him crying out—

"A dozen men! name of God, I have fifty thousand, and will never budge till this place is taken! Sound the charge!"

Which he did, and over the walls we went, and the fortress was ours. Old D'Aulon thought her mind was wandering; but all she meant was, that she felt the might of fifty thousand men surging in her heart. It was a fanciful expression; but, to my thinking, truer word was never said.

Then there was the affair near Lagny, where we charged the intrenched Burgundians through the open field four times, the last time victoriously; the best prize of it Franquet d'Arras, the freebooter and pitiless scourge of the region round-about.

Now and then other such affairs; and at last, away toward the end of May, 1430, we were in the neighborhood of Compiègne, and Joan resolved to go to the help of that place, which was being besieged by the Duke of Burgundy.

I had been wounded lately, and was not able to ride without help; but the good Dwarf took me on behind him, and I held on to him and was safe enough. We started at midnight, in a sullen downpour of warm rain, and went slowly and softly and in dead silence, for we had to slip through the enemy's lines. We were challenged only once; we made no answer, but held our breath and crept steadily and stealthily along, and got through without any accident. About three or half past we reached Compiègne, just as the gray dawn was breaking in the east.

Joan set to work at once, and concerted a plan with Guillaume de Flavy, captain of the city—a plan for a sortie toward evening against the enemy, who was posted in three bodies on the other side of the Oise, in the level plain. From our side one of the city gates communicated with a bridge. The end of this bridge was defended on the other side of the river by one of those fortresses called a boulevard; and this boulevard also commanded a raised road, which stretched from its front across the plain to the village of Marguy. A force of Burgundians occupied Marguy; another was camped at Clairoix, a couple of miles *above* the raised road; and a body of English was holding Venette, a mile and a half *below* it. A kind of bow-and-arrow arrangement, you see: the causeway the arrow, the boulevard at the feather-end of it, Marguy at the barb, Venette at one end of the bow, Clairoix at the other.

Joan's plan was to go straight per causeway against Marguy, carry it by assault, then turn swiftly upon Clairoix, up to the right, and capture that camp in the same way, then face to the rear and be ready for heavy work, for the Duke of Burgundy lay behind Clairoix with a reserve. Flavy's lieutenant, with archers and the artillery of the boulevard, was to keep the English troops from coming up from below and seizing the causeway and cutting off Joan's retreat in case she should have to make one. Also, a fleet of covered boats was to be stationed near the boulevard as an additional help in case a retreat should become necessary.

It was the 24th of May. At four in the afternoon Joan

moved out at the head of six hundred cavalry—on her last march in this life!

It breaks my heart. I had got myself helped up on to the walls, and from there I saw much that happened, the rest was told me long afterwards by our two knights and other eye-witnesses. Joan crossed the bridge, and soon left the boulevard behind her and went skimming away over the raised road with her horsemen clattering at her heels. She had on a brilliant silver-gilt cape over her armor, and I could see it flap and flare and rise and fall like a little patch of white flame.

It was a bright day, and one could see far and wide over that plain. Soon we saw the English force advancing, swiftly and in handsome order, the sunlight flashing from its arms.

Joan crashed into the Burgundians at Marguy and was repulsed. Then we saw the other Burgundians moving down from Clairoix. Joan rallied her men and charged again, and was again rolled back. Two assaults occupy a good deal of time—and time was precious here. The English were approaching the road, now, from Venette, but the boulevard opened fire on them and they were checked. Joan heartened her men with inspiring words and led them to the charge again in great style. This time she carried Marguy with a hurrah. Then she turned at once to the right and plunged into the plain and struck the Clairoix force, which was just arriving; then there was heavy work, and plenty of it, the two armies hurling each other backward turn about and about, and victory inclining first to the one, then to the other. Now all of a sudden there was a panic on our side. Some say one thing caused it, some another. Some say the cannonade made our front ranks think retreat was being cut off by the English, some say the rear ranks got the idea that Joan was killed. Anyway our men broke, and went flying in a wild rout for the causeway. Joan tried to rally them and face them around, crying to them that victory was sure, but it did no good, they divided and swept by her like a wave. Old D'Aulon begged her to retreat while there was yet a chance for safety, but she refused; so he seized her horse's bridle and bore her along with the wreck and ruin in spite of herself. And so along the causeway they came swarming, that wild confusion of frenzied men and horses—and the artillery had to stop firing, of course; consequently the English and Burgundians closed in in safety, the former in front, the latter behind their prey. Clear to the boulevard the French were washed in this enveloping inundation; and there, cornered in an angle formed by the flank of the boulevard and the slope of the causeway, they bravely fought a hopeless fight, and sank down one by one.

Flavy, watching from the city wall, ordered the gate to be closed and the drawbridge raised. This shut Joan out.

The little personal guard around her thinned swiftly. Both of our good knights went down, disabled; Joan's two brothers fell wounded; then Noël Rainguesson—all wounded while loyally sheltering Joan from blows aimed at her. When only the Dwarf and the Paladin were left, they would not give up, but stood their ground stoutly, a pair of steel towers streaked and splashed with blood; and where the axe of the one fell, and the sword of the other, an enemy gasped and died. And so fighting, and loyal to their duty to the last, good simple souls, they came to their honorable end. Peace to their memories! they were very dear to me.

Then there was a cheer and a rush, and Joan, still defiant, still laying about her with her sword, was seized by her cape and dragged from her horse. She was borne away a prisoner to the Duke of Burgundy's camp, and after her followed the victorious army roaring its joy.

The awful news started instantly on its round; from lip to lip it flew; and wherever it came it struck the people as with a sort of paralysis; and they murmured over and over again, as if they were talking to themselves, or in their sleep, "The Maid of Orleans taken! . . . Joan of Arc a prisoner! . . . the Savior of France lost to us!"—and would keep saying that over, as if they couldn't understand how it could be, or how God could permit it, poor creatures!

You know what a city is like when it is hung from eaves to

pavement with rustling black? Then you know what Tours was like, and some other cities. But can any man tell you what the mourning in the hearts of the peasantry of France was like? No, nobody can tell you that; and, poor dumb things, they could not have told you themselves; but it was there—indeed yes. Why, it was the spirit of a whole nation hung with crape!

The 24th of May. We will draw down the curtain, now, upon the most strange, and pathetic, and wonderful military drama that has been played upon the stage of the world. Joan of Arc will march no more.

Book III

TRIAL AND MARTYRDOM

CHAPTER I

I CANNOT bear to dwell at great length upon the shameful history of the summer and winter following the capture. For a while I was not much troubled, for I was expecting every day to hear that Joan had been put to ransom, and that the King —no, not the King, but grateful France—had come eagerly forward to pay it. By the laws of war she could not be denied the privilege of ransom. She was not a rebel; she was a legitimately constituted soldier, head of the armies of France by her King's appointment, and guilty of no crime known to military law; therefore she could not be detained upon any pretext, if ransom were proffered.

But day after day dragged by and no ransom was offered! It seems incredible, but it is true. Was that reptile Tremouille busy at the King's ear? All we know is, that the King was silent, and made no offer and no effort in behalf of this poor girl who had done so much for him.

But unhappily there was alacrity enough in another quarter. The news of the capture reached Paris the day after it happened, and the glad English and Burgundians deafened the world all the day and all the night with the clamor of their joy-bells and the thankful thunder of their artillery; and the next day the Vicar General of the Inquisition sent a message to the Duke of Burgundy requiring the delivery of the prisoner into the hands of the Church to be tried as an idolater.

The English had seen their opportunity, and it was the English power that was really acting, not the Church. The Church was being used as a blind, a disguise; and for a forcible reason: the Church was not only able to take the life of

Joan of Arc, but to blight her influence and the valor-breeding inspiration of her name, whereas the English power could but kill her body; that would not diminish or destroy the influence of her name; it would magnify it and make it permanent. Joan of Arc was the only power in France that the English did not despise, the only power in France that they considered formidable. If the Church could be brought to take her life, or to proclaim her an idolater, a heretic, a witch, sent from Satan, not from heaven, it was believed that the English supremacy could be at once reinstated.

The Duke of Burgundy listened—but waited. He could not doubt that the French King or the French people would come forward presently and pay a higher price than the English. He kept Joan a close prisoner in a strong fortress, and continued to wait, week after week. He was a French Prince, and was at heart ashamed to sell her to the English. Yet with all his waiting no offer came to him from the French side.

One day Joan played a cunning trick on her jailer, and not only slipped out of her prison, but locked him up in it. But as she fled away she was seen by a sentinel, and was caught and brought back.

Then she was sent to Beaurevoir, a stronger castle. This was early in August, and she had been in captivity more than two months, now. Here she was shut up in the top of a tower which was sixty feet high. She ate her heart there for another long stretch—about three months and a half. And she was aware, all these weary five months of captivity, that the English, under cover of the Church, were dickering for her as one would dicker for a horse or a slave, and that France was silent, the King silent, all her friends the same. Yes, it was pitiful.

And yet when she heard at last that Compiègne was being closely besieged and likely to be captured, and that the enemy had declared that no inhabitant of it should escape massacre, not even children of seven years of age, she was in a fever at once to fly to our rescue. So she tore her bed-

clothes to strips and tied them together and descended this
frail rope in the night, and it broke and she fell and was badly bruised, and remained three days insensible, meantime neither eating nor drinking.

And now came relief to us, led by the Count of Vendôme,
and Compiègne was saved and the siege raised. This was a
disaster to the Duke of Burgundy. He had to have money,
now. It was a good time for a new bid to be made for Joan
of Arc. The English at once sent a French Bishop — that
forever infamous Pierre· Cauchon of Beauvais. He was
partly promised the Archbishopric of Rouen, which was vacant, if he should succeed. He claimed the right to preside
over Joan's ecclesiastical trial because the battle-ground
where she was taken was within his diocese.

By the military usage of the time the ransom of a royal
Prince was 10,000 livres of gold, which is 61,125 francs—a
fixed sum, you see. It must be accepted, when offered; it
could not be refused.

Cauchon brought the offer of this very sum from the English—a royal Prince's ransom for the poor little peasant girl
of Domremy. It shows in a striking way the English idea of
her formidable importance. It was accepted. For that sum
Joan of Arc the Savior of France was sold; sold to her enemies; to the enemies of her country; enemies who had lashed
and thrashed and thumped and trounced France for a century and made holiday sport of it; enemies who had forgotten,
years and years ago, what a Frenchman's face was like, so
used were they to seeing nothing but his back; enemies
whom she had whipped, whom she had cowed, whom she
had taught to respect French valor, new-born in her nation by the breath of her spirit; enemies who hungered for
her life as being the only puissance able to stand between
English triumph and French degradation. Sold to a French
priest by a French Prince, with the French King and the
French nation standing thankless by and saying nothing.

And she—what did she say? Nothing. Not a reproach

passed her lips. She was too great for that—she was Joan of Arc; and when that is said, all is said.

As a soldier, her record was spotless. She could not be called to account for anything under that head. A subterfuge must be found, and, as we have seen, was found. She must be tried by priests for crimes against religion. If none could be discovered, some must be invented. Let the miscreant Cauchon alone to contrive those.

Rouen was chosen as the scene of the trial. It was in the heart of the English power; its population had been under English dominion so many generations that they were hardly French now, save in language. The place was strongly garrisoned. Joan was taken there near the end of December, 1430, and flung into a dungeon. Yes, and clothed in chains, that free spirit!

Still France made no move. How do I account for this? I think there is only one way. You will remember that whenever Joan was not at the front, the French held back and ventured nothing; that whenever she led, they swept everything before them, so long as they could see her white armor or her banner; that every time she fell wounded or was reported killed—as at Compiègne—they broke in panic and fled like sheep. I argue from this that they had undergone no real transformation as yet; that at bottom they were still under the spell of a timorousness born of generations of unsuccess, and a lack of confidence in each other and in their leaders born of old and bitter experience in the way of treacheries of all sorts—for their kings had been treacherous to their great vassals and to their generals, and these in turn were treacherous to the head of the state and to each other. The soldiery found that they could depend utterly on Joan, and upon her alone. With her gone, everything was gone. She was the sun that melted the frozen torrents and set them boiling; with that sun removed, they froze again, and the army and all France became what they had been before, mere dead corpses—that and nothing more; incapable of thought, hope, ambition, or motion.

JACQUES D'ARC AND UNCLE LAXART WATCHING THE PROCESSION

CHAPTER II

My wound gave me a great deal of trouble clear into the
first part of October; then the fresher weather renewed my
life and strength. All this time there were reports drifting
about that the King was going to ransom Joan. I believed
these, for I was young and had not yet found out the little-
ness and meanness of our poor human race, which brags
about itself so much, and thinks it is better and higher than
the other animals.

In October I was well enough to go out with two sorties,
and in the second one, on the 23d, I was wounded again.
My luck had turned, you see. On the night of the 25th the
besiegers decamped, and in the disorder and confusion one
of their prisoners escaped and got safe into Compiègne, and
hobbled into my room as pallid and pathetic an object as you
would wish to see.

"What? Alive? Nöel Rainguesson!"

It was indeed he. It was a most joyful meeting, that you
will easily know; and also as sad as it was joyful. We could
not speak Joan's name. One's voice would have broken
down. We knew who was meant when she was mentioned;
we could say "she" and "her," but we could not speak the
name.

We talked of the personal staff. Old D'Aulon, wounded
and a prisoner, was still with Joan and serving her, by permission of the Duke of Burgundy. Joan was being treated
with the respect due to her rank and to her character as a
prisoner of war taken in honorable conflict. And this was
continued—as we learned later—until she fell into the hands
of that bastard of Satan, Pierre Cauchon, Bishop of Beauvais.

Noël was full of noble and affectionate praises and appreciations of our old boastful big Standard-bearer, now gone silent forever, his real and imaginary battles all fought, his work done, his life honorably closed and completed.

"And think of his luck!" burst out Noël, with his eyes full of tears. "Always the pet child of luck! See how it followed him and stayed by him, from his first step all through, in the field or out of it; always a splendid figure in the public eye, courted and envied everywhere; always having a chance to do fine things and always doing them; in the beginning called the Paladin in joke, and called it afterwards in earnest because he magnificently made the title good; and at last—supremest luck of all—died in the field! died with his harness on; died faithful to his charge, the Standard in his hand; died—oh, think of it—with the approving eye of Joan of Arc upon him! He drained the cup of glory to the last drop, and went jubilant to his peace, blessedly spared all part in the disaster which was to follow. What luck, what luck! And we? What was our sin that we are still here, we who have also earned our place with the happy dead?"

And presently he said:

"They tore the sacred Standard from his dead hand and carried it away, their most precious prize after its captured owner. But they haven't it now. A month ago we put our lives upon the risk—our two good knights, my fellow-prisoners, and I—and stole it, and got it smuggled by trusty hands to Orleans, and there it is now, safe for all time in the Treasury."

I was glad and grateful to learn that. I have seen it often since, when I have gone to Orleans on the 8th of May to be the petted old guest of the city and hold the first place of honor at the banquets and in the processions—I mean since Joan's brothers passed from this life. It will still be there, sacredly guarded by French love, a thousand years from now —yes, as long as any shred of it hangs together.*

---

* It remained there three hundred and sixty years, and then was destroyed in a public bonfire, together with two swords, a plumed cap, several

Two or three weeks after this talk came the tremendous news like a thunder-clap, and we were aghast—Joan of Arc sold to the English!

Not for a moment had we ever dreamed of such a thing. We were young, you see, and did not know the human race, as I have said before. We had been so proud of our country, so sure of her nobleness, her magnanimity, her gratitude. We had expected little of the King, but of France we had expected everything. Everybody knew that in various towns patriot priests had been marching in procession urging the people to sacrifice money, property, everything, and buy the freedom of their heaven-sent deliverer. That the money would be raised we had not thought of doubting.

But it was all over, now, all over. It was a bitter time for us. The heavens seemed hung with black; all cheer went out from our hearts. Was this comrade here at my bedside really Noël Rainguesson, that light-hearted creature whose whole life was but one long joke, and who used up more breath in laughter than in keeping his body alive? No, no; *that* Noël I was to see no more. This one's heart was broken. He moved grieving about, and absently, like one in a dream; the stream of his laughter was dried at its source.

Well, that was best. It was my own mood. We were company for each other. He nursed me patiently through the dull long weeks, and at last, in January, I was strong enough to go about again. Then he said:

"Shall we go, now?"

suits of state apparel, and other relics of the Maid, by a mob in the time of the Revolution. Nothing which the hand of Joan of Arc is known to have touched now remains in existence except a few preciously guarded military and state papers which she signed, her pen being guided by a clerk or her secretary Louis de Conte. A bowlder exists from which she is known to have mounted her horse when she was once setting out upon a campaign. Up-to a quarter of a century ago there was a single hair from her head still in existence. It was drawn through the wax of a seal attached to the parchment of a state document. It was surreptitiously snipped out, seal and all, by some vandal relic-hunter, and carried off. Doubtless it still exists, but only the thief knows where.—TRANSLATOR.

"Yes."

There was no need to explain. Our hearts were in Rouen, we would carry our bodies there. All that we cared for in this life was shut up in that fortress. We could not help her, but it would be some solace to us to be near her, to breathe the air that she breathed, and look daily upon the stone walls that hid her. What if we should be made prisoners there? Well, we could but do our best, and let luck and fate decide what should happen.

And so we started. We could not realize the change which had come upon the country. We seemed able to choose our own route and go wherever we pleased, unchallenged and unmolested. When Joan of Arc was in the field, there was a sort of panic of fear everywhere; but now that she was out of the way, fear had vanished. Nobody was troubled about you or afraid of you, nobody was curious about you or your business, everybody was indifferent.

We presently saw that we could take to the Seine, and not weary ourselves out with land travel. So we did it, and were carried in a boat to within a league of Rouen. Then we got ashore; not on the hilly side, but on the other, where it is as level as a floor. Nobody could enter or leave the city without explaining himself. It was because they feared attempts at a rescue of Joan.

We had no trouble. We stopped in the plain with a family of peasants and stayed a week, helping them with their work for board and lodging, and making friends of them. We got clothes like theirs, and wore them. When we had worked our way through their reserves and gotten their confidence, we found that they secretly harbored French hearts in their bodies. Then we came out frankly and told them everything, and found them ready to do anything they could to help us. Our plan was soon made, and was quite simple. It was to help them drive a flock of sheep to the market of the city. One morning early we made the venture in a melancholy drizzle of rain, and passed through the frowning gates unmolested. Our friends had friends living over a humble wine-shop in a quaint tall building situated in one of the narrow lanes that run down from the cathedral to the river, and with these they bestowed us; and the next day they smuggled our own proper clothing and other belongings to us. The family that lodged us—the Pierrons—were French in sympathy, and we needed to have no secrets from them.

CHAPTER III

IT was necessary for me to have some way to gain bread for Noël and myself; and when the Pierrons found that I knew how to write, they applied to their confessor in my behalf, and he got a place for me with a good priest named Manchon, who was to be the chief recorder in the Great Trial of Joan of Arc now approaching. It was a strange position for me—clerk to the recorder—and dangerous if my sympathies and late employment should be found out. But there was not much danger. Manchon was at bottom friendly to Joan and would not betray me; and my name would not, for I had discarded my surname and retained only my given one, like a person of low degree.

I attended Manchon constantly straight along, out of January and into February, and was often in the citadel with him —in the very fortress where Joan was imprisoned, though not in the dungeon where she was confined, and so did not see her, of course.

Manchon told me everything that had been happening before my coming. Ever since the purchase of Joan, Cauchon had been busy packing his jury for the destruction of the Maid —weeks and weeks he had spent in this bad industry. The University of Paris had sent him a number of learned and able and trusty ecclesiastics of the stripe he wanted; and he had scraped together a clergyman of like stripe and great fame here and there and yonder, until he was able to construct a formidable court numbering half a hundred distinguished names. French names they were, but their interests and sympathies were English.

A great officer of the Inquisition was also sent from Paris, for the accused must be tried by the forms of the Inquisition; but this was a brave and righteous man, and he said squarely that this court had no power to try the case, wherefore he refused to act; and the same honest talk was uttered by two or three others.

The Inquisitor was right. The case as here resurrected against Joan had already been tried long ago at Poitiers, and decided in her favor. Yes, and by a higher tribunal than this one, for at the head of it was an Archbishop—he of Rheims—Couchon's own metropolitan. So here, you see, a lower court was impudently preparing to re-try and re-decide a cause which had already been decided by its superior, a court of higher authority. Imagine it! No, the case could not properly be tried again. Cauchon could not properly preside in this new court, for more than one reason: Rouen was not in his diocese; Joan had not been arrested in her domicile, which was still Domremy; and finally this proposed judge was the prisoner's outspoken enemy, and therefore he was incompetent to try her. Yet all these large difficulties were gotten rid of. The territorial Chapter of Rouen finally granted territorial letters to Cauchon—though only after a struggle and under compulsion. Force was also applied to the Inquisitor, and he was obliged to submit.

So, then, the little English King, by his representative, formally delivered Joan into the hands of the court, but with this reservation: *if the court failed to condemn her, he was to have her back again!*

Ah, dear, what chance was there for that forsaken and friendless child? Friendless indeed—it is the right word. For she was in a black dungeon, with half a dozen brutal common soldiers keeping guard night and day in the room where her cage was—for she was in a cage; an iron cage, and chained to her bed by neck and hands and feet. Never a person near her whom she had ever seen before; never a woman at all. Yes, this was indeed friendlessness.

Now it was a vassal of Jean de Luxembourg who captured Joan at Compiègne, and it was Jean who sold her to the Duke of Burgundy. Yet this very De Luxembourg was shameless enough to go and show his face to Joan in her cage. He came with two English earls, Warwick and Stafford. He was a poor reptile. He told her he would get her set free if she would promise not to fight the English any more. She had been in that cage a long time now, but not long enough to break her spirit. She retorted scornfully—

"Name of God, you but mock me. I know that you have neither the power nor the will to do it."

He insisted. Then the pride and dignity of the soldier rose in Joan, and she lifted her chained hands and let them fall with a clash, saying—

"See these! They know more than you, and can prophesy better. I know that the English are going to kill me, for they think that when I am dead they can get the Kingdom of France. It is not so. Though there were a hundred thousand of them they would never get it."

This defiance infuriated Stafford, and he—now think of it —he a free, strong man, she a chained and helpless girl—he drew his dagger and flung himself at her to stab her. But Warwick seized him and held him back. Warwick was wise. Take her life in that way? Send her to Heaven stainless and undisgraced? It would make her the idol of France, and the whole nation would rise and march to victory and emancipation under the inspiration of her spirit. No, she must be saved for another fate than that.

Well, the time was approaching for the Great Trial. For more than two months Cauchon had been raking and scraping everywhere for any odds and ends of evidence or suspicion or conjecture that might be made usable against Joan, and carefully suppressing all evidence that came to hand in her favor. He had limitless ways and means and powers at his disposal for preparing and strengthening the case for the prosecution, and he used them all.

But Joan had no one to prepare her case for her, and she was shut up in those stone walls and had no friend to appeal to for help. And as for witnesses, she could not call a single one in her defence; they were all far away, under the French flag, and this was an English court; they would have been seized and hanged if they had shown their faces at the gates of Rouen. No, the prisoner must be the *sole* witness— witness for the prosecution, witness for the defence; and with a verdict of death resolved upon before the doors were opened for the court's first sitting.

When she learned that the court was made up of ecclesiastics in the interest of the English, she begged that in fairness an equal number of priests of the French party should be added to these. Cauchon scoffed at her message, and would not even deign to answer it.

By the law of the Church—she being a minor under twenty-one—it was her right to have counsel to conduct her case, advise her how to answer when questioned, and protect her from falling into traps set by cunning devices of the prosecution. She probably did not know that this was her right, and that she could demand it and require it, for there was none to tell her that; but she begged for this help at any rate. Cauchon refused it. She urged and implored, pleading her youth and her ignorance of the complexities and intricacies of the law and of legal procedure. Cauchon refused again, and said she must get along with her case as best she might by herself. Ah, his heart was a stone.

Cauchon prepared the *proces verbal*. I will simplify that by calling it the Bill of Particulars. It was a detailed list of the charges against her, and formed the basis of the trial. Charges? It was a list of *suspicions and public rumors*— those were the words used. It was merely charged that she was suspected of having been guilty of heresies, witchcraft, and other such offences against religion.

Now by law of the Church, a trial of that sort could not be begun until a searching inquiry had been made into the history and character of the accused, and it was essential that the result of this inquiry be added to the *proces verbal* and form a part of it. You remember that that was the first thing they did before the trial at Poitiers. They did it again, now. An ecclesiastic was sent to Domremy. There and all about the neighborhood he made an exhaustive search into Joan's history and character, and came back with his verdict. It was very clear. The searcher reported that he found Joan's character to be in every way what he " would like his own sister's character to be." Just about the same report that was brought back to Poitiers, you see. Joan's was a character which could endure the minutest examination.

This verdict was a strong point for Joan, you will say. Yes, it *would* have been if it could have seen the light; but Cauchon was awake, and it disappeared from the *proces verbal* before the trial. People were prudent enough not to inquire what became of it.

One would imagine that Cauchon was ready to begin the trial by this time. But no, he devised one more scheme for poor Joan's destruction, and it promised to be a deadly one.

One of the great personages picked out and sent down by the University of Paris was an ecclesiastic named Nicolas Loyseleur. He was tall, handsome, grave, of smooth soft speech and courteous and winning manners. There was no seeming of treachery or hypocrisy about him, yet he was full of both. He was admitted to Joan's prison by night, disguised as a cobbler; he pretended to be from her own country; he professed to be secretly a patriot; he revealed the fact that he was a priest. She was filled with gladness to see one from the hills and plains that were so dear to her; happier still to look upon a priest and disburden her heart in confession, for the offices of the Church were the bread of life, the breath of her nostrils to her, and she had been long forced to pine for them in vain. She opened her whole innocent heart to this creature, and in return he gave her advice concerning her trial which could have destroyed her if her deep native wisdom had not protected her against following it.

You will ask, what value could this scheme have, since the secrets of the confessional are sacred and cannot be revealed? True — but suppose another person should overhear them?

That person is not bound to keep the secret. Well, that is what happened. Cauchon had previously caused a hole to be bored through the wall; and he stood with his ear to that hole and heard all. It is pitiful to think of these things. One wonders how they could treat that poor child so. She had not done them any harm.

CHAPTER IV

On Tuesday the 20th of February, whilst I sat at my master's work in the evening, he came in, looking sad, and said it had been decided to begin the trial at eight o'clock the next morning, and I must get ready to assist him.

Of course I had been expecting such news every day for many days; but no matter, the shock of it almost took my breath away and set me trembling like a leaf. I suppose that without knowing it I had been half imagining that at the last moment something would happen, something that would stop this fatal trial: maybe that La Hire would burst in at the gates with his hellions at his back; maybe that God would have pity and stretch forth His mighty hand. But now—now there was no hope.

The trial was to begin in the chapel of the fortress and would be public. So I went sorrowing away and told Noël, so that he might be there early and secure a place. It would give him a chance to look again upon the face which we so revered and which was so precious to us. All the way, both going and coming, I ploughed through chattering and rejoicing multitudes of English soldiery and English-hearted French citizens. There was no talk but of the coming event. Many times I heard the remark, accompanied by a pitiless laugh—

"The fat Bishop has got things as he wants them at last, and says he will lead the vile witch a merry dance and a short one."

But here and there I glimpsed compassion and distress in a face, and it was not always a French one. English soldiers feared Joan, but they admired her for her great deeds and her unconquerable spirit.

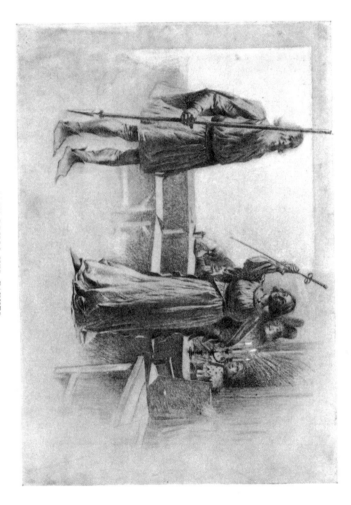

JOAN DRILLS HER FATHER

In the morning Manchon and I went early, yet as we approached the vast fortress we found crowds of men already there and still others gathering. The chapel was already full and the way barred against further admissions of unofficial persons. We took our appointed places. Throned on high sat the president, Cauchon, Bishop of Beauvais, in his grand robes, and before him in rows sat his robed court—fifty distinguished ecclesiastics, men of high degree in the Church, of clear-cut intellectual faces, men of deep learning, veteran adepts in strategy and casuistry, practised setters of traps for ignorant minds and unwary feet. When I looked around upon this army of masters of legal fence, gathered here to find just one verdict and no other, and remembered that Joan must fight for her good name and her life single-handed against them, I asked myself what chance an ignorant poor country girl of nineteen could have in such an unequal conflict; and my heart sank down low, very low. When I looked again at that obese president, puffing and wheezing there, his great belly distending and receding with each breath, and noted his three chins, fold above fold, and his knobby and knotty face, and his purple and splotchy complexion, and his repulsive cauliflower nose, and his cold and malignant eyes—a brute, every detail of him—my heart sank lower still. And when I noted that all were afraid of this man, and shrank and fidgeted in their seats when his eye smote theirs, my last poor ray of hope dissolved away and wholly disappeared.

There was one unoccupied seat in this place, and only one. It was over against the wall, in view of every one. It was a little wooden bench without a back, and it stood apart and solitary on a sort of dais. Tall men-at-arms in morion, breast-plate, and steel gauntlets stood as stiff as their own halberds on each side of this dais, but no other creature was near by it. A pathetic little bench to me it was, for I knew whom it was for; and the sight of it carried my mind back to the great court at Poitiers, where Joan sat upon one like it and calmly fought her cunning fight with the astonished doctors of the Church and Parliament, and rose from it victorious and applauded by all, and went forth to fill the world with the glory of her name.

What a dainty little figure she was, and how gentle and innocent, how winning and beautiful in the fresh bloom of her seventeen years! Those were grand days. And so recent—for she was but just nineteen now—and how much she had seen since, and what wonders she had accomplished!

But now—oh, all was changed, now. She had been languishing in dungeons, away from light and air and the cheer of friendly faces, for nearly three-quarters of a year—she, born child of the sun, natural comrade of the birds and of all happy free creatures. She would be weary, now, and worn with this long captivity, her forces impaired; despondent, perhaps, as knowing there was no hope. Yes, all was changed.

All this time there had been a muffled hum of conversation, and rustling of robes and scraping of feet on the floor, a combination of dull noises which filled all the place. Suddenly—

"Produce the accused!"

It made me catch my breath. My heart began to thump like a hammer. But there was silence, now—silence absolute. All those noises ceased, and it was as if they had never been. Not a sound; the stillness grew oppressive; it was like a weight upon one. All faces were turned towards the door; and one could properly expect that, for most of the people there suddenly realized, no doubt, that they were about to see, in actual flesh and blood, what had been to them before only an embodied prodigy, a word, a phrase, a world-girdling Name.

The stillness continued. Then, far down the stone-paved corridors, one heard a vague slow sound approaching: *clank* . . . . . . clink . . clank—Joan of Arc, Deliverer of France, in chains!

My head swam; all things whirled and spun about me. Ah, *I* was realizing, too.

CHAPTER V

I GIVE you my honor, now, that I am not going to distort or discolor the facts of this miserable trial. No, I will give them to you honestly, detail by detail, just as Manchon and I set them down daily in the official record of the court, and just as one may read them in the printed histories. There will be only this difference: that in talking familiarly with you I shall use my right to comment upon the proceedings and explain them as I go along, so that you can understand them better; also, I shall throw in trifles which came under our eyes and have a certain interest for you and me, but were not important enough to go into the official record.*

To take up my story, now, where I left off. We heard the clanking of Joan's chains down the corridors; she was approaching.

Presently she appeared; a thrill swept the house, and one heard deep breaths drawn. Two guardsmen followed her at a short distance to the rear. Her head was bowed a little, and she moved slowly, she being weak and her irons heavy. She had on men's attire—all black; a soft woollen stuff, intensely black, funereally black, not a speck of relieving color in it from her throat to the floor. A wide collar of this same black stuff lay in radiating folds upon her shoulders and breast; the sleeves of her doublet were full, down to the elbows, and tight thence to her manacled wrists; below the doublet, tight black hose down to the chains on her ankles.

---

* He kept his word. His account of the Great Trial will be found to be in strict and detailed accordance with the sworn facts of history.—TRANS-LATOR.

Half-way to her bench she stopped, just where a wide shaft of light fell slanting from a window, and slowly lifted her face. Another thrill!—it was totally colorless, white as snow; a face of gleaming snow set in vivid contrast upon that slender statue of sombre unmitigated black. It was smooth and pure and girlish, beautiful beyond belief, infinitely sad and sweet. But, dear, dear! when the challenge of those untamed eyes fell upon that judge, and the droop vanished from her form and it straightened up soldierly and noble, my heart leaped for joy; and I said, all is well, all is well—they have not broken her, they have not conquered her, she is Joan of Arc still! Yes, it was plain to me, now, that there was one spirit there which this dreaded judge could not quell nor make afraid.

She moved to her place and mounted the dais and seated herself upon her bench, gathering her chains into her lap and nestling her little white hands there. Then she waited in tranquil dignity, the only person there who seemed unmoved and unexcited. A bronzed and brawny English soldier, standing at martial ease in the front rank of the citizen spectators, did now most gallantly and respectfully put up his great hand and give her the military salute; and she, smiling friendly, put up hers and returned it; whereat there was a sympathetic little break of applause, which the judge sternly silenced.

Now the memorable inquisition called in history the Great Trial began. Fifty experts against a novice, and no one to help the novice!

The judge summarized the circumstances of the case and the public reports and suspicions upon which it was based; then he required Joan to kneel and make oath that she would answer with exact truthfulness to all questions asked her.

Joan's mind was not asleep. It suspected that dangerous possibilities might lie hidden under this apparently fair and reasonable demand. She answered with the simplicity which so often spoiled the enemy's best-laid plans in the trial at Poitiers, and said,

"No; for I do not know what you are going to ask me; you might ask of me things which I would not tell you."

This incensed the Court, and brought out a brisk flurry of angry exclamations. Joan was not disturbed. Cauchon raised his voice and began to speak in the midst of this noise, but he was so angry that he could hardly get his words out. He said—

"With the divine assistance of our Lord we require you to expedite these proceedings for the welfare of your conscience. Swear, with your hands upon the Gospels, that you will answer true to the questions which shall be asked you!" and he brought down his fat hand with a crash upon his official table.

Joan said, with composure—

"As concerning my father and mother, and the faith, and what things I have done since my coming into France, I will gladly answer; but as regards the revelations which I have received from God, my Voices have forbidden me to confide them to any save my King—"

Here there was another angry outburst of threats and expletives, and much movement and confusion; so she had to stop, and wait for the noise to subside; then her waxen face flushed a little and she straightened up and fixed her eye on the judge, and finished her sentence in a voice that had the old ring in it—

"— and I will never reveal these things though you cut my head off!"

Well, maybe you know what a deliberative body of Frenchmen is like. The judge and half the court were on their feet in a moment, and all shaking their fists at the prisoner and all storming and vituperating at once, so that you could hardly hear yourself think. They kept this up several minutes; and because Joan sat untroubled and indifferent, they grew madder and noisier all the time. Once she said, with a fleeting trace of the old-time mischief in her eye and manner—

"Prithee speak one at a time, fair lords, then I will answer all of you."

At the end of three whole hours of furious debating over the oath, the situation had not changed a jot. The Bishop was still requiring an unmodified oath, Joan was refusing for the twentieth time to take any except the one which she had herself proposed. There was a physical change apparent, but it was confined to court and judge; they were hoarse, droopy, exhausted by their long frenzy, and had a sort of haggard look in their faces, poor men, whereas Joan was still placid and reposeful and did not seem noticeably tired.

The noise quieted down; there was a waiting pause of some moments' duration. Then the judge surrendered to the prisoner, and with bitterness in his voice told her to take the oath after her own fashion. Joan sunk at once to her knees; and as she laid her hands upon the Gospels, that big English soldier set free his mind:

"By God if she were but English, she were not in this place another half a second!"

It was the soldier in him responding to the soldier in her. But what a stinging rebuke it was, what an arraignment of French character and French royalty! Would that he could have uttered just that one phrase in the hearing of Orleans! I know that that grateful city, that adoring city, would have risen, to the last man and the last woman, and marched upon Rouen. Some speeches—speeches that shame a man and humble him—burn themselves into the memory and remain there. That one is burnt into mine.

After Joan had made oath, Cauchon asked her her name, and where she was born, and some questions about her family; also what her age was. She answered these. Then he asked her how much education she had.

"I have learned from my mother the Pater Noster, the Ave Maria, and the Belief. All that I know was taught me by my mother."

Questions of this unessential sort dribbled on for a considerable time. Everybody was tired out by now, except Joan. The tribunal prepared to rise. At this point Cauchon forbade Joan to try to escape from prison, upon pain of being held guilty of the crime of heresy—singular logic! She answered simply—

"I am not bound by this prohibition. If I could escape I would not reproach myself, for I have given no promise, and I shall not."

Then she complained of the burden of her chains, and asked that they might be removed, for she was strongly guarded in that dungeon and there was no need of them. But the Bishop refused, and reminded her that she had broken out of prison twice before. Joan of Arc was too proud to insist. She only said, as she rose to go with the guard—

"It is true I have wanted to escape, and I do want to escape." Then she added, in a way that would touch the pity of anybody, I think, "It is the right of every prisoner."

And so she went from the place in the midst of an impressive stillness, which made the sharper and more distressful to me the clank of those pathetic chains.

What presence of mind she had! One could never surprise her out of it. She saw Noël and me there when she first took her seat on her bench; and we flushed to the forehead with excitement and emotion, but her face showed nothing, betrayed nothing. Her eyes sought us fifty times that day, but they passed on and there was never any ray of recognition in them. Another would have started upon seeing us, and then —why then there could have been trouble for us, of course.

We walked slowly home together, each busy with his own grief and saying not a word.

CHAPTER VI

THAT night Manchon told me that all through the day's proceedings Cauchon had had some clerks concealed in the embrasure of a window who were to make a special report garbling Jean's answers and twisting them from their right meaning. Ah, that was surely the cruelest man and the most shameless that has lived in this world. But his scheme failed. Those clerks had human hearts in them, and their base work revolted them, and they turned to and boldly made a straight report, whereupon Cauchon cursed them and ordered them out of his presence with a threat of drowning, which was his favorite and most frequent menace. The matter had gotten abroad and was making great and unpleasant talk, and Cauchon would not try to repeat this shabby game right away. It comforted me to hear that.

When we arrived at the citadel next morning, we found that a change had been made. The chapel had been found too small. The court had now removed to a noble chamber situated at the end of the great hall of the castle. The number of judges was increased to sixty-two—one ignorant girl against such odds, and none to help her.

The prisoner was brought in. She was as white as ever, but she was looking no whit worse than she looked when she had first appeared the day before. Isn't it a strange thing? Yesterday she had sat five hours on that backless bench with her chains in her lap, baited, badgered, persecuted by that unholy crew, without even the refreshment of a cup of water—for she was never offered anything, and if I have made you know her by this time you will know without my telling you that she was not a person likely to ask favors of those people. And she had spent the night caged in her wintry dungeon with her chains upon her; yet here she was, as I say, collected, unworn, and ready for the conflict; yes, and the only person there who showed no signs of the wear and worry of yesterday. And her eyes—ah, you should have seen them and broken your hearts. Have you seen that veiled deep glow, that pathetic hurt dignity, that unsubdued and unsubduable spirit that burns and smoulders in the eye of a caged eagle and makes you feel mean and shabby under the burden of its mute reproach? Her eyes were like that. How capable they were, and how wonderful! Yes, at all times and in all circumstances they could express as by print every shade of the wide range of her moods. In them were hidden floods of gay sunshine, the softest and peacefulest twilights, and devastating storms and lightnings. Not in this world have there been others that were comparable to them. Such is my opinion, and none that had the privilege to see them would say otherwise than this which I have said concerning them.

The seance began. And how did it begin, should you think? Exactly as it began before—with that same tedious thing which had been settled once, after so much wrangling. The Bishop opened thus:

"You are required, now, to take the oath pure and simple, to answer truly all questions asked you."

Joan replied placidly—

"I have made oath yesterday, my lord; let that suffice."

The Bishop insisted and insisted, with rising temper; Joan but shook her head and remained silent. At last she said—

"I made oath yesterday; it is sufficient." Then she sighed and said, "Of a truth, you do burden me too much."

The Bishop still insisted, still commanded, but he could not move her. At last he gave it up and turned her over for the day's inquest to an old hand at tricks and traps and deceptive plausibilities—Beaupere, a doctor of theology. Now notice the form of this sleek strategist's first remark—flung out in an easy, off-hand way that would have thrown any un-watchful person off his guard—

"Now, Joan, the matter is very simple; just speak up and frankly and truly answer the questions which I am going to ask you, as you have sworn to do."

It was a failure. Joan was not asleep. She saw the arti-fice. She said—

"No. You could ask me things which I could not tell you — and would not." Then, reflecting upon how profane and out of character it was for these ministers of God to be prying into matters which had proceeded from His hands under the awful seal of His secrecy, she added, with a warn-ing note in her tone, "If you were well informed concerning me you would wish me out of your hands. I have done noth-ing but by revelation."

Beaupere changed his attack, and began an approach from another quarter. He would slip upon her, you see, under cover of innocent and unimportant questions.

"Did you learn any trade at home?"

"Yes, to sew and to spin." Then the invincible soldier, victor of Patay, conqueror of the lion Talbot, deliverer of Or-leans, restorer of a king's crown, commander-in-chief of a nation's armies, straightened herself proudly up, gave her head a little toss, and said with naïve complacency, "And when it comes to that, I am not afraid to be matched against any woman in Rouen!"

The crowd of spectators broke out with applause—which pleased Joan — and there was many a friendly and petting smile to be seen. But Cauchon stormed at the people and warned them to keep still and mind their manners.

Beaupere asked other questions. Then—

"Had you other occupations at home?"

"Yes. I helped my mother in the household work and went to the pastures with the sheep and the cattle."

Her voice trembled a little, but one could hardly notice it. As for me, it brought those old enchanted days flooding back to me, and I could not see what I was writing for a little while.

Beaupere cautiously edged along up with other questions toward the forbidden ground, and finally repeated a question which she had refused to answer a little while back—as to whether she had received the Eucharist in those days at other festivals than that of Easter. Joan merely said—

"*Passez outre.*" Or, as one might say, "Pass on to matters which you are privileged to pry into."

I heard a member of the court say to a neighbor—

"As a rule, witnesses are but dull creatures, and an easy prey—yes, and easily embarrassed, easily frightened—but truly one can neither scare this child nor find her dozing."

Presently the house pricked up its ears and began to listen eagerly, for Beaupere began to touch upon Joan's Voices, a matter of consuming interest and curiosity to everybody. His purpose was, to trick her into heedless sayings that could indicate that the Voices had sometimes given her evil advice—hence that they had come from Satan, you see. To have dealings with the devil—well, that would send her to the stake in brief order, and that was the deliberate end and aim of this trial.

"When did you first hear these Voices?"

"I was thirteen when I first heard a Voice coming from God to help me to live well. I was frightened. It came at mid-day, in my father's garden in the summer."

"Had you been fasting?"

"Yes."

"The day before?"

"No."

"From what direction did it come?"

"From the right—from toward the church."

"Did it come with a bright light?"

"Oh, indeed yes. It was brilliant. When I came into France I often heard the Voices very loud."

"What did the Voice sound like?"

"It was a noble Voice, and I thought it was sent to me from God. The third time I heard it I recognized it as being an angel's."

"You could understand it?"

"Quite easily. It was always clear."

"What advice did it give you as to the salvation of your soul?"

"It told me to live rightly, and be regular in attendance upon the services of the Church. And it told me that I must go to France."

"In what species of form did the Voice appear?"

Joan looked suspiciously at the priest a moment, then said, tranquilly—

"As to that, I will not tell you."

"Did the Voice seek you often?"

"Yes. Twice or three times a week, saying, 'Leave your village and go to France.'"

"Did your father know about your departure?"

"No. The Voice said, 'Go to France'; therefore I could not abide at home any longer."

"What else did it say?"

"That I should raise the siege of Orleans."

"Was that all?"

"No, I was to go to Vaucouleurs, and Robert de Baudricourt would give me soldiers to go with me to France; and I answered, saying that I was a poor girl who did not know how to ride, neither how to fight."

Then she told how she was baulked and interrupted at Vaucouleurs, but finally got her soldiers, and began her march.

"How were you dressed?"

The court of Poitiers had distinctly decided and decreed that as God had appointed her to do a man's work, it was meet and no scandal to religion that she should dress as a man; but no matter, this court was ready to use any and all weapons against Joan, even broken and discredited ones, and much was going to be made of this one before this trial should end.

"I wore a man's dress, also a sword which Robert de Baudricourt gave me, but no other weapon."

THE PALADIN TELLS HOW HE WON PATAY

"Who was it that advised you to wear the dress of a man?"

Joan was suspicious again. She would not answer.

The question was repeated.

She refused again.

"Answer. It is a command!"

"*Passez outre*," was all she said.

So Beaupere gave up the matter for the present.

"What did Baudricourt say to you when you left?"

"He made them that were to go with me promise to take charge of me, and to me he said, 'Go, and let happen what may!'" (*Advienne que pourra!*)

After a good deal of questioning upon other matters she was asked again about her attire. She said it was necessary for her to dress as a man.

"Did your Voice advise it?"

Joan merely answered placidly—

"I believe my Voice gave me good advice."

It was all that could be got out of her, so the questions wandered to other matters, and finally to her first meeting with the King at Chinon. She said she chose out the King, who was unknown to her, by the revelation of her Voices. All that happened at that time was gone over. Finally—

"Do you still hear those Voices?"

"They come to me every day."

"What do you ask of them?"

"I have never asked of them any recompense but the salvation of my soul."

"Did the Voice always urge you to follow the army?"

He is creeping upon her again. She answered—

"It required me to remain behind at St. Denis. I would have obeyed if I had been free, but I was helpless by my wound, and the knights carried me away by force."

"When were you wounded?"

"I was wounded in the moat before Paris, in the assault."

The next question reveals what Beaupere had been leading up to—

"Was it a feast day?"

You see? The suggestion is that a voice coming from God would hardly advise or permit the violation, by war and bloodshed, of a sacred day.

Joan was troubled a moment, then she answered yes, it was a feast day.

"Now then, tell me this : did you hold it right to make the attack on such a day?"

This was a shot which might make the first breach in a wall which had suffered no damage thus far. There was immediate silence in the court and intense expectancy noticeable all about. But Joan disappointed the house. She merely made a slight little motion with her hand, as when one brushes away a fly, and said with reposeful indifference—

"*Passez outre.*"

Smiles danced for a moment in some of the sternest faces there, and several even laughed outright. The trap had been long and laboriously prepared; it fell, and was empty.

The court rose. It had sat for hours, and was cruelly fatigued. Most of the time had been taken up with apparently idle and purposeless inquiries about the Chinon events, the exiled Duke of Orleans, Joan's first proclamation, and so on, but all this seemingly random stuff had really been sown thick with hidden traps. But Joan had fortunately escaped them all, some by the protecting luck which attends upon ignorance and innocence, some by happy accident, the others by force of her best and surest helper, the clear vision and lightning intuitions of her extraordinary mind.

Now then, this daily baiting and badgering of this friendless girl, a captive in chains, was to continue a long, long time —dignified sport, a kennel of mastiffs and blood-hounds harassing a kitten!—and I may as well tell you, upon sworn testimony, what it was like from the first day to the last. When poor Joan had been in her grave a quarter of a century, the Pope called together that great court which was to re-examine her history, and whose just verdict cleared her illustrious name from every spot and stain, and laid upon the verdict and conduct of our Rouen tribunal the blight of its everlasting execrations. Manchon and several of the judges who had been members of our court were among the witnesses who appeared before that Tribunal of Rehabilitation. Recalling these miserable proceedings which I have been telling you about, Manchon testified thus:—here you have it, all in fair print in the official history:

When Joan spoke of her apparitions she was interrupted at almost every word. They wearied her with long and multiplied interrogatories upon all sorts of things. Almost every day the interrogatories of the morning lasted *three or four hours;* then from these morning-interrogatories they extracted the particularly difficult and subtle points, and these served as material for the afternoon-interrogatories, which lasted *two or three hours.* Moment by moment they skipped from one subject to another; yet in spite of this *she always responded with an astonishing wisdom and memory.* She often corrected the judges, saying, " But I have already answered that once before—ask the recorder," referring them to me.

And here is the testimony of one of Joan's judges. Remember, these witnesses are not talking about two or three days, they are talking about a tedious long *procession* of days:

They asked her profound questions, but she extricated herself quite well. Sometimes the questioners changed suddenly and passed to another subject *to see if she would not contradict herself.* They burdened her with long interrogatories of two or three hours, from which *the judges themselves went forth fatigued.* From the snares with which she was beset *the expertest man in the world could not have extricated himself but with difficulty.* She gave her responses with great prudence ; indeed to such a degree that during *three weeks I believed she was inspired.*

Ah, had she a mind such as I have described ? You see what these priests say under oath—picked men, men chosen for their places in that terrible court on account of their learning, their experience, their keen and practised intellects and their strong bias against the prisoner. They make that poor young country girl out the match, and more than the match, of the sixty-two trained adepts. Isn't it so ? They from the University of Paris, she from the sheepfold and the cow-stable ! Ah yes, she was great, she was wonderful. It took six thousand years to produce her ; her like will not be seen in the earth again in fifty thousand. Such is my opinion.

CHAPTER VII

THE third meeting of the court was in that same spacious
chamber, next day, 24th of February.

How did it begin work? In just the same old way.
When the preparations were ended, the robed sixty-two
massed in their chairs and the guards and order-keepers dis-
tributed to their stations, Cauchon spoke from his throne
and commanded Joan to lay her hands upon the Gospels and
swear to tell the truth concerning everything asked her!

Joan's eyes kindled, and she rose; rose and stood, fine and
noble, and faced toward the Bishop and said—

"Take care what you do, my Lord, you who are my judge,
for you take a terrible responsibility on yourself and you
presume too far."

It made a great stir, and Cauchon burst out upon her with
an awful threat—the threat of instant condemnation unless
she obeyed. That made the very bones in my body turn
cold, and I saw cheeks about me blanch—for it meant fire and
the stake! But Joan, still standing, answered him back, proud
and undismayed—

"Not all the clergy in Paris and Rouen could condemn me,
lacking the right!"

This made a great tumult, and part of it was applause from
the spectators. Joan resumed her seat. The Bishop still insisted. Joan said—

"I have already made oath. It is enough."

The Bishop shouted—

"In refusing to swear, you place yourself under suspicion!"

"Let be. I have sworn already. It is enough."

The Bishop continued to insist. Joan answered that "she would tell what she knew—but not all that she knew."

The Bishop plagued her straight along, till at last she said, in a weary tone—

"I came from God; I have nothing more to do here. Return me to God, from whom I came."

It was piteous to hear; it was the same as saying, "You only want my life; take it and let me be at peace."

The Bishop stormed out again—

"Once more I command you to—"

Joan cut in with a nonchalant "*Passez outre*," and Cauchon retired from the struggle; but he retired with some credit this time, for he offered a compromise, and Joan, always clear-headed, saw protection for herself in it and promptly and willingly accepted it. She was to swear to tell the truth "as touching the matters set down in the *proces verbal*." They could not sail her outside of definite limits, now; her course was over a charted sea, henceforth. The Bishop had granted more than he had intended, and more than he would honestly try to abide by.

By command, Beaupere resumed his examination of the accused. It being Lent, there might be a chance to catch her neglecting some detail of her religious duties. I could have told him he would fail there. Why, religion was her life!

"Since when have you eaten or drunk?"

If the least thing had passed her lips in the nature of sustenance, neither her youth nor the fact that she was being half starved in her prison could save her from dangerous suspicion of contempt for the commandments of the Church.

"I have done neither since yesterday at noon."

The priest shifted to the Voices again.

"When have you heard your Voice?"

"Yesterday and to-day."

"At what time?"

"Yesterday it was in the morning."

"What were you doing, then?"

"I was asleep and it woke me."

"By touching your arm?"

"No; without touching me."

"Did you thank it? Did you kneel?"

He had Satan in his mind, you see; and was hoping, perhaps, that by-and-by it could be shown that she had rendered homage to the archenemy of God and man.

"Yes, I thanked it; and knelt in my bed where I was chained, and joined my hands and begged it to implore God's help for me so that I might have light and instruction as touching the answers I should give here."

"Then what did the Voice say?"

"It told me to answer boldly, and God would help me." Then she turned toward Cauchon and said, "You say that you are my judge; now I tell you again, take care what you do, for in truth I am sent of God and you are putting yourself in great danger."

Beaupere asked her if the Voice's counsels were not fickle and variable.

"No. It never contradicts itself. This very day it has told me again to answer boldly."

"Has it forbidden you to answer only part of what is asked you?"

"I will tell you nothing as to that. I have revelations touching the King my master, and those I will not tell you." Then she was stirred by a great emotion, and the tears sprang to her eyes and she spoke out as with strong conviction, saying—

"I believe wholly—as wholly as I believe the Christian faith and that God has redeemed us from the fires of hell, that God speaks to me by that Voice!"

Being questioned further concerning the Voice, she said she was not at liberty to tell all she knew.

"Do you think God would be displeased at your telling the whole truth?"

"The Voice has commanded me to tell the King certain things, and not you—and some very lately—even last night;

things which I would he knew. He would be more easy at his dinner."

"Why doesn't the Voice speak to the King itself, as it did when you were with him? Would it not if you asked it?"

"I do not know if it be the wish of God." She was pensive, a moment or two, busy with her thoughts and far away, no doubt; then she added a remark in which Beaupere, always watchful, always alert, detected a possible opening—a chance to set a trap. Do you think he jumped at it instantly, betraying the joy he had in his find, as a young hand at craft and artifice would do? No, oh, no, you could not tell that he had noticed the remark at all. He slid indifferently away from it at once, and began to ask idle questions about other things, so as to slip around and spring on it from behind, so to speak: tedious and empty questions as to whether the Voice had told her she would escape from this prison; and if it had furnished answers to be used by her in to-day's seance; if it was accompanied with a glory of light; if it had eyes, etc. That risky remark of Joan's was this:

"Without the Grace of God I could do nothing."

The court saw the priest's game, and watched his play with a cruel eagerness. Poor Joan was grown dreamy and absent; possibly she was tired. Her life was in imminent danger, and she did not suspect it. The time was ripe now, and Beaupere quietly and stealthily sprung his trap:

"Are you in a state of Grace?"

Ah, we had two or three honorable brave men in that pack of judges; and Jean Lefevre was one of them. He sprang to his feet and cried out—

"It is a terrible question! The accused is not obliged to answer it!"

Cauchon's face flushed black with anger to see this plank flung to the perishing child, and he shouted—

"Silence! and take your seat. The accused will answer the question!"

There was no hope, no way out of the dilemma; for whether she said yes or whether she said no, it would be all the same—

23

a disastrous answer, for the Scriptures had said one *cannot know* this thing. Think what hard hearts they were to set this fatal snare for that ignorant young girl and be proud of such work and happy in it. It was a miserable moment for me while we waited; it seemed a year. All the house showed excitement; and mainly it was glad excitement. Joan looked out upon these hungering faces with innocent untroubled eyes, and then humbly and gently she brought out that immortal answer which brushed the formidable snare away as it had been but a cobweb:

"*If I be not in a state of Grace, I pray God place me in it; if I be in it, I pray God keep me so.*"

Ah, you will never see an effect like that; no, not while you live. For a space there was the silence of the grave. Men looked wondering into each other's faces, and some were awed and crossed themselves; and I heard Lefevre mutter—

"It was beyond the wisdom of man to devise that answer. *Whence* come this child's amazing inspirations?"

Beaupere presently took up his work again, but the humiliation of his defeat weighed upon him, and he made but a rambling and dreary business of it, he not being able to put any heart in it.

He asked Joan a thousand questions about her childhood and about the oak wood, and the fairies, and the children's games and romps under our dear *Arbre Fée de Bourlemont,* and this stirring up of old memories broke her voice and made her cry a little, but she bore up as well as she could, and answered everything.

Then the priest finished by touching again upon the matter of her apparel—a matter which was never to be lost sight of in this still-hunt for this innocent creature's life, but kept always hanging over her, a menace charged with mournful possibilities:

"Would you like a woman's dress?"

"Indeed yes, if I may go out from this prison—but here, no."

CHAPTER VIII

THE court met next on Monday the 27th. Would you believe it? The Bishop ignored the contract limiting the examination to matters set down in the *proces verbal* and again commanded Joan to take the oath without reservations. She said—

"You should be content; I have sworn enough."

She stood her ground, and Cauchon had to yield.

The examination was resumed, concerning Joan's Voices.

"You have said that you recognized them as being the voices of angels the third time that you heard them. What angels were they?"

"St. Catherine and St. Marguerite."

"How did you know that it was those two saints? How could you tell the one from the other?"

"I know it was they; and I know how to distinguish them."

"By what sign?"

"By their manner of saluting me. I have been these seven years under their direction, and I knew who they were because they told me."

"Whose was the first Voice that came to you when you were thirteen years old?"

"It was the Voice of St. Michael. I saw him before my eyes; and he was not alone, but attended by a cloud of angels."

"Did you see the archangel and the attendant angels in the body, or in the spirit?"

"I saw them with the eyes of my body, just as I see you; and when they went away I cried because they did not take me with them."

It made me see that awful shadow again that fell dazzling white upon her that day under *l'Arbre Fée de Bourlemont,* and it made me shiver again, though it was so long ago. It was really not very long gone by, but it seemed so, because so much had happened since.

"In what shape and form did St. Michael appear?"

"As to that, I have not received permission to speak."

"What did the archangel say to you that first time?"

"I cannot answer you to-day."

Meaning, I think, that she would have to get permission of her Voices first.

Presently, after some more questions as to the revelations which had been conveyed through her to the King, she complained of the unnecessity of all this, and said—

"I will say again, as I have said before, many times in these sittings, that I answered all questions of this sort before the court at Poitiers, and I would that you would bring here the record of that court and read from that. Prithee send for that book."

There was no answer. It was a subject that had to be got around and put aside. That book had wisely been gotten out of the way, for it contained things which would be very awkward here. Among them was a decision that Joan's mission was from God, whereas it was the intention of this inferior court to show that it was from the devil; also a decision permitting Joan to wear male attire, whereas it was the purpose of this court to make the male attire do hurtful work against her.

"How was it that you were moved to come into France— by your own desire?"

"Yes; and by command of God. But that it was his will I would not have come. I would sooner have had my body torn in sunder by horses than come, lacking that."

Beaupere shifted once more to the matter of the male attire, now, and proceeded to make a solemn talk about it. That tried Joan's patience; and presently she interrupted and said—

"It is a trifling thing and of no consequence. And I did not put it on by counsel of any man, but by command of God."

"Robert de Baudricourt did not order you to wear it?"

"No."

"Do you think you did well in taking the dress of a man?"

"I did well. to do whatsoever thing God commanded me to do."

"But in this particular case do you think you did well in taking the dress of a man?"

"I have done nothing but by command of God."

Beaupere made various attempts to lead her into contradictions of herself; also to put her words and acts in disaccord with the Scriptures. But it was lost time. He did not succeed. He returned to her visions, the light which shone about them, her relations with the King, and so on.

"Was there an angel above the King's head the first time you saw him?"

"By the Blessed Mary!—"

She forced her impatience down, and finished her sentence with tranquillity: "If there was one I did not see it."

"Was there light?"

"There were more than three hundred soldiers there, and five hundred torches, without taking account of spiritual light."

"What made the King believe in the revelations which you brought him?"

"He had signs; also the counsel of the clergy."

"What revelations were made to the King?"

"You will not get that out of me this year." Presently she added: "During three weeks I was questioned by the clergy at Chinon and Poitiers. The King had a sign before he would believe; and the clergy were of opinion that my acts were good and not evil."

The subject was dropped now for a while, and Beaupere took up the matter of the miraculous sword of Fierbois to see if he could not find a chance there to fix the crime of sorcery upon Joan.

"How did you know that there was an ancient sword buried in the ground under the rear of the altar of the church of St. Catherine of Fierbois?"

Joan had no concealments to make as to this:

"I knew the sword was there because my Voices told me so; and I sent to ask that it be given to me to carry in the wars. It seemed to me that it was not very deep in the ground. The clergy of the church caused it to be sought for and dug up; and they polished it, and the rust fell easily off from it."

"Were you wearing it when you were taken in battle at Compiègne?"

"No. But I wore it constantly until I left St. Denis after the attack upon Paris."

This sword, so mysteriously discovered and so long and so constantly victorious, was suspected of being under the protection of enchantment.

"Was that sword blest? What blessing had been invoked upon it?"

"None. I loved it because it was found in the church of St. Catherine, for I loved that church very dearly."

She loved it because it had been built in honor of one of her angels.

"Didn't you lay it upon the altar, to the end that it might be lucky?" (The altar of St. Denis)

"No."

"Didn't you pray that it might be made lucky?"

"Truly it were no harm to wish that my harness might be fortunate."

"Then it was not that sword which you wore in the field of Compiègne? What sword did you wear there?"

The sword of the Burgundian Franquet d'Arras, whom I took prisoner in the engagement at Lagny. I kept it because it was a good war-sword—good to lay on stout thumps and blows with."

She said that quite simply; and the contrast between her delicate little self and the grim soldier-words which she

THE CAPTURE OF JOAN OF ARC AT COMPIÈGNE

dropped with such easy familiarity from her lips made many spectators smile.

"What is become of the other sword? Where is it now?"

"Is that in the *proces verbal?*"

Beaupere did not answer.

"Which do you love best, your banner or your sword?"

Her eye lighted gladly at the mention of her banner, and she cried out—

"I love my banner best—oh, forty times more than the sword! Sometimes I carried it myself when I charged the enemy, to avoid killing any one." Then she added, naïvely, and with again that curious contrast between her girlish little personality and her subject, "I have never killed any one."

It made a great many smile; and no wonder, when you consider what a gentle and innocent little thing she looked. One could hardly believe she had ever even seen men slaughtered, she looked so little fitted for such things.

"In the final assault at Orleans did you tell your soldiers that the arrows shot by the enemy and the stones discharged from their catapults and cannon would not strike any one but you?"

"No. And the proof is, that more than a hundred of my men were struck. I told them to have no doubts and no fears; that they would raise the siege. I was wounded in the neck by an arrow in the assault upon the bastille that commanded the bridge, but St. Catherine comforted me and I was cured in fifteen days without having to quit the saddle and leave my work."

"Did you know that you were going to be wounded?"

"Yes; and I had told it to the King beforehand. I had it from my Voices."

"When you took Jargeau, why did you not put its commandant to ransom?"

"I offered him leave to go out unhurt from the place, with all his garrison; and if he would not I would take it by storm."

"And you did, I believe."

" Yes."

" Had your Voices counselled you to take it by storm ?"

" As to that, I do not remember."

Thus closed a weary long sitting, without result. Every device that could be contrived to trap Joan into wrong thinking, wrong doing, or disloyalty to the Church, or sinfulness as a little child at home or later had been tried, and none of them had succeeded. She had come unscathed through the ordeal.

Was the court discouraged? No. Naturally it was very much surprised, very much astonished, to find its work baffling and difficult instead of simple and easy, but it had powerful allies in the shape of hunger, cold, fatigue, persecution, deception, and treachery; and opposed to this array nothing but a defenceless and ignorant girl who must some time or other surrender to bodily and mental exhaustion or get caught in one of the thousand traps set for her.

And had the court made no progress during these seemingly resultless sittings? Yes. It had been feeling its way, groping here, groping there, and had found one or two vague trails which might freshen by-and-by and lead to something. The male attire, for instance, and the visions and Voices. Of course no one doubted that she had seen supernatural beings and been spoken to and advised by them. And of course no one doubted that by supernatural help miracles had been done by Joan, such as choosing out the King in a crowd when she had never seen him before, and her discovery of the sword buried under the altar. It would have been foolish to doubt these things, for we all know that the air is full of devils and angels that are visible to traffickers in magic on the one hand and to the stainlessly holy on the other; but what many and perhaps most did doubt was, that Joan's visions, voices, and miracles came from God. It was hoped that in time they could be proven to have been of satanic origin. Therefore, as you see, the court's persistent fashion of coming back to that subject every little while and spooking around it and prying into it was not to pass the time—it had a strictly business end in view.

CHAPTER IX

THE next sitting opened on Thursday the first of March. Fifty-eight judges present—the others resting.

As usual, Joan was required to take an oath without reservations. She showed no temper this time. She considered herself well buttressed by the *proces verbal* compromise which Cauchon was so anxious to repudiate and creep out of ; so she merely refused, distinctly and decidedly ; and added, in a spirit of fairness and candor—

"But as to matters set down in the *proces verbal*, I will freely tell the whole truth—yes, as freely and fully as if I were before the Pope."

Here was a chance ! We had two or three Popes, then ; only one of them could be the true Pope, of course. Everybody judiciously shirked the question of *which* was the true Pope and refrained from naming him, it being clearly dangerous to go into particulars in this matter. Here was an opportunity to trick an unadvised girl into bringing herself into peril, and the unfair judge lost no time in taking advantage of it. He asked, in a plausibly indolent and absent way—

"Which one do you consider to be the true Pope ?"

The house took an attitude of deep attention, and so waited to hear the answer and see the prey walk into the trap. But when the answer came it covered the judge with confusion, and you could see many people covertly chuckling. For Joan asked in a voice and manner which almost deceived even me, so innocent it seemed—

"Are there two ?"

One of the ablest priests in that body and one of the best

swearers there, spoke right out so that half the house heard him, and said—

" By God it was a master stroke !"

As soon as the judge was better of his embarrassment he came back to the charge, but was prudent and passed by Joan's question—

" Is it true that you received a letter from the Count of Armagnac asking you which of the three Popes he ought to obey?"

" Yes, and answered it."

Copies of both letters were produced and read. Joan said that hers had not been quite strictly copied. She said she had received the Count's letter when she was just mounting her horse; and added—

" So, in dictating a word or two of reply I said I would try to answer him from Paris or somewhere where I could be at rest."

She was asked again which Pope she had considered the right one.

" I was not able to instruct the Count of Armagnac as to which one he ought to obey "; then she added, with a frank fearlessness which sounded fresh and wholesome in that den of trimmers and shufflers, " but as for me, I hold that we are bound to obey our Lord the Pope who is at Rome."

The matter was dropped. Then they produced and read a copy of Joan's first effort at dictating—her proclamation summoning the English to retire from the siege of Orleans and vacate France—truly a great and fine production for an unpractised girl of seventeen.

" Do you acknowledge as your own the document which has just been read ?"

" Yes, except that there are errors in it—words which make me give myself too much importance." I saw what was coming; I was troubled and ashamed. " For instance, I did not say ' Deliver up to the Maid' (*rendez a la Pucelle*); I said 'Deliver up to the King' (*rendez au Roi*); and I did not call myself 'Commander - in - Chief' (*chef de guerre*). All those are words which my secretary substituted ; or mayhap he mis-heard me or forgot what I said."

She did not look at me when she said it ; she spared me that embarrassment. I hadn't misheard her at all, and hadn't forgotten. I changed her language purposely, for she *was* Commander-in-Chief and entitled to call herself so, and it was becoming and proper, too ; and who was going to surrender anything to the King ?—at that time a stick, a cipher ? If any surrendering was done, it would be to the noble Maid of Vaucouleurs, already famed and formidable though she had not yet struck a blow.

Ah, there would have been a fine and disagreeable episode (for me) there, if that pitiless court had discovered that the very scribbler of that piece of dictation, secretary to Joan of Arc, was present—and not only present, but helping build the record ; and not only that, but destined at a far distant day to testify against lies and perversions smuggled into it by Cauchon and deliver them over to eternal infamy !

" Do you acknowledge that you dictated this proclamation?"

" I do."

" Have you repented of it ? Do you retract it ?"

Ah, then she was indignant !

" No ! Not even these chains "— and she shook them— " not even these chains can chill the hopes that I uttered there. And more !"— she rose, and stood a moment with a divine strange light kindling in her face, then her words burst forth as in a flood—" I warn you now that before seven years a disaster will smite the English, oh, many fold greater than the fall of Orleans ! and—"

" Silence ! Sit down !"

" —and then, soon after, they will lose all France !"

Now consider these things. The French armies no longer existed. The French cause was standing still, our King was standing still, there was no hint that by-and-by the Constable Richemont would come forward and take up the great work of Joan of Arc and finish it. In face of all this, Joan made that prophecy—made it with perfect confidence—*and it came true.*

For within five years Paris fell — 1436 — and our King marched into it flying the victor's flag. So the first part of the prophecy was then fulfilled—in fact, almost the entire prophecy; for, with Paris in our hands, the fulfilment of the rest of it was assured.

Twenty years later all France was ours excepting a single town—Calais.

Now that will remind you of an earlier prophecy of Joan's. At the time that she wanted to take Paris and could have done it with ease if our King had but consented, she said that that was the golden time; that with Paris ours, all France would be ours in six months. But if this golden opportunity to recover France was wasted, said she, "*I give you twenty years to do it in.*"

She was right. After Paris fell, in 1436, the rest of the work had to be done city by city, castle by castle, and it took twenty years to finish it.

Yes, it was the first day of March, 1431, there in the court, that she stood in the view of everybody and uttered that strange and incredible prediction. Now and then, in this world, somebody's prophecy turns up correct, but when you come to look into it there is sure to be considerable room for suspicion that the prophecy was made after the fact. But here the matter is different. There in that court Joan's prophecy was set down in the official record at the hour and moment of its utterance, years before the fulfilment, and there you may read it to this day. Twenty-five years after Joan's death the record was produced in the great Court of the Rehabilitation and verified under oath by Manchon and me, and surviving judges of our court confirmed the exactness of the record in their testimony.

Joan's startling utterance on that now so celebrated first of March stirred up a great turmoil, and it was some time before it quieted down again. Naturally everybody was troubled, for a prophecy is a grisly and awful thing, whether one thinks it ascends from hell or comes down from heaven. All that these people felt sure of was, that the inspiration back of it was genuine and puissant. They would have given their right hands to know the source of it.

At last the questions began again.

"How do you know that those things are going to happen?"

"I know it by revelation. And I know it as surely as I know that you sit here before me."

This sort of answer was not going to allay the spreading uneasiness. Therefore, after some further dallying the judge got the subject out of the way and took up one which he could enjoy more.

"What language do your Voices speak?"

"French."

"St. Marguerite, too?"

"Verily; why not? She is on our side, not on the English?"

Saints and angels who did not condescend to speak English! a grave affront. They could not be brought into court and punished for contempt, but the tribunal could take silent note of Joan's remark and remember it against her; which they did. It might be useful by-and-by.

"Do your saints and angels wear jewelry?—crowns, rings, ear-rings?"

To Joan, questions like this were profane frivolities and not worthy of serious notice; she answered indifferently. But the question brought to her mind another matter, and she turned upon Cauchon and said—

"I had two rings. They have been taken away from me during my captivity. You have one of them. It is the gift of my brother. Give it back to me. If not to me, then I pray that it be given to the Church."

The judges conceived the idea that maybe these rings were for the working of enchantments. Perhaps they could be made to do Joan a damage.

"Where is the other ring?"

"The Burgundians have it."

"Where did you get it?"

"My father and mother gave it to me."

" Describe it."

"It is plain and simple and has '*Jesus and Mary*' engraved upon it."

Everybody could see that that was not a valuable equipment to do devil's work with. So that trail was not worth following. Still, to make sure, one of the judges asked Joan if she had ever cured sick people by touching them with the ring. She said no.

" Now as concerning the fairies, that were used to abide near by Domremy whereof there are many reports and traditions. It is said that your godmother surprised these creatures on a summer's night dancing under the tree called *L'Arbre Fée de Bourlemont*. Is it not possible that your pretended saints and angels are but those fairies ?"

" Is that in your *proces* ?"

She made no other answer.

" Have you not conversed with St. Marguerite and St. Catherine under that tree ?"

" I do not know."

" Or by the fountain near the tree ?"

" Yes, sometimes."

" What promises did they make you ?"

" None but such as they had God's warrant for."

" But what promises *did* they make ?"

" That is not in your *proces*; yet I will say this much : they told me that the King would become master of his kingdom in spite of his enemies."

" And what else ?"

There was a pause ; then she said humbly—

" They promised to lead me to Paradise."

If faces do really betray what is passing in men's minds, a fear came upon many in that house, at this time, that maybe, after all, a chosen servant and herald of God was here being hunted to her death. The interest deepened. Movements and whisperings ceased : the stillness became almost painful.

Have you noticed that almost from the beginning the nature of the questions asked Joan showed that in some way or other the questioner very often already knew his fact before he asked his question? Have you noticed that somehow or other the questioners usually knew just how and where to search for Joan's secrets; that they really knew the bulk of her privacies—a fact not suspected by her—and that they had no task before them but to trick her into exposing those secrets?

Do you remember Loyseleur the hypocrite, the treacherous priest, tool of Cauchon? Do you remember that under the sacred seal of the confessional Joan freely and trustingly revealed to him everything concerning her history save only a few things regarding her supernatural revelations which her Voices had forbidden her to tell to any one — and that the unjust judge, Cauchon, was a hidden listener all the time?

Now you understand how the inquisitors were able to devise that long array of minutely prying questions; questions whose subtlety and ingenuity and penetration are astonishing until we come to remember Loyseleur's performance and recognize their source. Ah, Bishop of Beauvais, you are now lamenting this cruel iniquity these many years in hell! Yes verily, unless one has come to your help. There is but one among the redeemed that would do it; and it is futile to hope that that one has not already done it—Joan of Arc.

We will return to the court and the questionings.

"Did they make you still another promise?"

"Yes, but that is not in your *proces*. I will not tell it now, but before three months I will tell it you."

The judge seems to know the matter he is asking about, already; one gets this idea from his next question.

"Did your Voices tell you that you would be liberated before three months?"

Joan often showed a little flash of surprise at the good guessing of the judges, and she showed one this time. I was frequently in terror to find my mind (which *I* could not control) criticising the Voices and saying, "They counsel her to speak boldly—a thing which she would do without any suggestion from *them* or anybody else—but when it comes to telling her any useful thing, such as how these conspirators manage to guess their way so skilfully into her affairs, they are always off attending to some other business." I am reverent by nature; and when such thoughts swept through my head they made me cold with fear, and if there was a storm and thunder at the time, I was so ill that I could but with difficulty abide at my post and do my work.

Joan answered—

"That is not in your *proces*. I do not know when I shall be set free, but some who wish me out of this world will go from it before me."

It made some of them shiver.

"Have your Voices told you that you will be delivered from this prison?"

Without a doubt they had, and the judge knew it before he asked the question.

"Ask me again in three months and I will tell you."

She said it with such a happy look, the tired prisoner! And I? And Noël Rainguesson, drooping yonder?—why, the floods of joy went streaming through us from crown to sole! It was all that we could do to hold still and keep from making fatal exposure of our feelings.

She was to be set free in three months. That was what she meant; we saw it. The Voices had told her so, and told her true—true to the very day—May 30. But we know, now, that they had mercifully hidden from her *how* she was to be set free, but left her in ignorance. Home again! That was our understanding of it—Noël's and mine; that was our dream; and now we would count the days, the hours, the minutes. They would fly lightly along; they would soon be over. Yes, we would carry our idol home; and there, far from the pomps and tumults of the world, we would take up our happy life again and live it out as we had begun it, in the free air and the sunshine, with the friendly sheep and the friendly people for comrades, and the grace and charm of the meadows, the woods, and the river always before our eyes and their deep peace in our hearts. Yes, that was our dream, the dream that carried us bravely through that three months to an exact and awful fulfilment, the thought of which would have killed us, I think, if we had foreknown it and been obliged to bear the burden of it upon our hearts the half of those heavy days.

Our reading of the prophecy was this: We believed the King's soul was going to be smitten with remorse; and that he would privately plan a rescue with Joan's old lieutenants, D'Alençon and the Bastard and La Hire, and that this rescue would take place at the end of the three months. So we made up our minds to be ready and take a hand in it.

In the present and also in later sittings Joan was urged to name the exact day of her deliverance; but she could not do that. She had not the permission of her Voices. Moreover, the Voices themselves did not name the precise day. Ever since the fulfilment of the prophecy, I have believed that Joan had the idea that her deliverance was going to come in the form of death. But not *that* death! Divine as she was, dauntless as she was in battle, she was human also. She was not solely a saint, an angel, she was a clay-made girl also —as human a girl as any in the world, and full of a human girl's sensitivenesses and tendernesses and delicacies. And so, *that* death! No, she could not have lived the three months with that one before her, I think. You remember that the first time she was wounded she was frightened, and cried, just as any other girl of seventeen would have done, although she had known for eighteen days that she was going to be wounded on that very day. No, she was not afraid of any ordinary death, and an ordinary death was what she believed the prophecy of deliverance meant, I think, for her face showed happiness, not horror, when she uttered it.

Now I will explain why I think as I do. Five weeks before she was captured in the battle of Compiègne, her Voices told her what was coming. They did not tell her the day or the place, but said she would be taken prisoner and that it would be before the feast of St. John. She begged that death, certain and swift, should be her fate, and the captivity

24

brief; for she was a free spirit, and dreaded the confinement. The Voices made no promise, but only told her to bear whatever came. Now as they did not *refuse* the swift death, a hopeful young thing like Joan would naturally cherish that fact and make the most of it, allowing it to grow and establish itself in her mind. And so now that she was told she was to be " delivered " in three months, I think she believed it meant that she would die in her bed in the prison, and that that was why she looked happy and content—the gates of Paradise standing open for her, the time so short, you see, her troubles so soon to be over, her reward so close at hand. Yes, that would make her look happy, that would make her patient and bold, and able to fight her fight out like a soldier. Save herself if she could, of course, and try her best, for that was the way she was made; but die with her face to the front if die she must.

Then later, when she charged Cauchon with trying to kill her with a poisoned fish, her notion that she was to be " delivered " by death in the prison—if she had it, and I believe she had—would naturally be greatly strengthened, you see.

But I am wandering from the trial. Joan was asked to definitely name the time that she would be delivered from prison.

" I have always said that I was not permitted to tell you everything. I am to be set free, and I desire to ask leave of my Voices to tell you the day. This is why I wish for delay."

" Do your Voices forbid you to tell the truth ?"

" Is it that you wish to know matters concerning the King of France ? I tell you again that he will regain his kingdom, and that I know it as well as I know that you sit here before me in this tribunal." She sighed and, after a little pause, added: " I should be dead but for this revelation, which comforts me always."

Some trivial questions were asked her about St. Michael's dress and appearance. She answered them with dignity, but one saw that they gave her pain. After a little she said—

" I have great joy in seeing him, for when I see him I have

THE MAID OF ORLEANS
(From the portrait, by an unknown painter, in the Hôtel de Ville at Rouen)

the feeling that I am not in mortal sin." She added, "Sometimes St. Marguerite and St. Catherine have allowed me to confess myself to them."

Here was a possible chance to set a successful snare for her innocence.

"When you confessed were you in mortal sin, do you think?"

But her reply did her no hurt. So the inquiry was shifted once more to the revelations made to the King — secrets which the court had tried again and again to force out of Joan, but without success.

"Now as to the sign given to the King—"

"I have already told you that I will tell you nothing about it."

"Do you know what the sign was?"

"As to that, you will not find out from me."

All this refers to Joan's secret interview with the King— held apart, though two or three others were present. It was known—through Loyseleur, of course—that this sign was a crown and was a pledge of the verity of Joan's mission. But that is all a mystery until this day—the nature of the crown, I mean—and will remain a mystery to the end of time. We can never know whether a real crown descended upon the King's head, or only a symbol, the mystic fabric of a vision.

"Did you see a crown upon the King's head when he received the revelation?"

"I cannot tell you as to that, without perjury."

"Did the King have that crown at Rheims?"

"I think the King put upon his head a crown which he found there; but a much richer one was brought him afterwards."

"Have you seen that one?"

"I cannot tell you, without perjury. But whether I have seen it or not, I have heard say that it was rich and magnificent."

They went on and pestered her to weariness about that mysterious crown, but they got nothing more out of her. The sitting closed. A long, hard day for all of us.

CHAPTER X

THE court rested a day, then took up work again on Saturday the third of March.

This was one of our stormiest sessions. The whole court was out of patience; and with good reason. These three-score distinguished churchmen, illustrious tacticians, veteran legal gladiators, had left important posts where their supervision was needed, to journey hither from various regions and accomplish a most simple and easy matter—condemn and send to death a country lass of nineteen who could neither read nor write, knew nothing of the wiles and perplexities of legal procedure, could call not a single witness in her defence, was allowed no advocate or adviser, and must conduct her case by herself against a hostile judge and a packed jury. In two hours she would be hopelessly entangled, routed, defeated, convicted. Nothing could be more certain than this—so they thought. But it was a mistake. The two hours had strung out into days; what promised to be a skirmish had expanded into a siege; the thing which had looked so easy had proven to be surprisingly difficult; the light victim who was to have been puffed away like a feather remained planted like a rock; and on top of all this, if anybody had a right to laugh it was the country lass and not the court.

She was not doing that, for that was not her spirit; but others were doing it. The whole town was laughing in its sleeve, and the court knew it, and its dignity was deeply hurt. The members could not hide their annoyance.

And so, as I have said, the session was stormy. It was easy to see that these men had made up their minds to force words from Joan to-day which should shorten up her case and bring it to a prompt conclusion. It shows that after all their experience with her they did not know her yet. They went into the battle with energy. They did not leave the questioning to a particular member; no, everybody helped. They volleyed questions at Joan from all over the house, and sometimes so many were talking at once that she had to ask them to deliver their fire one at a time and not by platoons. The beginning was as usual:

"You are once more required to take the oath pure and simple."

"I will answer to what is in the *proces verbal*. When I do more, I will choose the occasion for myself."

That old ground was debated and fought over inch by inch with great bitterness and many threats. But Joan remained steadfast, and the questionings had to shift to other matters. Half an hour was spent over Joan's apparitions—their dress, hair, general appearance, and so on—in the hope of fishing something of a damaging sort out of the replies; but with no result.

Next, the male attire was reverted to, of course. After many well-worn questions had been re-asked, one or two new ones were put forward.

"Did not the King or the Queen sometimes ask you to quit the male dress?"

"That is not in your *proces*."

"Do you think you would have sinned if you had taken the dress of your sex?"

"I have done best to serve and obey my sovereign Lord and Master."

After a while the matter of Joan's Standard was taken up, in the hope of connecting magic and witchcraft with it.

"Did not your men copy your banner in their pennons?"

"The lancers of my guard did it. It was to distinguish them from the rest of the forces. It was their own idea."

"Were they often renewed?"

"Yes. When the lances were broken they were renewed."

The purpose of the questions unveils itself in the next one.

"Did you not say to your men that pennons made like your banner would be lucky?"

The soldier-spirit in Joan was offended at this puerility. She drew herself up, and said with dignity and fire: "What I said to them was, 'Ride these English down!' and I did it myself."

Whenever she flung out a scornful speech like that at these French menials in English livery it lashed them into a rage; and that is what happened this time. There were ten, twenty, sometimes even thirty of them on their feet at a time, storming at the prisoner minute after minute, but Joan was not disturbed.

By-and-by there was peace, and the inquiry was resumed.

It was now sought to turn against Joan the thousand loving honors which had been done her when she was raising France out of the dirt and shame of a century of slavery and castigation.

"Did you not cause paintings and images of yourself to be made?"

"No. At Arras I saw a painting of myself kneeling in armor before the King and delivering him a letter; but I caused no such things to be made."

"Were not masses and prayers said in your honor?"

"If it was done it was not by my command. But if any prayed for me I think it was no harm."

"Did the French people believe you were sent of God?"

"As to that, I know not: but whether they believed it or not, I was not the less sent of God."

"If they thought you were sent of God do you think it was well thought?"

"If they believed it, their trust was not abused."

"What impulse was it, think you, that moved the people to kiss your hands, your feet, and your vestments?"

"They were glad to see me, and so they did those things; and I could not have prevented them if I had had the heart. Those poor people came lovingly to me because I had not done them any hurt, but had done the best I could for them according to my strength."

See what modest little words she uses to describe that touching spectacle, her marches about France walled in on both sides by the adoring multitudes: "They were glad to see me." Glad? Why, they were transported with joy to see her. When they could not kiss her hands or her feet, they knelt in the mire and kissed the hoof-prints of her horse. They worshipped her; and that is what these priests were trying to prove. It was nothing to them that she was not to blame for what other people did. No, if she was worshipped, it was enough; she was guilty of mortal sin. Curious logic, one must say.

"Did you not stand sponsor for some children baptized at Rheims?"

"At Troyes I did, and at St. Denis; and I named the boys Charles, in honor of the King, and the girls I named Joan."

"Did not women touch their rings to those which you wore?"

"Yes, many did, but I did not know their reason for it."

"At Rheims was your Standard carried into the church? Did you stand at the altar with it in your hand at the Coronation?"

"Yes."

"In passing through the country did you confess yourself in the churches and receive the sacrament?"

"Yes."

"In the dress of a man?"

"Yes. But I do not remember that I was in armor."

It was almost a concession! almost a half-surrender of the permission granted her by the Church at Poitiers to dress as a man. The wily court shifted to another matter: to pursue this one at this time might call Joan's attention to her small mistake, and by her native cleverness she might recover her lost ground. The tempestuous session had worn her and drowsed her alertness.

"It is reported that you brought a dead child to life in the church at Lagny. Was that in answer to your prayers?"

"As to that, I have no knowledge. Other young girls were

praying for the child, and I joined them and prayed also, do-ing no more than they."

" Continue."

" While we prayed it came to life, and cried. It had been dead three days, and was as black as my doublet. It was straightway baptized, then it passed from life again and was buried in holy ground."

" Why did you jump from the tower of Beaurevoir by night and try to escape ?"

" I would go to the succor of Compiègne."

It was insinuated that this was an attempt to commit the deep crime of suicide to avoid falling into the hands of the English.

" Did you not say that you would rather die than be deliv-ered into the power of the English ?"

Joan answered frankly, without perceiving the trap—

" Yes ; my words were, that I would rather that my soul be returned unto God than that I should fall into the hands of the English."

It was now insinuated that when she came to, after jumping from the tower, she was angry and blasphemed the name of God ; and that she did it again when she heard of the defec-tion of the Commandant of Soissons. She was hurt and in-dignant at this, and said—

" It is not true. I have never cursed. It is not my custom to swear."

CHAPTER XI

A HALT was called. It was time. Cauchon was losing ground in the fight, Joan was gaining it. There were signs that here and there in the court a judge was being softened toward Joan by her courage, her presence of mind, her fortitude, her constancy, her piety, her simplicity and candor, her manifest purity, the nobility of her character, her fine intelligence, and the good brave fight she was making, all friendless and alone against unfair odds, and there was grave room for fear that this softening process would spread further and presently bring Cauchon's plans in danger.

Something must be done, and it was done. Cauchon was not distinguished for compassion, but he now gave proof that he had it in his character. He thought it pity to subject so many judges to the prostrating fatigues of this trial when it could be conducted plenty well enough by a handful of them. Oh, gentle Judge! But he did not remember to modify the fatigues for the little captive.

He would let all the judges but a handful go, but he would select the handful himself, and he did. He chose tigers. If a lamb or two got in, it was by oversight, not intention; and he knew what to do with lambs when discovered.

He called a small council, now, and during five days they sifted the huge bulk of answers thus far gathered from Joan. They winnowed it of all chaff, all useless matter—that is, all matter favorable to Joan; they saved up all matter which could be twisted to her hurt, and out of this they constructed a basis for a new trial which should have the semblance of a continuation of the old one. Another change. It was plain that the public trial had wrought damage: its proceedings had been discussed all over the town and had moved many to pity the abused prisoner. There should be no more of that. The sittings should be secret hereafter, and no spectators admitted. So Noël could come no more. I sent this news to him. I had not the heart to carry it myself. I would give the pain a chance to modify before I should see him in the evening.

On the tenth of March the secret trial began. A week had passed since I had seen Joan. Her appearance gave me a great shock. She looked tired and weak. She was listless and far away, and her answers showed that she was dazed and not able to keep perfect run of all that was done and said. Another court would not have taken advantage of her state, seeing that her life was at stake here, but would have adjourned and spared her. Did this one? No; it worried her for hours, and with a glad and eager ferocity, making all it could out of this great chance, the first one it had had.

She was tortured into confusing herself concerning the "sign" which had been given the King, and the next day this was continued hour after hour. As a result, she made partial revealments of particulars forbidden by her Voices; and seemed to me to state as facts things which were but allegories and visions mixed with facts.

The third day she was brighter, and looked less worn. She was almost her normal self again, and did her work well. Many attempts were made to beguile her into saying indiscreet things, but she saw the purpose in view and answered with tact and wisdom.

"Do you know if St. Catherine and St. Marguerite hate the English?"

"They love whom Our Lord loves, and hate whom He hates."

"Does God hate the English?"

"Of the love or the hatred of God toward the English I know nothing." Then she spoke up with the old martial ring in her voice and the old audacity in her words, and added, "But I know this—that God will send victory to the

French, and that all the English will be flung out of France
but the dead ones!"

"Was God on the side of the English when they were
prosperous in France?"

"I do not know if God hates the French, but I think that
he allowed them to be chastised for their sins."

It was a sufficiently naïve way to account for a chastise-
ment which had now strung out for ninety - six years. But
nobody found fault with it. There was nobody there who
would not punish a sinner ninety-six years if he could, nor
anybody there who would ever dream of such a thing as the
Lord's being any shade less stringent than men.

"Have you ever embraced St. Marguerite and St. Catherine?"

"Yes, both of them."

The evil face of Cauchon betrayed satisfaction when she
said that.

"When you hung garlands upon *L'Arbre Fée de Bourlemont*, did you do it in honor of your apparitions?"

"No."

Satisfaction again. No doubt Cauchon would take it for
granted that she hung them there out of sinful love for the
fairies.

"When the saints appeared to you did you bow, did you
make reverence, did you kneel?"

"Yes; I did them the most honor and the most reverence
that I could."

A good point for Cauchon if he could eventually make it
appear that these were no saints to whom she had done reverence, but devils in disguise.

Now there was the matter of Joan's keeping her supernatural commerce a secret from her parents. Much might be
made of that. In fact, particular emphasis had been given to
it in a private remark written in the margin of the *proces*: "*She
concealed her visions from her parents and from every one.*"
Possibly this disloyalty to her parents might itself be the
sign of the satanic source of her mission.

"Do you think it was right to go away to the wars without getting your parents' leave? It is written one must honor his father and his mother."

"I have obeyed them in all things but that. And for that I have begged their forgiveness in a letter and gotten it."

"Ah, you asked their pardon? So you *knew* you were guilty of sin in going without their leave!"

Joan was stirred. Her eyes flashed, and she exclaimed—

"I was commanded of God, and it was right to go! If I had had a hundred fathers and mothers and been a king's daughter to boot I would have gone."

"Did you never ask your Voices if you might tell your parents?"

"They were willing that I should tell them, but I would not for anything have given my parents that pain."

To the minds of the questioners this headstrong conduct savored of pride. That sort of pride would move one to seek sacrilegious adorations.

"Did not your Voices call you Daughter of God?"

Joan answered with simplicity, and unsuspiciously—

"Yes; before the siege of Orleans and since, they have several times called me Daughter of God."

Further indications of pride and vanity were sought.

"What horse were you riding when you were captured? Who gave it you?"

"The King."

"You had other things—riches—of the King?"

"For myself I had horses and arms, and money to pay the service in my household."

"Had you not a treasury?"

"Yes. Ten or twelve thousand crowns." Then she said with naïveté, "It was not a great sum to carry on a war with."

"You have it yet?"

"No. It is the King's money. My brothers hold it for him."

"What were the arms which you left as an offering in the church of St. Denis?"

"My suit of silver mail and a sword."

"Did you put them there in order that they might be adored?"

"No. It was but an act of devotion. And it is the custom of men of war who have been wounded to make such offering there. I had been wounded before Paris."

Nothing appealed to those stony hearts, those dull imaginations—not even this pretty picture, so simply drawn, of the wounded girl-soldier hanging her toy harness there in curious companionship with the grim and dusty iron mail of the historic defenders of France. No, there was nothing in it for them; nothing, unless evil and injury for that innocent creature could be gotten out of it somehow.

"Which aided most — you the Standard, or the Standard you?"

"Whether it was the Standard or whether it was I, is nothing—the victories came from God."

"But did you base your hopes of victory in yourself or in your Standard?"

"In neither. In God, and not otherwhere."

"Was not your Standard waved around the King's head at the Coronation?"

"No. It was not."

"Why was it that *your* Standard had place at the crowning of the King in the Cathedral of Rheims, rather than those of the other captains?"

Then, soft and low, came that touching speech which will live as long as language lives, and pass into all tongues, and move all gentle hearts wheresoever it shall come, down to the latest day:

"It had borne the burden, it had earned the honor." *

---

* What she said has been many times translated, but never with success. There is a haunting pathos about the original which eludes all efforts to convey it into our tongue. It is as subtle as an odor, and escapes in the transmission. Her words were these:

"Il avait été a la peine, c'etait bien raison qu'il fut a l'honneur."

Monseigneur Ricard, Honorary Vicar-General to the Archbishop of Aix,

How simple it is, and how beautiful. And how it beggars
the studied eloquence of the masters of oratory. Eloquence
was a native gift of Joan of Arc; it came from her lips without effort and without preparation. Her words were as sublime as her deeds, as sublime as her character; they had their
source in a great heart and were coined in a great brain.

finely speaks of it ("*Jeanne d'Arc la Vénérable*," page 197) as "that
sublime reply, enduring in the history of celebrated sayings like the cry of
a French and Christian soul wounded unto death in its patriotism and its
faith."—TRANSLATOR.

CHAPTER XII

Now as a next move, this small secret court of holy assassins did a thing so base that even at this day, in my old age, it is hard to speak of it with patience.

In the beginning of her commerce with her Voices there at Domremy, the child Joan solemnly devoted her life to God, vowing her pure body and her pure soul to his service. You will remember that her parents tried to stop her from going to the wars by haling her to the court at Toul to compel her to make a marriage which she had never promised to make—a marriage with our poor, good, windy, big, hard-fighting and most dear and lamented comrade the Standard-bearer, who fell in honorable battle and sleeps in God these sixty years, peace to his ashes! And you will remember how Joan, sixteen years old, stood up in that venerable court and conducted her case all by herself, and tore the poor Paladin's case to rags and blew it away with a breath; and how the astonished old judge on the bench spoke of her as "this marvellous child."

You remember all that. Then think what I felt, to see these false priests here in the tribunal wherein Joan had fought a fourth lone fight in three years, deliberately twist that matter entirely around and try to make out that Joan haled the Paladin into court and pretended that he had promised to marry her, and was bent on making him do it.

Certainly there was no baseness that those people were ashamed to stoop to in their hunt for that friendless girl's life. What they wanted to show was this—that she had committed the sin of relapsing from her vow and trying to violate it.

Joan detailed the true history of the case, but lost her temper as she went along, and finished with some words for Cauchon which he remembers yet, whether he is fanning himself in the world he belongs in or has swindled his way into the other.

The rest of this day and part of the next the court labored upon the old theme—the male attire. It was shabby work for those grave men to be engaged in; for they well knew one of Joan's reasons for clinging to the male dress was, that soldiers of the guard were always present in her room whether she was asleep or awake, and that the male dress was a better protection for her modesty than the other.

The court knew that one of Joan's purposes had been the deliverance of the exiled Duke of Orleans, and they were curious to know how she had intended to manage it. Her plan was characteristically business-like, and her statement of it as characteristically simple and straightforward:

"I would have taken English prisoners enough in France for his ransom; and failing that, I would have invaded England and brought him out by force."

That was just her way. If a thing was to be done, it was love first, and hammer and tongs to follow; but no shilly-shallying between. She added with a little sigh—

"If I had had my freedom three years, I would have delivered him."

"Have you the permission of your Voices to break out of prison whenever you can?"

"I have asked their leave several times, but they have not given it."

I think it is as I have said, she expected the deliverance of death, and within the prison walls, before the three months should expire.

"Would you escape if you saw the doors open?"

She spoke up frankly and said—

"Yes—for I should see in that the permission of Our Lord. God helps who help themselves, the proverb says. But except I thought I had permission, I would not go."

RAINGUESSON AND DE CONTE MAKING THEIR WAY TO ROUEN

Now, then, at this point, something occurred which convinces me, every time I think of it—and it struck me so at the time—that for a moment, at least, her hopes wandered to the King, and put into her mind the same notion about her deliverance which Noël and I had settled upon—a rescue by her old soldiers. I think the idea of the rescue did occur to her, but only as a passing thought, and that it quickly passed away.

Some remark of the Bishop of Beauvais moved her to remind him once more that he was an unfair judge, and had no right to preside there, and that he was putting himself in great danger.

"What danger?" he asked.

"I do not know. St. Catherine has promised me help, but I do not know the form of it. I do not know whether I am to be delivered from this prison or whether when you send me to the scaffold there will happen a trouble by which I shall be set free. Without much thought as to this matter, I am of the opinion that it may be one or the other." After a pause she added these words, memorable forever — words whose meaning she may have miscaught, misunderstood, as to that we can never know; words which she may have rightly understood; as to that also, we can never know; but words whose mystery fell away from them many a year ago and revealed their real meaning to all the world:

"But what my Voices have said clearest is, that I shall be delivered *by a great victory*." She paused, my heart was beating fast, for to me that great victory meant the sudden bursting in of our old soldiers with war-cry and clash of steel at the last moment and the carrying off of Joan of Arc in triumph. But oh, that thought had such a short life! For now she raised her head and finished, with those solemn words which men still so often quote and dwell upon—words which filled me with fear, they sounded so like a prediction. "And always they say 'Submit to whatever comes; *do not grieve for your martyrdom;* from it you will ascend into the Kingdom of Paradise.'"

25

Was she thinking of fire and the stake? I think not. I thought of it myself, but I believe she was only thinking of this slow and cruel martyrdom of chains and captivity and insult. Surely martyrdom was the right name for it.

It was Jean de la Fontaine who was asking the questions. He was willing to make the most he could out of what she had said:

"As the Voices have told you you are going to Paradise, you feel certain that that will happen and that you will not be damned in hell. Is that so?"

"I believe what they told me. I know that I shall be saved."

"It is a weighty answer."

"To me the knowledge that I shall be saved is a great treasure."

"Do you think that after that revelation you could be able to commit mortal sin?"

"As to that, I do not know. My hope for salvation is in holding fast to my oath to keep my body and my soul pure."

"Since you know you are to be saved do you think it necessary to go to confession?"

The snare was ingeniously devised, but Joan's simple and humble answer left it empty—

"One cannot keep his conscience too clean."

We were now arriving at the last day of this new trial. Joan had come through the ordeal well. It had been a long and wearisome struggle for all concerned. All ways had been tried to convict the accused, and all had failed, thus far. The inquisitors were thoroughly vexed and dissatisfied. However, they resolved to make one more effort, put in one more day's work. This was done—March 17th. Early in the sitting a notable trap was set for Joan:

"Will you submit to the determination of the Church all your words and deeds, whether good or bad?"

That was well planned. Joan was in imminent peril now. If she should heedlessly say yes, it would put her mission *itself* upon trial, and one would know how to decide its source and character promptly. If she should say no, she would render herself chargeable with the crime of heresy.

But she was equal to the occasion. She drew a distinct line of separation between the Church's authority over her as a subject member, and the matter of her *mission*. She said she loved the Church and was ready to support the Christian faith with all her strength; but as to the works done under her mission, those must be judged by God alone, who had commanded them to be done.

The judge still insisted that she submit them to the decision of the Church. She said—

"I will submit them to Our Lord who sent me. It would seem to me that He and His Church are one, and that there should be no difficulty about this matter." Then she turned upon the judge and said, "Why do you make a difficulty where there is no room for any?"

Then Jean de la Fontaine corrected her notion that there was but one Church. There were two—the Church Triumphant, which is God, the saints, the angels, and the redeemed, and has its seat in heaven; and the Church Militant, which is our Holy Father the Pope, Vicar of God, the prelates, the clergy and all good christians and catholics, the which Church has its seat in the earth, is governed by the Holy Spirit, and cannot err. "Will you not submit those matters to the Church Militant?"

"I am come to the King of France from the Church Triumphant on high by its commandant, and to that Church I will submit all those things which I have done. For the Church Militant I have no other answer now."

The court took note of this straightly worded refusal, and would hope to get profit out of it; but the matter was dropped for the present, and a long chase was then made over the old hunting-ground—the fairies, the visions, the male attire, and all that.

In the afternoon the satanic Bishop himself took the chair and presided over the closing scenes of the trial. Along toward the finish, this question was asked by one of the judges:

"You have said to my lord the Bishop that you would answer him as you would answer before our Holy Father the Pope, and yet there are several questions which you continually refuse to answer. Would you not answer the Pope more fully than you have answered before my lord of Beauvais? Would you not feel obliged to answer the Pope, who is the Vicar of God, more fully?"

Now fell a thunder-clap out of a clear sky—

"*Take me to the Pope.* I will answer to everything that I ought to."

It made the Bishop's purple face fairly blanch with consternation. If Joan had only known, if she had only known! She had lodged a mine under this black conspiracy able to blow the Bishop's schemes to the four winds of heaven, and she didn't know it. She had made that speech by mere instinct, not suspecting what tremendous forces were hidden in it, and there was none to tell her what she had done. I knew, and Manchon knew; and if she had known how to read writing we could have hoped to get the knowledge to her somehow; but speech was the only way, and none was allowed to approach her near enough for that. So there she sat, once more Joan of Arc the Victorious, but all unconscious of it. She was miserably worn and tired, by the long day's struggle and by illness, or she must have noticed the effect of that speech and divined the reason of it.

She had made many master-strokes, but this was *the* master-stroke. It was *an appeal to Rome*. It was her clear right; and if she had persisted in it Cauchon's plot would have tumbled about his ears like a house of cards, and he would have gone from that place the worst beaten man of the century. He was daring, but he was not daring enough to stand up against that demand if Joan had urged it. But no, she was ignorant, poor thing, and did not know what a blow she had struck for life and liberty.

France was not the Church. Rome had no interest in the destruction of this messenger of God. Rome would have given her a fair trial, and that was all that her cause needed.

From that trial she would have gone forth free and honored and blest.

But it was not so fated. Cauchon at once diverted the questions to other matters and hurried the trial quickly to an end.

As Joan moved feebly away, dragging her chains, I felt stunned and dazed, and kept saying to myself, " Such a little while ago she said the saving word and could have gone free; and now, there she goes to her death; yes, it is to her death, I know it, I feel it. They will double the guards; they will never let any come near her now between this and her condemnation, lest she get a hint and speak that word again. This is the bitterest day that has come to me in all this miserable time."

CHAPTER XIII

So the second trial in the prison was over. Over, and no definite result. The character of it I have described to you. It was baser in one particular than the previous one; for this time the charges had not been communicated to Joan, therefore she had been obliged to fight in the dark. There was no opportunity to do any thinking beforehand; there was no foreseeing what traps might be set, and no way to prepare for them. Truly it was a shabby advantage to take of a girl situated as this one was. One day, during the course of it, an able lawyer of Normandy, Maître Lohier, happened to be in Rouen, and I will give you his opinion of that trial, so that you may see that I have been honest with you, and that my partisanship has not made me deceive you as to its unfair and illegal character. Cauchon showed Lohier the *proces* and asked his opinion about the trial. Now this was the opinion which he gave to Cauchon. He said that the whole thing was null and void; for these reasons: 1, because the trial was secret, and full freedom of speech and action on the part of those present not possible; 2, because the trial touched the honor of the King of France, yet he was not summoned to defend himself, nor any one appointed to represent him; 3, because the charges against the prisoner were not communicated to her; 4, because the accused, although young and simple, had been forced to defend her cause without help of counsel, notwithstanding she had so much at stake.

Did that please Bishop Cauchon? It did not. He burst out upon Lohier with the most savage cursings, and swore he would have him drowned. Lohier escaped from Rouen and got out of France with all speed, and so saved his life.

Well, as I have said, the second trial was over, without definite result. But Cauchon did not give up. He could trump up another. And still another and another, if necessary. He had the half-promise of an enormous prize—the Archbishopric of Rouen—if he should succeed in burning the body and damning to hell the soul of this young girl who had never done him any harm; and such a prize as that, to a man like the Bishop of Beauvais, was worth the burning and damning of fifty harmless girls, let alone one.

So he set to work again straight off, next day; and with high confidence, too, intimating with brutal cheerfulness that he should succeed this time. It took him and the other scavengers nine days to dig matter enough out of Joan's testimony and their own inventions to build up the new mass of charges. And it was a formidable mass indeed, for it numbered sixty-six articles!

This huge document was carried to the castle the next day, March 27th; and there, before a dozen carefully selected judges, the new trial was begun.

Opinions were taken, and the tribunal decided that Joan should hear the articles read, this time. Maybe that was on account of Lohier's remark upon that head; or maybe it was hoped that the reading would kill the prisoner with fatigue—for, as it turned out, this reading occupied several days. It was also decided that Joan should be required to answer squarely to every article, and that if she refused she should be considered *convicted*. You see, Cauchon was managing to narrow her chances more and more all the time; he was drawing the toils closer and closer.

Joan was brought in, and the Bishop of Beauvais opened with a speech to her which ought to have made even himself blush, so laden it was with hypocrisy and lies. He said that this court was composed of holy and pious churchmen whose hearts were full of benevolence and compassion toward her, and that they had no wish to hurt her body, but only a desire to instruct her and lead her into the way of truth and salvation.

Why, this man was born a devil; now think of his describing himself and those hardened slaves of his in such language as that.

And yet, worse was to come. For now, having in mind another of Lohier's hints, he had the cold effrontery to make to Joan a proposition which I think will surprise you when you hear it. He said that this court, recognizing her untaught estate and her inability to deal with the complex and difficult matters which were about to be considered, had determined, out of their pity and their mercifulness, to allow her to choose one or more persons *out of their own number* to help her with counsel and advice!

Think of that — a court made up of Loyseleur and his breed of reptiles. It was granting leave to a lamb to ask help of a wolf. Joan looked up to see if he was serious, and perceiving that he was at least pretending to be, she declined, of course.

The Bishop was not expecting any other reply. He had made a show of fairness and could have it entered on the minutes, therefore he was satisfied.

Then he commanded Joan to answer straightly to every accusation; and threatened to cut her off from the Church if she failed to do that or delayed her answers beyond a given length of time. Yes, he was narrowing her chances down, step by step.

Thomas de Courcelles began the reading of that interminable document, article by article. Joan answered to each article in its turn; sometimes merely denying its truth, sometimes by saying her answer would be found in the records of the previous trials.

What a strange document that was, and what an exhibition and exposure of the heart of man, the one creature authorized to boast that he is made in the image of God. To know Joan of Arc was to know one who was wholly noble, pure, truthful, brave, compassionate, generous, pious, unselfish, modest, blameless as the very flowers in the fields—a nature fine and beautiful, a character supremely great. To know her from

that document would be to know her as the exact reverse
of all that. Nothing that she *was* appears in it, everything
that she was *not* appears there in detail.

Consider some of the things it charges against her, and remember who it is it is speaking of. It calls her a sorceress,
a false prophet, an invoker and companion of evil spirits, a
dealer in magic, a person ignorant of the Catholic faith, a
schismatic; she is sacrilegious, an idolator, an apostate, a
blasphemer of God and his saints, scandalous, seditious, a dis-
turber of the peace; she incites men to war, and to the spill-
ing of human blood; she discards the decencies and proprieties of her sex, irreverently assuming the dress of a man and
the vocation of a soldier; she beguiles both princes and people; she usurps divine honors, and has caused herself to be
adored and venerated, offering her hands and her vestments
to be kissed.

There it is—every fact of her life distorted, perverted, reversed. As a child she had loved the fairies, she had spoken
a pitying word for them when they were banished from their
home, she had played under their tree and around their fountain—hence she was a comrade of evil spirits. She had lifted
France out of the mud and moved her to strike for freedom,
and led her to victory after victory—hence she was a dis-
turber of the peace—as indeed she was, and a provoker of
war—as indeed she was again! and France will be proud of
it and grateful for it for many a century to come. And she
had been adored—as if she could help that, poor thing, or was
in any way to blame for it. The cowed veteran and the wavering recruit had drunk the spirit of war from her eyes and
touched her sword with theirs and moved forward invincible
—hence she was a sorceress.

And so the document went on, detail by detail, turning
these waters of life to poison, this gold to dross, these proofs
of a noble and beautiful life to evidences of a foul and odious
one.

Of course the sixty-six articles were just a rehash of the
things which had come up in the course of the previous trials,

so I will touch upon this new trial but lightly. In fact Joan went but little into detail herself, usually merely saying "That is not true—*passez outre*"; or, "I have answered that before —let the clerk read it in his record"; or saying some other brief thing.

She refused to have her mission examined and tried by the earthly Church. The refusal was taken note of.

She denied the accusation of idolatry and that she had sought men's homage. She said—

"If any kissed my hands and my vestments it was not by my desire, and I did what I could to prevent it."

She had the pluck to say to that deadly tribunal that she did not know the fairies to be evil beings. She knew it was a perilous thing to say, but it was not in her nature to speak anything but the truth when she spoke at all. Danger had no weight with her in such things. Note was taken of her remark.

She refused, as always before, when asked if she would put off the male attire if she were given permission to commune. And she added this:

"When one receives the sacrament, the manner of his dress is a small thing and of no value in the eyes of Our Lord."

She was charged with being so stubborn in clinging to her male dress that she would not lay it off even to get the blessed privilege of hearing mass. She spoke out with spirit and said:

"I would rather die than be untrue to my oath to God."

She was reproached with doing man's work in the wars and thus deserting the industries proper to her sex. She answered, with some little touch of soldierly disdain—

"As to the matter of women's work, there's plenty to do it."

It was always a comfort to me to see the soldier-spirit crop up in her. While that remained in her she would be Joan of Arc, and able to look trouble and fate in the face.

"It appears that this mission of yours which you claim

you had from God, was to make war and pour out human blood."

Joan replied quite simply, contenting herself with explaining that war was not her first move, but her second:

"To begin with, I demanded that peace should be made. If it was refused, then I would fight."

The judge mixed the Burgundians and English together in speaking of the enemy which Joan had come to make war upon. But she showed that she made a distinction between them by act and word, the Burgundians being Frenchmen and therefore entitled to less brusque treatment than the English. She said:

"As to the Duke of Burgundy, I required of him, both by letters and by his ambassadors, that he make peace with the King. As to the English, the only peace for them was that they leave the country and go home."

Then she said that even with the English she had shown a pacific disposition, since she had warned them away by proclamation before attacking them.

"If they had listened to me," said she, "they would have done wisely." At this point she uttered her prophecy again, saying with emphasis, "Before seven years they will see it themselves."

Then they presently began to pester her again about her male costume, and tried to persuade her to voluntarily promise to discard it. I was never deep, so I think it no wonder that I was puzzled by their persistency in what seemed a thing of no consequence, and could not make out what their reason could be. But we all know, now. We all know now that it was another of their treacherous projects. Yes, if they could but succeed in getting her to formally discard it they could play a game upon her which would quickly destroy her. So they kept at their evil work until at last she broke out and said—

"Peace! Without the permission of God I will not lay it off though you cut off my head!"

At one point she corrected the *proces verbal*, saying—

"It makes me say that everything which I have done was done by the counsel of Our Lord. I did not say that. I said 'all which I have *well* done.'"

Doubt was cast upon the authenticity of her mission because of the ignorance and simplicity of the messenger chosen. Joan smiled at that. She could have reminded these people that Our Lord, who is no respecter of persons, had chosen the lowly for his high purposes even oftener than he had chosen bishops and cardinals; but she phrased her rebuke in simpler terms:

"It is the prerogative of Our Lord to choose His instruments where He will."

She was asked what form of prayer she used in invoking counsel from on high. She said the form was brief and simple; then she lifted her pallid face and repeated it, clasping her chained hands:

"Most dear God, in honor of your holy passion I beseech you, if you love me, that you will reveal to me what I am to answer to these churchmen. As concerns my dress, I know by what command I have put it on, but I know not in what manner I am to lay it off. I pray you tell me what to do."

She was charged with having dared, against the precepts of God and His saints, to assume empire over men and make herself Commander-in-Chief. That touched the soldier in her. She had a deep reverence for priests, but the soldier in her had but small reverence for a priest's opinions about war; so, in her answer to this charge she did not condescend to go into any explanations or excuses, but delivered herself with bland indifference and military brevity.

"If I was Commander-in-Chief, it was to thrash the English!"

Death was staring her in the face here, all the time, but no matter: she dearly loved to make these English-hearted Frenchmen squirm, and whenever they gave her an opening she was prompt to jab her sting into it. She got great refreshment out of these little episodes. Her days were a desert; these were the oases in it.

THE TRIAL OF JOAN OF ARC

Her being in the wars with men was charged against her as an indelicacy. She said—

"I had a woman with me when I could—in towns and lodgings. In the field I always slept in my armor."

That she and her family had been ennobled by the King was charged against her as evidence that the source of her deeds were sordid self-seeking. She answered that she had not asked this grace of the King, it was his own act.

This third trial was ended at last. And once again there was no definite result.

Possibly a fourth trial might succeed in defeating this apparently unconquerable girl. So the malignant Bishop set himself to work to plan it.

He appointed a commission to reduce the substance of the sixty-six articles to twelve compact lies, as a basis for the new attempt. This was done. It took several days.

Meantime Cauchon went to Joan's cell one day, with Manchon and two of the judges, Isambard de la Pierre and Martin Ladvenue, to see if he could not manage somehow to beguile Joan into submitting her mission to the examination and decision of the church militant—that is to say, to that part of the church militant which was represented by himself and his creatures.

Joan once more positively refused. Isambard de la Pierre had a heart in his body, and he so pitied this persecuted poor girl that he ventured to do a very daring thing; for he asked her if she would be willing to have her case go before the Council of Basel, and said it contained as many priests of her party as of the English party.

Joan cried out that she would gladly go before so fairly constructed a tribunal as that; but before Isambard could say another word, Cauchon turned savagely upon him and exclaimed—

"Shut up, in the devil's name!"

Then Manchon ventured to do a brave thing, too, though he did it in great fear for his life. He asked Cauchon if he should enter Joan's submission to the Council of Basel upon the minutes.

"No! It is not necessary."

"Ah," said poor Joan, reproachfully, "you set down everything that is against me, but you will not set down what is for me."

It was piteous. It would have touched the heart of a brute. But Cauchon was more than that.

CHAPTER XIV

WE were now in the first days of April. Joan was ill. She had fallen ill the 29th of March, the day after the close of the third trial, and was growing worse when the scene which I have just described occurred in her cell. It was just like Cauchon to go there and try to get some advantage out of her weakened state.

Let us note some of the particulars in the new indictment—the Twelve Lies.

Part of the first one says Joan asserts that she has found her salvation. She never said anything of the kind. It also says she refuses to submit herself to the Church. Not true. She was willing to submit all her acts to this Rouen tribunal except those done by command of God in fulfilment of her mission. Those she reserved for the judgment of God. She refused to recognize Cauchon and his serfs as the Church, but was willing to go before the Pope or the Council of Basel.

A clause of another of the Twelve says she admits having threatened with death those who would not obey her. Distinctly false. Another clause says she declares that all she has done has been done by command of God. What she really said was, all that she had done *well*—a correction made by herself as you have already seen.

Another of the Twelve says she claims that she has never committed any sin. She never made any such claim.

Another makes the wearing of the male dress a sin. If it was, she had high Catholic authority for committing it—that of the Archbishop of Rheims and the tribunal of Poitiers.

The Tenth Article was resentful against her for "pretending" that St. Catherine and St. Marguerite spoke French and not English, and were French in their politics.

The Twelve were to be submitted first to the learned doctors of theology of the University of Paris for approval. They were copied out and ready by the night of April 4th. Then Manchon did another bold thing: he wrote in the margin that many of the Twelve put statements in Joan's mouth which were the exact opposite of what she had said. *That* fact would not be considered important by the University of Paris, and would not influence its decision or stir its humanity, in case it had any—which it hadn't when acting in a political capacity, as at present—but it was a brave thing for that good Manchon to do, all the same.

The Twelve were sent to Paris next day, April 5th. That afternoon there was a great tumult in Rouen, and excited crowds were flocking through all the chief streets, chattering and seeking for news; for a report had gone abroad that Joan of Arc was sick unto death. In truth these long seances had worn her out, and she was ill indeed. The heads of the English party were in a state of consternation; for if Joan should die uncondemned by the Church and go to the grave unsmirched, the pity and the love of the people would turn her wrongs and sufferings and death into a holy martyrdom, and she would be even a mightier power in France dead, than she had been when alive.

The Earl of Warwick and the English Cardinal (Winchester) hurried to the castle and sent messengers flying for physicians. Warwick was a hard man, a rude coarse man, a man without compassion. There lay the sick girl stretched in her chains in her iron cage—not an object to move man to ungentle speech, one would think; yet Warwick spoke right out in her hearing and said to the physicians—

"Mind you take good care of her. The King of England has no mind to have her die a natural death. She is dear to him, for he bought her dear, and he does not want her to die, save at the stake. Now then, mind you cure her."

The doctors asked Joan what had made her ill. She said

the Bishop of Beauvais had sent her a fish and she thought it was that.

Then Jean d'Estivet burst out on her, and called her names and abused her. He understood Joan to be charging the Bishop with poisoning her, you see; and that was not pleasing to him, for he was one of Cauchon's most loving and conscienceless slaves, and it outraged him to have Joan injure his master in the eyes of these great English chiefs, these being men who could ruin Cauchon and would promptly do it if they got the conviction that he was capable of saving Joan from the stake by poisoning her and thus cheating the English out of all the real value gainable by her purchase from the Duke of Burgundy.

Joan had a high fever, and the doctors proposed to bleed her. Warwick said—

"Be careful about that; she is smart and is capable of killing herself."

He meant that to escape the stake she might undo the bandage and let herself bleed to death.

But the doctors bled her anyway, and then she was better.

Not for long, though. Jean d'Estivet could not hold still, he was so worried and angry about the suspicion of poisoning which Joan had hinted at; so he came back in the evening and stormed at her till he brought the fever all back again.

When Warwick heard of this he was in a fine temper, you may be sure, for here was his prey threatening to escape again, and all through the over-zeal of this meddling fool. Warwick gave D'Estivet a quite admirable cursing—admirable as to strength, I mean, for it was said by persons of culture that the art of it was not good—and after that the meddler kept still.

Joan remained ill more than two weeks; then she grew better. She was still very weak, but she could bear a little persecution now without much danger to her life. It seemed to Cauchon a good time to furnish it. So he called together some of his doctors of theology and went to her dungeon. Manchon and I went along to keep the record—that

26

is, to set down what might be useful to Cauchon, and leave out the rest.

The sight of Joan gave me a shock. Why, she was but a shadow! It was difficult for me to realize that this frail little creature with the sad face and drooping form was the same Joan of Arc that I had so often seen, all fire and enthusiasm, charging through a hail of death and the lightning and thunder of the guns at the head of her battalions. It wrung my heart to see her looking like this.

But Cauchon was not touched. He made another of those conscienceless speeches of his, all dripping with hypocrisy and guile. He told Joan that among her answers had been some which had seemed to endanger religion; and as she was ignorant and without knowledge of the Scriptures, he had brought some good and wise men to instruct her, if she desired it. Said he, " We are churchmen, and disposed by our good will as well as by our vocation to procure for you the salvation of your soul and your body, in every way in our power, just as we would do the like for our nearest kin or for ourselves. In this we but follow the example of Holy Church, who never closes the refuge of her bosom against any that are willing to return."

Joan thanked him for these sayings and said :

" I seem to be in danger of death from this malady; if it be the pleasure of God that I die here, I beg that I may be heard in confession and also receive my Saviour; and that I may be buried in consecrated ground."

Cauchon thought he saw his opportunity at last; this weakened body had the fear of an unblessed death before it and the pains of hell to follow. This stubborn spirit would surrender now. So he spoke out and said—

" Then if you want the Sacraments, you must do as all good Catholics do, and submit to the Church."

He was eager for her answer; but when it came there was no surrender in it, she still stood to her guns. She turned her head away and said wearily—

" I have nothing more to say."

Cauchon's temper was stirred, and he raised his voice threat-
eningly and said that the more she was in danger of death the
more she ought to amend her life; and again he refused the
things she begged for unless she would submit to the Church.
Joan said—

"If I die in this prison I beg you to have me buried in
holy ground; if you will not, I cast myself upon my Saviour."

There was some more conversation of the like sort, then
Cauchon demanded again, and imperiously, that she submit
herself and all her deeds to the Church. His threatening and
storming went for nothing. That body was weak, but the
spirit in it was the spirit of Joan of Arc; and out of that came
the steadfast answer which these people were already so familiar with and detested so sincerely—

"Let come what may, I will neither do nor say any otherwise than I have said already in your tribunals."

Then the good theologians took turn about and worried her
with reasonings and arguments and Scriptures; and always
they held the lure of the Sacraments before her famishing soul,
and tried to bribe her with them to surrender her mission to
the Church's judgment—that is to *their* judgment—as if *they*
were the Church! But it availed nothing. I could have told
them that beforehand, if they had asked me. But they never
asked me anything; I was too humble a creature for their notice.

Then the interview closed with a threat; a threat of fearful import; a threat calculated to make a Catholic Christian
feel as if the ground were sinking from under him—

"The Church calls upon you to submit; disobey, and she
will abandon you as if you were a pagan!"

Think of being abandoned by the Church!—that august
Power in whose hands is lodged the fate of the human race;
whose sceptre stretches beyond the furthest constellation that
twinkles in the sky; whose authority is over the millions that
live and over the billions that wait trembling in purgatory for
ransom or doom; whose smile opens the gates of Heaven to
you, whose frown delivers you to the fires of everlasting hell;

a Power whose dominion overshadows and belittles earthly empire as earthly empire overshadows and belittles the pomps and shows of a village. To be abandoned by one's King— yes, that is death, and death is much; but to be abandoned by Rome, to be abandoned by the Church! Ah, death is nothing to that, for that is consignment to endless *life*—and such a life!

I could see the red waves tossing in that shoreless lake of fire, I could see the black myriads of the damned rise out of them and struggle and sink and rise again; and I knew that Joan was seeing what I saw, while she paused musing; and I believed that she must yield now, and in truth I hoped she would, for these men were able to make the threat good and deliver her over to eternal suffering, and I knew that it was in their natures to do it.

But I was foolish to think that thought and hope that hope. Joan of Arc was not made as others are made. Fidelity to principle, fidelity to truth, fidelity to her word, all these were in her bone and in her flesh—they were parts of her. She could not change, she could not cast them out. She was the very genius of Fidelity, she was Steadfastness incarnated. Where she had taken her stand and planted her foot, there she would abide; hell itself could not move her from that place.

Her Voices had not given her permission to make the sort of submission that was required, therefore she would stand fast. She would wait, in perfect obedience, let come what might.

My heart was like lead in my body when I went out from that dungeon; but she—she was serene, she was not troubled. She had done what she believed to be her duty, and that was sufficient; the consequences were not her affair. The last thing she said, that time, was full of this serenity, full of contented repose—

"I am a good Christian born and baptized, and a good Christian I will die."

CHAPTER XV

Two weeks went by; the second of May was come, the chill was departed out of the air, the wild flowers were springing in the glades and glens, the birds were piping in the woods, all nature was brilliant with sunshine, all spirits were renewed and refreshed, all hearts glad, the world was alive with hope and cheer, the plain beyond the Seine stretched away soft and rich and green, the river was limpid and lovely, the leafy islands were dainty to see, and flung still daintier reflections of themselves upon the shining water; and from the tall bluffs above the bridge Rouen was become again a delight to the eye, the most exquisite and satisfying picture of a town that nestles under the arch of heaven anywhere.

When I say that all hearts were glad and hopeful, I mean it in a general sense. There were exceptions—we who were the friends of Joan of Arc, also Joan of Arc herself, that poor girl shut up there in that frowning stretch of mighty walls and towers: brooding in darkness, so close to the flooding downpour of sunshine yet so impossibly far away from it; so longing for any little glimpse of it, yet so implacably denied it by those wolves in the black gowns who were plotting her death and the blackening of her good name.

Cauchon was ready to go on with his miserable work. He had a new scheme to try, now. He would see what persuasion could do—argument, eloquence, poured out upon the incorrigible captive from the mouth of a trained expert. That was his plan. But the reading of the Twelve Articles to her was not a part of it. No, even Cauchon was ashamed to lay that monstrosity before her; even he had a remnant of shame in him, away down deep, a million fathoms deep, and that remnant asserted itself now and prevailed.

On this fair second of May, then, the black company gathered itself together in the spacious chamber at the end of the great hall of the castle—the Bishop of Beauvais on his throne, and sixty-two minor judges massed before him, with the guards and recorders at their stations and the orator at his desk.

Then we heard the far clank of chains, and presently Joan entered with her keepers and took her seat upon her isolated bench. She was looking well, now, and most fair and beautiful after her fortnight's rest from wordy persecution.

She glanced about and noted the orator. Doubtless she divined the situation.

The orator had written his speech all out, and had it in his hand, though he held it back of him out of sight. It was so thick that it resembled a book. He began flowingly, but in the midst of a flowery period his memory failed him and he had to snatch a furtive glance at his manuscript—which much injured the effect. Again this happened, and then a third time. The poor man's face was red with embarrassment, the whole great house was pitying him, which made the matter worse; then Joan dropped in a remark which completed his trouble. She said:

"*Read your book*—and then I will answer you!"

Why, it was almost cruel the way those mouldy veterans laughed; and as for the orator, he looked so flustered and helpless that almost anybody would have pitied him, and I had difficulty to keep from doing it myself. Yes, Joan was feeling very well after her rest, and the native mischief that was in her lay near the surface. It did not show, when she made the remark, but I knew it was close in there back of the words.

When the orator had gotten back his composure he did a wise thing; for he followed Joan's advice: he made no more attempts at sham impromptu oratory, but read his speech straight from his "book." In the speech he compressed the Twelve Articles into six and made these his text.

Every now and then he stopped and asked questions, and Joan replied. The nature of the church militant was explained, and once more Joan was asked to submit herself to it.

She gave her usual answer.

Then she was asked—

"Do you believe the Church can err?"

"I believe it cannot err; but for those deeds and words of mine which were done and uttered by command of God, I will answer to him alone."

"Will you say that you have no judge upon earth? Is not our Holy Father the Pope your judge?"

"I will say nothing to you about it. I have a good Master who is our Lord and to Him I will submit all."

Then came these terrible words:

"If you do not submit to the Church you will be pronounced a heretic by these judges here present and burned at the stake!"

Ah, that would have smitten you or me dead with fright, but it only roused the lion heart of Joan of Arc, and in her answer rang that martial note which had used to stir her soldiers like a bugle-call—

"I will not say otherwise than I have said already; and if I saw the fire before me I would say it again!"

It was uplifting to hear her battle-voice once more and see the battle-light burn in her eye. Many there were stirred; every man that was a *man* was stirred, whether friend or foe; and Manchon risked his life again, good soul, for he wrote in the margin of the record in good plain letters these brave words: "*Superba responsio!*" and there they have remained these sixty years, and there you may read them to this day.

"*Superba responsio!*" Yes, it was just that. For this "superb answer" came from the lips of a girl of nineteen with death and hell staring her in the face.

Of course the matter of the male attire was gone over again; and as usual at wearisome length; also, as usual, the customary bribe was offered: if she would discard that dress voluntarily they would let her hear mass. But she answered as she had often answered before—

"I will go in a woman's robe to all services of the church if I may be permitted, but I will resume the other dress when I return to my cell."

They set several traps for her in a tentative form ; that is to say, they placed supposititious propositions before her and cunningly tried to commit her to one end of the propositions without committing themselves to the other. But she always saw the game and spoiled it. The trap was in this form—

"Would you be willing to do so and so if we should give you leave?"

Her answer was always in this form or to this effect:

"When you give me leave, then you will know."

Yes, Joan was at her best, that second of May. She had all her wits about her, and they could not catch her anywhere. It was a long, long session, and all the old ground was fought over again, foot by foot, and the orator-expert worked all his persuasions, all his eloquence ; but the result was the familiar one—a drawn battle, the sixty-two retiring upon their base, the solitary enemy holding her original position within her original lines.

CHAPTER XVI

THE brilliant weather, the heavenly weather, the bewitching weather made everybody's heart to sing, as I have told you; yes, Rouen was feeling light-hearted and gay, and most willing and ready to break out and laugh upon the least occasion; and so when the news went around that the young girl in the tower had scored another defeat against Bishop Cauchon there was abundant laughter—abundant laughter among the citizens of both parties, for they all hated the Bishop. It is true, the English-hearted majority of the people wanted Joan burned, but that did not keep them from laughing at the man they hated. It would have been perilous for anybody to laugh at the English chiefs or at the majority of Cauchon's assistant judges, but to laugh at Cauchon or D'Estivet and Loyseleur was safe—nobody would report it.

The difference between Cauchon and *cochon** was not noticeable in speech, and so there was plenty of opportunity for puns: the opportunities were not thrown away.

Some of the jokes got well worn in the course of two or three months, from repeated use; for every time Cauchon started a new trial the folk said "The sow has littered† again"; and every time the trial failed they said it over again, with its other meaning, "The hog has made a mess of it."

And so, on the third of May, Noël and I, drifting about the town, heard many a wide-mouthed lout let go his joke and his laugh, and then move to the next group, proud of his wit and happy, to work it off again—

---

* Hog, pig.
† *Cochonner*, to litter, to farrow; also, "to make a mess of!"

"'Ods blood, the sow has littered five times, and five times has made a mess of it!"

And now and then one was bold enough to say—but he said it softly—

"Sixty-three and the might of England against a girl, and she camps on the field five times!"

Cauchon lived in the great palace of the Archbishop, and it was guarded by English soldiery; but no matter, there was never a dark night but the walls showed, next morning, that the rude joker had been there with his paint and brush. Yes, he had been there, and had smeared the sacred walls with pictures of hogs in all attitudes except flattering ones; hogs clothed in a Bishop's vestments and wearing a Bishop's mitre irreverently cocked on the side of their heads.

Cauchon raged and cursed over his defeats and his impotence during seven days, then he conceived a new scheme. You shall see what it was; for you have not cruel hearts, and you would never guess it.

On the ninth of May there was a summons, and Manchon and I got our materials together and started. But this time we were to go to one of the other towers—not the one which was Joan's prison. It was round and grim and massive, and built of the plainest and thickest and solidest masonry—a dismal and forbidding structure.*

We entered the circular room on the ground floor, and I saw what turned me sick—the instruments of torture and the executioners standing ready! Here you have the black heart of Cauchon at the blackest, here you have the proof that in his nature there was no such thing as pity. One wonders if he ever knew his mother or ever had a sister.

Cauchon was there, and the Vice-Inquisitor and the Abbot of St. Corneille; also six others, among them that false Loyseleur. The guards were in their places, the rack was there, and by it stood the executioner and his aids in their crimson

---

* The lower half of it remains to-day just as it was then; the upper half is of a later date.—TRANSLATOR.

EXECUTION OF JOAN OF ARC
(From the mural painting by J. E. Lenepveu in the Panthéon at Paris)

hose and doubtlets, meet color for their bloody trade. The picture of Joan rose before me stretched upon the rack, her feet tied to one end of it, her wrists to the other, and those red giants turning the windlass and pulling her limbs out of their sockets. It seemed to me that I could hear the bones snap and the flesh tear apart, and I did not see how that body of anointed servants of the merciful Jesus could sit there and look so placid and indifferent.

After a little, Joan arrived and was brought in. She saw the rack, she saw its attendants, and the same picture which I had been seeing must have risen in her mind; but do you think she quailed, do you think she shuddered? No, there was no sign of that sort. She straightened herself up, and there was a slight curl of scorn about her lip; but as for fear, she showed not a vestige of it.

This was a memorable session, but it was the shortest one of all the list. When Joan had taken her seat a résumé of her "crimes" was read to her. Then Cauchon made a solemn speech. In it he said that in the course of her several trials Joan had refused to answer some of the questions and had answered others with lies, but that now he was going to have the truth out of her, and the whole of it.

His manner was full of confidence this time; he was sure he had found a way at last to break this child's stubborn spirit and make her beg and cry. He would score a victory this time and stop the mouths of the jokers of Rouen. You see, he was only just a man, after all, and couldn't stand ridicule any better than other people. He talked high, and his splotchy face lighted itself up with all the shifting tints and signs of evil pleasure and promised triumph—purple, yellow, red, green—they were all there, with sometimes the dull and spongy blue of a drowned man, the uncanniest of them all. And finally he burst out in a great passion and said—

"There is the rack, and there are its ministers! You will reveal all, now, or be put to the torture. Speak."

Then she made that great answer, which will live forever;

made it without fuss or bravado, and yet how fine and noble was the sound of it—

"I will tell you nothing more than I have told you; no, not even if you tear the limbs from my body. And even if in my pain I *did* say something otherwise, I would always say afterwards that it was the torture that spoke and not I."

There was no crushing that spirit. You should have seen Cauchon. Defeated again, and he had not dreamed of such a thing. I heard it said next day, around the town, that he had a full confession, all written out, in his pocket and all ready for Joan to sign. I do not know that that was true, but it probably was, for her mark signed at the bottom of a confession would be the kind of evidence (for effect with the public) which Cauchon and his people would particularly value, you know.

No, there was no crushing that spirit, and no beclouding that clear mind. Consider the depth, the wisdom of that answer, coming from an ignorant girl. Why, there were not six men in the world who had ever reflected that words forced out of a person by horrible tortures were not necessarily words of verity and truth, yet this unlettered peasant girl put her finger upon that flaw with an unerring instinct. I had always supposed that torture brought out the truth—everybody supposed it; and when Joan came out with those simple common-sense words they seemed to flood the place with light. It was like a lightning-flash at midnight which suddenly reveals a fair valley sprinkled over with silver streams and gleaming villages and farmsteads where was only an impenetrable world of darkness before. Manchon stole a sidewise look at me, and his face was full of surprise; and there was the like to be seen in other faces there. Consider—they were old, and deeply cultured, yet here was a village maid able to teach them something which they had not known before. I heard one of them mutter—

"Verily it is a wonderful creature. She has laid her hand upon an accepted truth that is as old as the world, and it has crumbled to dust and rubbish under her touch. Now whence got she that marvellous insight?"

The judges laid their heads together and began to talk low. It was plain, from chance words which one caught now and then, that Cauchon and Loyseleur were insisting upon the application of the torture, and that most of the others were urgently objecting.

Finally Cauchon broke out with a good deal of asperity in his voice and ordered Joan back to her dungeon. That was a happy surprise for me. I was not expecting that the Bishop would yield.

When Manchon came home that night he said he had found out why the torture was not applied. There were two reasons. One was, a fear that Joan might die under the torture, which would not suit the English at all; the other was, that the torture would effect nothing if Joan was going to take back everything she said under its pains; and as to putting her mark to a confession, it was believed that not even the rack could ever make her do that.

So all Rouen laughed again, and kept it up for three days, saying—

"The sow has littered six times, and made six messes of it."

And the palace walls got a new decoration—a mitred hog carrying a discarded rack home on its shoulder, and Loyseleur weeping in its wake. Many rewards were offered for the capture of these painters, but nobody applied. Even the English guard feigned blindness and would not see the artists at work.

The Bishop's anger was very high now. He could not reconcile himself to the idea of giving up the torture. It was the pleasantest idea he had invented yet, and he would not cast it by. So he called in some of his satellites on the twelfth, and urged the torture again. But it was a failure. With some, Joan's speech had wrought an effect; others feared she might die under the torture; others did not believe that any amount of suffering could make her put her mark to a lying confession. There were fourteen men present, including the Bishop. Eleven of them voted dead against the torture, and stood their ground in spite of Cauchon's abuse. Two voted with the Bishop and insisted upon the torture. These two were Loyseleur and the orator — the man whom Joan had bidden to " read his book "—Thomas de Courcelles, the renowned pleader, and master of eloquence.

Age has taught me charity of speech ; but it fails me when I think of those three names—Cauchon, Courcelles, Loyseleur.

CHAPTER XVII

ANOTHER ten days' wait. The great theologians of that treasury of all valuable knowledge and all wisdom, the University of Paris, were still weighing and considering and discussing the Twelve Lies.

I had but little to do, these ten days, so I spent them mainly in walks about the town with Noël. But there was no pleasure in them, our spirits being so burdened with cares, and the outlook for Joan growing so steadily darker and darker all the time. And then we naturally contrasted our circumstances with hers: this freedom and sunshine, with her darkness and chains; our comradeship, with her lonely estate; our alleviations of one sort and another, with her destitution in all. She was used to liberty, but now she had none; she was an out-of-door creature by nature and habit, but now she was shut up day and night in a steel cage like an animal; she was used to the light, but now she was always in a gloom where all objects about her were dim and spectral; she was used to the thousand various sounds which are the cheer and music of a busy life, but now she heard only the monotonous footfall of the sentry pacing his watch; she had been fond of talking with her mates, but now there was no one to talk to; she had had an easy laugh, but it was gone dumb, now; she had been born for comradeship, and blithe and busy work, and all manner of joyous activities, but here were only dreariness, and leaden hours, and weary inaction, and brooding stillness, and thoughts that travel day and night and night and day round and round in the same circle, and wear the brain and break the heart with weariness.

It was death in life; yes, death in life, that is what it must have been. And there was another hard thing about it all. A young girl in trouble needs the soothing solace and support and sympathy of persons of her own sex, and the delicate offices and gentle ministries which only these can furnish; yet in all these months of gloomy captivity in her dungeon Joan never saw the face of a girl or a woman. Think how her heart would have leaped to see such a face.

Consider. If you would realize how great Joan of Arc was, remember that it was out of such a place and such circumstances that she came week after week and month after month and confronted the master intellects of France singlehanded, and baffled their cunningest schemes, defeated their ablest plans, detected and avoided their secretest traps and pitfalls, broke their lines, repelled their assaults, and camped on the field after every engagement; steadfast always, true to her faith and her ideals; defying torture, defying the stake, and answering threats of eternal death and the pains of hell with a simple "Let come what may, here I take my stand and will abide."

Yes, if you would realize how great was the soul, how profound the wisdom, and how luminous the intellect of Joan of Arc, you must study her there, where she fought out that long fight all alone — and not merely against the subtlest brains and deepest learning of France, but against the ignoblest deceits, the meanest treacheries, and the hardest hearts to be found in any land, pagan or Christian.

She was great in battle—we all know that; great in foresight; great in loyalty and patriotism; great in persuading discontented chiefs and reconciling conflicting interests and passions, great in the ability to discover merit and genius wherever it lay hidden; great in picturesque and eloquent speech; supremely great in the gift of firing the hearts of hopeless men with noble enthusiasms, the gift of turning hares into heroes, slaves and skulkers into battalions that march to death with songs upon their lips. But all these are exalting activities; they keep hand and heart and brain keyed up to their work : there is the joy of achievement, the inspiration of stir and movement, the applause which hails success; the soul is overflowing with life and energy, the faculties are at white heat; weariness, despondency, inertia— these do not exist.

Yes, Joan of Arc was great always, great everywhere, but she was greatest in the Rouen trials. There she rose above the limitations and infirmities of our human nature, and accomplished under blighting and unnerving and hopeless conditions all that her splendid equipment of moral and intellectual forces could have accomplished if they had been supplemented by the mighty helps of hope and cheer and light, the presence of friendly faces, and a fair and equal fight, with the great world looking on and wondering.

CHAPTER XVIII

TOWARD the end of the ten-day interval the University of Paris rendered its decision concerning the Twelve Articles. By this finding, Joan was guilty upon all the counts: she must renounce her errors and make satisfaction, or be abandoned to the secular arm for punishment.

The University's mind was probably already made up before the Articles were laid before it; yet it took it from the fifth to the eighteenth to produce its verdict. I think the delay may have been caused by temporary difficulties concerning two points:

1, As to who the fiends were who were represented in Joan's Voices;

2, As to whether her saints spoke French only.

You understand, the University decided emphatically that it was fiends who spoke in those Voices; it would need to prove that, and it did. It found out who the fiends were, and named them in the verdict: Belial, Satan, and Behemoth. This has always seemed a doubtful thing to me, and not entitled to much credit. I think so for this reason: if the University had actually known it was those three, it would for very consistency's sake have told *how* it knew it, and not stopped with the mere assertion, since it had made Joan explain how she knew they were *not* fiends. Does not that seem reasonable? To my mind the University's position was weak, and I will tell you why. It had claimed that Joan's angels were devils in disguise, and we all know that devils do disguise themselves as angels; up to that point the University's position was strong; but you see yourself that it eats its own argument when it turns around and pretends that *it* can tell who such apparitions are, while denying the like ability to a person with as good a head on her shoulders as the best one the University could produce.

The doctors of the University had to see those creatures in order to know; and if Joan was deceived, it is argument that they in their turn could also be deceived, for their insight and judgment were surely not clearer than hers.

As to the other point which I have thought may have proved a difficulty and cost the University delay, I will touch but a moment upon that, and pass on. The University decided that it was blasphemy for Joan to say that her saints spoke French and not English, and were on the French side in political sympathies. I think that the thing which troubled the doctors of theology was this: they had decided that the three Voices were Satan and two other devils; but they had also decided that these Voices were *not* on the French side—thereby tacitly asserting that they were on the English side; and if on the English side, then they must be angels and not devils. Otherwise, the situation was embarrassing. You see, the University being the wisest and deepest and most erudite body in the world, it would like to be logical if it could, for the sake of its reputation; therefore it would study and study, days and days, trying to find some good common-sense reason for proving the Voices devils in Article No. 1 and proving them angels in Article No. 10. However, they had to give it up. They found no way out: and so, to this day the University's verdict remains just so—devils in No. 1, angels in No. 10; and no way to reconcile the discrepancy.

The envoys brought the verdict to Rouen, and with it a letter for Cauchon which was full of fervid praise. The University complimented him on his zeal in hunting down this woman "whose venom had infected the faithful of the whole West," and as recompense it as good as promised him "a crown of imperishable glory in heaven." Only *that!*—a crown in heaven; a promissory note and no indorser; always something away off yonder; not a word about the Archbishopric of Rouen, which was the thing Cauchon was destroying his soul for. A crown in heaven; it must have sounded like a sarcasm to him, after all his hard work. What should *he* do in heaven? he did not know anybody there.

On the nineteenth of May a court of fifty judges sat in the archiepiscopal palace to discuss Joan's fate. A few wanted her delivered over to the secular arm at once for punishment, but the rest insisted that she be once more "charitably admonished" first.

So the same court met in the castle on the twenty-third, and Joan was brought to the bar. Pierre Maurice, a canon of Rouen, made a speech to Joan in which he admonished her to save her life and her soul by renouncing her errors and surrendering to the Church. He finished with a stern threat: if she remained obstinate the damnation of her soul was certain, the destruction of her body probable. But Joan was immovable. She said—

"If I were under sentence, and saw the fire before me, and the executioner ready to light it—more, if I were in the fire itself, I would say none but the things which I have said in these trials; and I would abide by them till I died."

A deep silence followed, now, which endured some moments. It lay upon me like a weight. I knew it for an omen. Then Cauchon, grave and solemn, turned to Pierre Maurice—

"Have you anything further to say?"

The priest bowed low, and said—

"Nothing, my lord."

"Prisoner at the bar, have you anything further to say?"

"Nothing."

"Then the debate is closed. To-morrow, sentence will be pronounced. Remove the prisoner."

She seemed to go from the place erect and noble. But I do not know; my sight was dim with tears.

To-morrow—twenty-fourth of May! Exactly a year since I saw her go speeding across the plain at the head of her troops, her silver helmet shining, her silvery cape fluttering in the wind, her white plumes flowing, her sword held aloft; saw her charge the Burgundian camp three times, and carry it; saw her wheel to the right and spur for the Duke's reserves; saw her fling herself against it in the last assault she was ever to make. And now that fatal day was come again—and see what it was bringing!

CHAPTER XIX

JOAN had been adjudged guilty of heresy, sorcery, and all the other terrible crimes set forth in the Twelve Articles, and her life was in Cauchon's hands at last. He could send her to the stake at once. His work was finished now, you think? He was satisfied? Not at all. What would his Archbishopric be worth if the people should get the idea into their heads that this faction of interested priests, slaving under the English lash, had wrongly condemned and burned Joan of Arc, Deliverer of France? That would be to make of her a holy martyr. Then her spirit would rise from her body's ashes, a thousand-fold reinforced, and sweep the English domination into the sea, and Cauchon along with it. No, the victory was not complete yet. Joan's guilt must be established by evidence which would satisfy the people. Where was that evidence to be found? There was only one person in the world who could furnish it—*Joan of Arc herself*. She must condemn herself, and in public—at least she must *seem* to do it.

But how was this to be managed? Weeks had been spent already in trying to get her to surrender—time wholly wasted; what was to persuade her now? Torture had been threatened, the fire had been threatened; what was left? Illness, deadly fatigue, and the sight of the fire, the presence of the fire! That was left.

Now that was a shrewd thought. She was but a girl, after all, and, under illness and exhaustion, subject to a girl's weaknesses.

Yes, it was shrewdly thought. She had tacitly said, herself, that under the bitter pains of the rack they would be

THE MAID OF ORLEANS
(From a statue by Freimet in the Rue de Rivoli at Paris)

able to extort a false confession from her. It was a hint worth remembering, and it was remembered.

She had furnished another hint at the same time: that as soon as the pains were gone, she would retract the confession. That hint was also remembered.

She had herself taught them what to do, you see. First, they must wear out her strength, then frighten her with the fire. Second, while the fright was on her, she must be made to sign a paper.

But she would demand a reading of the paper. They could not venture to refuse this, with the public there to hear. Suppose that during the reading her courage should return? she would refuse to sign, then. Very well, even that difficulty could be got over. They could read a short paper of no importance, then slip a long and deadly one into its place and trick her into signing *that*.

Yet there was still one other difficulty. If they made her seem to abjure, that would free her from the death penalty. They could keep her in a prison of the Church, but they could not kill her. That would not answer; for only her death would content the English. Alive she was a terror, in a prison or out of it. She had escaped from two prisons already.

But even that difficulty could be managed. Cauchon would make promises to her; in return, she would promise to leave off the male dress. He would violate his promises, and that would so situate her that she would not be able to keep hers. Her lapse would condemn her to the stake, and the stake would be ready.

These were the several moves; there was nothing to do but to make them, each in its order, and the game was won. One might almost name the day that the betrayed girl, the most innocent creature in France, and the noblest, would go to her pitiful death.

And the time was favorable—cruelly favorable. Joan's spirit had as yet suffered no decay, it was as sublime and masterful as ever; but her body's forces had been steadily wasting away in those last ten days, and a strong mind needs a healthy body for its rightful support.

The world knows, now, that Cauchon's plan was as I have sketched it to you, but the world did not know it at that time. There are sufficient indications that Warwick and all the other English chiefs except the highest one—the Cardinal of Winchester—were not let into the secret; also, that only Loyseleur and Beaupère, on the French side, knew the scheme. Sometimes I have doubted if even Loyseleur and Beaupère knew the whole of it at first. However, if any did, it was these two.

It is usual to let the condemned pass their last night of life in peace, but this grace was denied to poor Joan, if one may credit the rumors of the time. Loyseleur was smuggled into her presence, and in the character of priest, friend, and secret partisan of France and hater of England, he spent some hours in beseeching her to do "the only right and righteous thing"—submit to the Church, as a good Christian should; and that then she would straightway get out of the clutches of the dreaded English and be transferred to the Church's prison, where she would be honorably used and have women about her for jailers. He knew where to touch her. He knew how odious to her was the presence of her rough and profane English guards; he knew that her Voices had vaguely promised something which she interpreted to be escape, rescue, release of some sort, and the chance to burst upon France once more and victoriously complete the great work which she had been commissioned of Heaven to do. Also there was that other thing: if her failing body could be further weakened by loss of rest and sleep, now, her tired mind would be dazed and drowsy on the morrow, and in ill condition to stand out against persuasions, threats, and the sight of the stake, and also be purblind to traps and snares which it would be swift to detect when in its normal estate.

I do not need to tell you that there was no rest for me that night. Nor for Noël. We went to the main gate of the city before nightfall, with a hope in our minds, based upon that vague prophecy of Joan's Voices which seemed to promise a rescue by force at the last moment. The immense news had flown swiftly far and wide that at last Joan of Arc was condemned, and would be sentenced and burned alive on the morrow; and so, crowds of people were flowing in at the gate, and other crowds were being refused admission by the soldiery; these being people who brought doubtful passes or none at all. We scanned these crowds eagerly, but there was nothing about them to indicate that they were our old war-comrades in disguise, and certainly there were no familiar faces among them. And so, when the gate was closed at last, we turned away grieved, and more disappointed than we cared to admit, either in speech or thought.

The streets were surging tides of excited men. It was difficult to make one's way. Toward midnight our aimless tramp brought us to the neighborhood of the beautiful church of St. Ouen, and there all was bustle and work. The square was a wilderness of torches and people; and through a guarded passage dividing the pack, laborers were carrying planks and timbers and disappearing with them through the gate of the churchyard. We asked what was going forward; the answer was—

"Scaffolds and the stake. Don't you know that the French witch is to be burnt in the morning?"

Then we went away. We had no heart for that place.

At dawn we were at the city gate again; this time with a hope which our wearied bodies and fevered minds magnified into a large probability. We had heard a report that the Abbot of Jumièges with all his monks was coming to witness the burning. Our desire, abetted by our imagination, turned those nine hundred monks into Joan's old campaigners, and their Abbot into La Hire or the Bastard or D'Alençon; and we watched them file in, unchallenged, the multitude respectfully dividing and uncovering while they passed, with our hearts in our throats and our eyes swimming with tears of joy and pride and exultation; and we tried to catch glimpses of the faces under the cowls, and were prepared to give signal to any recognized face that we were Joan's men and ready and eager to kill and be killed in the good cause. How foolish we were; but we were young, you know, and youth hopeth all things, believeth all things.

CHAPTER XX

In the morning I was at my official post. It was on a platform raised the height of a man, in the churchyard, under the eaves of St. Ouen. On this same platform was a crowd of priests and important citizens, and several lawyers. Abreast it, with a small space between, was another and larger platform, handsomely canopied against sun and rain, and richly carpeted; also it was furnished with comfortable chairs, and with two which were more sumptuous than the others, and raised above the general level. One of these two was occupied by a prince of the royal blood of England, his Eminence the Cardinal of Winchester; the other by Cauchon, Bishop of Beauvais. In the rest of the chairs sat three bishops, the Vice-Inquisitor, eight abbots, and the sixty-two friars and lawyers who had sat as Joan's judges in her late trials.

Twenty steps in front of the platforms was another—a table-topped pyramid of stone, built up in retreating courses, thus forming steps. Out of this rose that grisly thing the stake; about the stake bundles of fagots and firewood were piled. On the ground at the base of the pyramid stood three crimson figures, the executioner and his assistants. At their feet lay what had been a goodly heap of brands, but was now a smokeless nest of ruddy coals; a foot or two from this was a supplemental supply of wood and fagots compacted into a pile shoulder-high and containing as much as six pack-horse loads. Think of that. We seem so delicately made, so destructible, so insubstantial; yet it is easier to reduce a granite statue to ashes than it is to do that with a man's body.

The sight of the stake sent physical pains tingling down the nerves of my body; and yet, turn as I would, my eyes would keep coming back to it, such fascination has the grewsome and the terrible for us.

The space occupied by the platforms and the stake was kept open by a wall of English soldiery, standing elbow to elbow, erect and stalwart figures, fine and sightly in their polished steel; while from behind them on every hand stretched far away a level plain of human heads; and there was no window and no housetop within our view, howsoever distant, but was black with patches and masses of people.

But there was no noise, no stir; it was as if the world was dead. The impressiveness of this silence and solemnity was deepened by a leaden twilight, for the sky was hidden by a pall of low-hanging storm-clouds; and above the remote horizon faint winkings of heat-lightning played, and now and then one caught the dull mutterings and complainings of distant thunder.

At last the stillness was broken. From beyond the square rose an indistinct sound, but familiar—curt, crisp phrases of command; next I saw the plain of heads dividing, and the steady swing of a marching host was glimpsed between. My heart leaped, for a moment. Was it La Hire and his hellions? No—that was not their gait. No, it was the prisoner and her escort; it was Joan of Arc, under guard, that was coming; my spirits sank as low as they had been before. Weak as she was, they made her walk; they would increase her weakness all they could. The distance was not great—it was but a few hundred yards—but short as it was it was a heavy tax upon one who had been lying chained in one spot for months, and whose feet had lost their powers.from inaction. Yes, and for a year Joan had known only the cool damps of a dungeon, and now she was dragging herself through this sultry summer heat, this airless and suffocating void. As she entered the gate, drooping with exhaustion, there was that creature Loyseleur at her side with his head bent to her ear. We knew afterwards that he had been with her again this morning in the prison wearing her with his persuasions and enticing her with false promises, and that he was now still at the same work at the gate, imploring her to yield everything that would be required of her, and assuring her that if she would do this all would be well with her: she would be rid of the dreaded English and find safety in the powerful shelter and protection of the Church. A miserable man, a stony-hearted man!

The moment Joan was seated on the platform she closed her eyes and allowed her chin to fall; and so sat, with her hands nestling in her lap, indifferent to everything, caring for nothing but rest. And she was so white again; white as ala baster.

How the faces of that packed mass of humanity lighted up with interest, and with what intensity all eyes gazed upon this fragile girl! And how natural it was; for these people realized that at last they were looking upon that person whom they had so long hungered to see; a person whose name and fame filled all Europe, and made all other names and all other renowns insignificant by comparison: Joan of Arc, the wonder of the time, and destined to be the wonder of all times! And I could read as by print, in their marvelling countenances, the words that were drifting through their minds: "Can it be true ; is it believable, that it is this little creature, this girl, this child with the good face, the sweet face, the beautiful face, the dear and bonny face, that has carried fortresses by storm, charged at the head of victorious armies, blown the might of England out of her path with a breath, and fought a long campaign, solitary and alone, against the massed brains and learning of France—and had won it if the fight had been fair!"

Evidently Cauchon had grown afraid of Manchon because of his pretty apparent leanings toward Joan, for another recorder was in the chief place, here, which left my master and me nothing to do but sit idle and look on.

Well, I supposed that everything had been done which could be thought of to tire Joan's body and mind, but it was a mistake ; one more device had been invented. This was to preach a long sermon to her in that oppressive heat.

When the preacher began, she cast up one distressed and disappointed look, then dropped her head again. This preacher was Guillaume Erard, an oratorical celebrity. He got his text from the Twelve Lies. He emptied upon Joan all the calumnies, in detail, that had been bottled up in that mess of venom, and called her all the brutal names that the Twelve were labelled with, working himself into a whirlwind of fury as he went on; but his labors were wasted, she seemed lost in dreams, she made no sign, she did not seem to hear. At last he launched this apostrophe:

"O France, how hast thou been abused! Thou hast always been the home of Christianity; but now, Charles, who calls himself thy King and governor, indorses like the heretic and schismatic that he is, the words and deeds of a worthless and infamous woman!" Joan raised her head, and her eyes began to burn and flash. The preacher turned toward her: "It is to you, Joan, that I speak, and I tell you that your King is schismatic and a heretic!"

Ah, he might abuse *her* to his heart's content; she could endure that; but to her dying moment she could never hear in patience a word against that ingrate, that treacherous dog our King, whose proper place was here, at this moment, sword in hand, routing these reptiles and saving this most noble servant that ever King had in this world — and he *would* have been there if he had not been what I have called him. Joan's loyal soul was outraged, and she turned upon the preacher and flung out a few words with a spirit which the crowd recognized as being in accordance with the Joan of Arc traditions—

"By my faith, sir! I make bold to say and swear, on pain of death, that he is the most noble Christian of all Christians, and the best lover of the faith and the Church!"

There was an explosion of applause from the crowd—which angered the preacher, for he had been aching long to hear an expression like this, and now that it was come at last it had fallen to the wrong person: he had done all the work; the other had carried off all the spoil. He stamped his foot and shouted to the sheriff—

"Make her shut up!"

That made the crowd laugh.

A mob has small respect for a grown man who has to call on a sheriff to protect him from a sick girl.

Joan had damaged the preacher's cause more with one sentence than he had helped it with a hundred; so he was much put out, and had trouble to get a good start again. But he needn't have bothered; there was no occasion. It was mainly an English-feeling mob. It had but obeyed a law of our nature—an irresistible law—to enjoy and applaud a spirited and promptly delivered retort, no matter who makes it. The mob was with the preacher; it had been beguiled for a moment, but only that; it would soon return. It was there to see this girl burnt; so that it got that satisfaction—without too much delay—it would be content.

Presently the preacher formally summoned Joan to submit to the Church. He made the demand with confidence, for he had gotten the idea from Loyseleur and Beaupère that she was worn to the bone, exhausted, and would not be able to put forth any more resistance; and indeed, to look at her it seemed that they must be right. Nevertheless, she made one more effort to hold her ground, and said, wearily—

"As to that matter, I have answered my judges before. I have told them to report all that I have said and done to our holy Father the Pope—to whom, and to God first, I appeal."

Again, out of her native wisdom, she had brought those words of tremendous import, but was ignorant of their value. But they could have availed her nothing in any case, now, with the stake there and these thousands of enemies about her. Yet they made every churchman there blench, and the preacher changed the subject with all haste. Well might those criminals blench, for Joan's appeal of her case to the Pope stripped Cauchon at once of jurisdiction over it, and annulled all that he and his judges had already done in the matter and all that they should do in it thenceforth.

Joan went on presently to reiterate, after some further talk, that she had acted by command of God in her deeds and utterances ; then, when an attempt was made to implicate the King, and friends of hers and his, she stopped that. She said—

" I charge my deeds and words upon no one, neither upon my King nor any other. If there is any fault in them, I am responsible and no other."

She was asked if she would not recant those of her words and deeds which had been pronounced evil by her judges. Her answer made confusion and damage again :

" I submit them to God and the Pope."

The Pope once more ! It was very embarrassing. Here was a person who was asked to submit her case to the Church, and who frankly consents—offers to submit it to the very head of it. What more could any one require ? How was one to answer such a formidably unanswerable answer as that ?

The worried judges put their heads together and whispered and planned and discussed. Then they brought forth this sufficiently shambling conclusion—but it was the best they could do, in so close a place : they said the Pope was so far away ; and it was not necessary to go to him, anyway, because these present judges had sufficient power and authority to deal with the present case, and were in effect " the Church " to that extent. At another time they could have smiled at this conceit, but not now ; they were not comfortable enough, now.

The mob was getting impatient. It was beginning to put on a threatening aspect ; it was tired standing, tired of the scorching heat ; and the thunder was coming nearer, the lightning was flashing brighter. It was necessary to hurry this matter to a close. Erard showed Joan a written form, which had been prepared and made all ready beforehand, and asked her to abjure.

" Abjure ? What is abjure ?"

She did not know the word. It was explained to her by

Massieu. She tried to understand, but she was breaking, under exhaustion, and she could not gather the meaning. It was all a jumble and confusion of strange words. In her despair she sent out this beseeching cry—

"I appeal to the Church universal whether I ought to abjure or no!"

Erard exclaimed—

"You shall abjure instantly, or instantly be burnt!"

She glanced up, at those awful words, and for the first time she saw the stake and the mass of red coals — redder and angrier than ever, now, under the constantly deepening storm-gloom. She gasped and staggered up out of her seat muttering and mumbling incoherently, and gazed vacantly upon the people and the scene about her like one who is dazed, or thinks he dreams, and does not know where he is.

The priests crowded about her imploring her to sign the paper, there were many voices beseeching and urging her at once, there was great turmoil and shouting and excitement, amongst the populace and everywhere.

"Sign! sign!" from the priests; "sign—sign and be saved!" And Loyseleur was urging at her ear, "Do as I told you—do not destroy yourself!"

Joan said plaintively to these people—

"Ah, you do not do well to seduce me."

The judges joined their voices to the others. Yes, even the iron in *their* hearts melted, and they said—

"Oh, Joan, we pity you so! Take back what you have said, or we must deliver you up to punishment."

And now there was another voice—it was from the other platform—pealing solemnly above the din: Cauchon's—reading the sentence of death!

Joan's strength was all spent. She stood looking about her in a bewildered way a moment, then slowly she sank to her knees, and bowed her head and said—

"I submit."

They gave her no time to reconsider—they knew the peril of that. The moment the words were out of her mouth

28

Massieu was reading to her the abjuration, and she was repeating the words after him mechanically, unconsciously—and *smiling;* for her wandering mind was far away in some happier world.

Then this short paper of six lines was slipped aside and a long one of many pages was smuggled into its place, and she, noting nothing, put her mark to it, saying, in pathetic apology, that she did not know how to write. But a secretary of the King of England was there to take care of that defect; he guided her hand with his own, and wrote her name—*Jehanne.*

The great crime was accomplished. She had signed— what? She did not know—but the others knew. She had signed a paper confessing herself a sorceress, a dealer with devils, a liar, a blasphemer of God and His angels, a lover of blood, a promoter of sedition, cruel, wicked, commissioned of Satan; and this signature of hers bound her to resume the dress of a woman. There were other promises, but that one would answer, without the others; that one could be made to destroy her.

Loyseleur pressed forward and praised her for having done "such a good day's work."

But she was still dreamy, she hardly heard.

Then Cauchon pronounced the words which dissolved the excommunication and restored her to her beloved Church, with all the dear privileges of worship. Ah, she heard that! You could see it in the deep gratitude that rose in her face and transfigured it with joy.

But how transient was that happiness! For Cauchon, without a tremor of pity in his voice, added these crushing words—

"And that she may repent of her crimes and repeat them no more, she is sentenced to perpetual imprisonment, with the bread of affliction and the water of anguish!"

Perpetual imprisonment! She had never dreamed of that— such a thing had never been hinted to her by Loyseleur or by any other. Loyseleur had distinctly said and promised that "all would be well with her." And the very last words spoken

JOAN SIGNS THE LIST OF ACCUSATIONS

to her by Erard, on that very platform, when he was urging her to abjure, was a straight, unqualified promise—that if she would do it she should *go free from captivity.*

She stood stunned and speechless a moment; then she remembered, with such solacement as the thought could furnish, that by another clear promise—a promise made by Cauchon himself—she would at least be the Church's captive, and have women about her in place of a brutal foreign soldiery. So she turned to the body of priests and said, with a sad resignation—

" Now, you men of the Church, take me to your prison, and leave me no longer in the hands of the English "; and she gathered up her chains and prepared to move.

But alas, now came these shameful words from Cauchon—and with them a mocking laugh :

"Take her to the prison whence she came !"

Poor abused girl! She stood dumb, smitten, paralyzed. It was pitiful to see. She had been beguiled, lied to, betrayed; she saw it all, now.

The rumbling of a drum broke upon the stillness, and for just one moment she thought of the glorious deliverance promised by her Voices—I read it in the rapture that lit her face ; then she saw what it was—her prison escort—and that light faded, never to revive again. And now her head began a piteous rocking motion, swaying slowly, this way and that, as is the way when one is suffering unwordable pain, or when one's heart is broken; then drearily she went from us, with her face in her hands, and sobbing bitterly.

CHAPTER XXI

THERE is no certainty that any one in all Rouen was in the secret of the deep game which Cauchon was playing except the Cardinal of Winchester. Then you can imagine the astonishment and stupefaction of that vast mob gathered there and those crowds of churchmen assembled on the two platforms, when they saw Joan of Arc moving away, alive and whole—slipping out of their grip at last, after all this tedious waiting, all this tantalizing expectancy.

Nobody was able to stir or speak, for a while, so paralyzing was the universal astonishment, so unbelievable the fact that the stake was actually standing there unoccupied and its prey gone. Then suddenly everybody broke into a fury of rage; maledictions and charges of treachery began to fly freely; yes, and even stones : a stone came near killing the Cardinal of Winchester—it just missed his head. But the man who threw it was not to blame, for he was excited, and a person who is excited never can throw straight.

The tumult was very great indeed, for a while. In the midst of it a chaplain of the Cardinal even forgot the proprieties so far as to opprobriously assail the august Bishop of Beauvais himself, shaking his fist in his face and shouting:

" By God, you are a traitor !"

" You lie !" responded the Bishop.

He a traitor ! Oh, far from it; he certainly was the last Frenchman that any Briton had a right to bring that charge against.

The Earl of Warwick lost his temper, too. He was a doughty soldier, but when it came to the intellectuals—when it came to delicate chicane, and scheming, and trickery—he couldn't see any further through a millstone than another. So he burst out in his frank warrior fashion, and swore that the King of England was being treacherously used, and that Joan of Arc was going to be allowed to cheat the stake. But they whispered comfort into his ear—

"Give yourself no uneasiness, my lord; we shall soon have her again."

Perhaps the like tidings found their way all around, for good news travels fast as well as bad. At any rate the ragings presently quieted down, and the huge concourse crumbled apart and disappeared. And thus we reached the noon of that fearful Thursday.

We two youths were happy; happier than any words can tell—for we were not in the secret any more than the rest. Joan's life was saved. We knew that, and that was enough. France would hear of this day's infamous work—and then! Why, then her gallant sons would flock to her standard by thousands and thousands, multitudes upon multitudes, and their wrath would be like the wrath of the ocean when the storm-winds sweep it; and they would hurl themselves against this doomed city and overwhelm it like the resistless tides of that ocean, and Joan of Arc would march again! In six days— seven days—one short week—noble France, grateful France, indignant France, would be thundering at these gates—let us count the hours, let us count the minutes, let us count the seconds! Oh happy day, oh day of ecstasy, how our hearts sang in our bosoms!

For we were young, then; yes, we were very young.

Do you think the exhausted prisoner was allowed to rest and sleep after she had spent the small remnant of her strength in dragging her tired body back to the dungeon?

No; there was no rest for her, with those sleuth-hounds on her track. Cauchon and some of his people followed her to her lair, straightway; they found her dazed and dull, her mental and physical forces in a state of prostration. They told her she had abjured; that she had made certain promises— among them, to resume the apparel of her sex; and that if she relapsed, the Church would cast her out for good and all. She heard the words, but they had no meaning to her. She was like a person who has taken a narcotic and is dying for sleep, dying for rest from nagging, dying to be let alone, and who mechanically does everything the persecutor asks, taking but dull note of the things done, and but dully recording them in the memory. And so Joan put on the gown which Cauchon and his people had brought; and would come to herself by-and-by, and have at first but a dim idea as to when and how the change had come about.

Cauchon went away happy and content. Joan had resumed woman's dress without protest; also she had been formally warned against relapsing. He had witnesses to these facts. How could matters be better?

But suppose she should *not* relapse?

Why, then she must be forced to do it.

Did Cauchon hint to the English guards that thenceforth if they chose to make their prisoner's captivity crueler and bitterer than ever, no official notice would be taken of it? Perhaps so; since the guards did begin that policy at once, and no official notice *was* taken of it. Yes, from that moment Joan's life in that dungeon was made almost unendurable. Do not ask me to enlarge upon it. I will not do it.

CHAPTER XXII

FRIDAY and Saturday were happy days for Noël and me. Our minds were full of our splendid dream of France aroused—France shaking her mane—France on the march—France at the gates—Rouen in ashes, and Joan free! Our imagination was on fire; we were delirious with pride and joy. For we were very young, as I have said.

We knew nothing about what had been happening in the dungeon the yester-afternoon. We supposed that as Joan had abjured and been taken back into the forgiving bosom of the Church, she was being gently used, now, and her captivity made as pleasant and comfortable for her as the circumstances would allow. So, in high contentment, we planned out our share in the great rescue, and fought our part of the fight over and over again during those two happy days—as happy days as ever I have known.

Sunday morning came. I was awake, enjoying the balmy, lazy weather, and thinking. Thinking of the rescue—what else? I had no other thought now. I was absorbed in that, drunk with the happiness of it.

I heard a voice shouting, far down the street, and soon it came nearer, and I caught the words—

"*Joan of Arc has relapsed! The witch's time has come!*"

It stopped my heart, it turned my blood to ice. That was more than sixty years ago, but that triumphant note rings as clear in my memory to-day as it rang in my ear that long-vanished summer morning. We are so strangely made; the memories that could make us happy pass away; it is the memories that break our hearts that abide.

Soon other voices took up that cry—tens, scores, hundreds

of voices; all the world seemed filled with the brutal joy of it. And there were other clamors—the clatter of rushing feet, merry congratulations, bursts of coarse laughter, the rolling of drums, the boom and crash of distant bands profaning the sacred day with the music of victory and thanksgiving.

About the middle of the afternoon came a summons for Manchon and me to go to Joan's dungeon—a summons from Cauchon. But by that time distrust had already taken possession of the English and their soldiery again, and all Rouen was in an angry and threatening mood. We could see plenty evidences of this from our own windows—fist-shaking, black looks, tumultuous tides of furious men billowing by along the street.

And we learned that up at the castle things were going very badly indeed; that there was a great mob gathered there who considered the relapse a lie and a priestly trick, and among them many half-drunk English soldiers. Moreover, these people had gone beyond words. They had laid hands upon a number of churchmen who were trying to enter the castle, and it had been difficult work to rescue them and save their lives.

And so Manchon refused to go. He said he would not go a step without a safeguard from Warwick. So next morning Warwick sent an escort of soldiers, and then we went. Matters had not grown peacefuler meantime, but worse. The soldiers protected us from bodily damage, but as we passed through the great mob at the castle we were assailed with insults and shameful epithets. I bore it well enough, though, and said to myself, with secret satisfaction, "In three or four short days, my lads, you will be employing your tongues in a different sort from this — and I shall be there to hear."

To my mind these were as good as dead men. How many of them would still be alive after the rescue that was coming? Not more than enough to amuse the executioner a short half-hour, certainly.

It turned out that the report was true. Joan had relapsed.

She was sitting there in her chains, clothed again in her male attire.

She accused nobody. That was her way. It was not in her character to hold a servant to account for what his master had made him do, and her mind had cleared, now, and she knew that the advantage which had been taken of her the previous morning had its origin, not in the subordinate, but in the master—Cauchon.

Here is what had happened. While Joan slept, in the early morning of Sunday, one of the guards stole her female apparel and put her male attire in its place. When she woke she asked for the other dress, but the guards refused to give it back. She protested, and said she was forbidden to wear the male dress. But they continued to refuse. She had to have clothing, for modesty's sake; moreover, she saw that she could not save her life if she must fight for it against treacheries like this; so she put on the forbidden garments, knowing what the end would be. She was weary of the struggle, poor thing.

We had followed in the wake of Cauchon, the Vice-Inquisitor, and the others—six or eight—and when I saw Joan sitting there, despondent, forlorn, and still in chains, when I was expecting to find her situation so different, I did not know what to make of it. The shock was very great. I had doubted the relapse, perhaps; possibly I had believed in it, but had not realized it.

Cauchon's victory was complete. He had had a harassed and irritated and disgusted look for a long time, but that was all gone now, and contentment and serenity had taken its place. His purple face was full of tranquil and malicious happiness. He went trailing his robes and stood grandly in front of Joan, with his legs apart, and remained so more than a minute, gloating over her and enjoying the sight of this poor ruined creature, who had won so lofty a place for him in the service of the meek and merciful Jesus, Saviour of the World, Lord of the Universe — in case England kept her promise to him, who kept no promises himself.

Presently the judges began to question Joan. One of them, named Marguerie, who was a man with more insight than prudence, remarked upon Joan's change of clothing, and said—

"There is something suspicious about this. How could it have come about without connivance on the part of others? Perhaps even something worse?"

"Thousand devils!" screamed Cauchon, in a fury. "Will you shut your mouth?"

"Armagnac! Traitor!" shouted the soldiers on guard, and made a rush for Marguerie with their lances levelled. It was with the greatest difficulty that he was saved from being run through the body. He made no more attempts to help the inquiry, poor man. The other judges proceeded with the questionings.

"Why have you resumed this male habit?"

I did not quite catch her answer, for just then a soldier's halberd slipped from his fingers and fell on the stone floor with a crash; but I thought I understood Joan to say that she had resumed it of her own motion.

"But you have promised and sworn that you would not go back to it."

I was full of anxiety to hear her answer to that question; and when it came it was just what I was expecting. She said —quite quietly—

"I have never intended and never understood myself to swear I would not resume it."

There—I had been sure, all along, that she did not know what she was doing and saying on the platform Thursday, and this answer of hers was proof that I had not been mistaken. Then she went on to add this—

"But I had a right to resume it, because the promises made to me have not been kept—promises that I should be allowed to go to mass, and receive the communion, and that I should be freed from the bondage of these chains—but they are still upon me, as you see."

"Nevertheless, you have abjured, and have especially promised to return no more to the dress of a man."

Then Joan held out her fettered hands sorrowfully toward these unfeeling men and said—

"I would rather die than continue so. But if they may be taken off, and if I may hear mass, and be removed to a penitential prison, and have a woman about me, I will be good, and will do what shall seem good to you that I do."

Cauchon sniffed scoffingly at that. Honor the compact which he and his had made with her? Fulfil its conditions? What need of that? Conditions had been a good thing to concede, temporarily, and for advantage; but they had served their turn—let something of a fresher sort and of more consequence be considered. The resumption of the male dress was sufficient for all practical purposes, but perhaps Joan could be led to add something to that fatal crime. So Cauchon asked her if her Voices had spoken to her since Thursday— and he reminded her of her abjuration.

"Yes," she answered ; and then it came out that the Voices had talked with her about the abjuration—*told* her about it, I suppose. She guilelessly reasserted the heavenly origin of her mission, and did it with the untroubled mien of one who was not conscious that she had ever knowingly repudiated it. So I was convinced once more that she had had no notion of what she was doing that Thursday morning on the platform. Finally she said, "My Voices told me I did very wrong to confess that what I had done was not well." Then she sighed, and said with simplicity, "But it was the fear of the fire that made me do so."

That is, fear of the fire had made her sign a paper whose contents she had not understood then, but understood now by revelation of her Voices and by testimony of her persecutors.

She was sane now, and not exhausted ; her courage had come back, and with it her inborn loyalty to the truth. She was bravely and serenely speaking it again, knowing that it would deliver her body up to that very fire which had such terrors for her.

That answer of hers was quite long, quite frank, wholly free from concealments or palliations. It made me shudder; I knew she was pronouncing sentence of death upon herself. So did poor Manchon. And he wrote in the margin abreast of it—

RESPONSIO MORTIFERA.

Fatal answer. Yes, all present knew that it was indeed a fatal answer. Then there fell a silence such as falls in a sick-room when the watchers by the dying draw a deep breath and say softly one to another, " All is over."

Here, likewise, all was over; but after some moments Cauchon, wishing to clinch this matter and make it final, put this question—

" Do you still believe that your Voices are St. Marguerite and St. Catherine ?"

" Yes—and that they come from God."

" Yet you denied them on the scaffold ?"

Then she made direct and clear affirmation that she had never had any intention to deny them; and that if—I noted the *if*—" if she had made some retractions and revocations on the scaffold it was from fear of the fire, and was a violation of the truth."

There it is again, you see. She certainly never knew what it was she had done on the scaffold until she was told of it afterwards by these people and by her Voices.

And now she closed this most painful scene with these words; and there was a weary note in them that was pathetic—

" I would rather do my penance all at once; let me die. I cannot endure captivity any longer."

The spirit born for sunshine and liberty so longed for release that it would take it in any form, even that.

Several among the company of judges went from the place troubled and sorrowful, the others in another mood. In the court of the castle we found the Earl of Warwick and fifty English waiting, impatient for news. As soon as Cauchon saw them he shouted—*laughing*—think of a man destroying a friendless poor girl and then having the heart to laugh at it:

" Make yourselves comfortable—it's all over with her !"

CHAPTER XXIII

THE young can sink into abysses of despondency, and it was so with Noël and me, now; but the hopes of the young are quick to rise again, and it was so with ours. We called back that vague promise of the Voices, and said the one to the other that the glorious release was to happen at "the last moment"—"that other time was not the last moment, but this is; it will happen now; the King will come, La Hire will come, and with them our veterans, and behind them all France!" And so we were full of heart again, and could already hear, in fancy, that stirring music the clash of steel and the war-cries and the uproar of the onset, and in fancy see our prisoner free, her chains gone, her sword in her hand.

But this dream was to pass also, and come to nothing. Late at night, when Manchon came in, he said—

"I am come from the dungeon, and I have a message for you from that poor child."

A message to me! If he had been noticing I think he would have discovered me—discovered that my indifference concerning the prisoner was a pretence; for I was caught off my guard, and was so moved and so exalted to be so honored by her that I must have shown my feeling in my face and manner.

"A message for me, your reverence?"

"Yes. It is something she wishes done. She said she had noticed the young man who helps me, and that he had a good face; and did I think he would do a kindness for her? I said I knew you would, and asked her what it was, and she said a letter—would you write a letter to her mother? And I said you would. But I said I would do it myself, and gladly; but she said no, that my labors were heavy, and she thought the young man would not mind the doing of this service for one not able to do it for herself, she not knowing how to write. Then I would have sent for you, and at that the sadness vanished out of her face. Why, it was as if she was going to see a friend, poor friendless thing. But I was not permitted. I did my best, but the orders remain as strict as ever, the doors are closed against all but officials ; as before, none but officials may speak to her. So I went back and told her, and she sighed, and was sad again. Now this is what she begs you to write to her mother. It is partly a strange message, and to me means nothing, but she said her mother would understand. You will ' convey her adoring love to her family and her village friends, and say there will be no rescue, for that this night—and it is the third time in the twelvemonth, and is final—she has seen The Vision of the Tree.' "

" How strange !"

" Yes, it *is* strange, but that is what she said ; and said her parents would understand. And for a little time she was lost in dreams and thinkings, and her lips moved, and I caught in her mutterings these lines, which she said over two or three times, and they seemed to bring peace and contentment to her. I set them down, thinking they might have some connection with her letter and be useful; but it was not so; they were a mere memory, floating idly in a tired mind, and they have no meaning, at least no relevancy."

I took the piece of paper, and found what I knew I should find :

> "And when in exile wand'ring we
> Shall fainting yearn for glimpse of thee,
> O rise upon our sight !"

There was no hope any more. I knew it now. I knew that Joan's letter was a message to·Noël and me, as well as to her family, and that its object was to banish vain hopes from our minds and tell us from her own mouth of the blow that was going to fall upon us, so that we, being her soldiers,

CAUCHON ACCUSES JOAN OF VIOLATING HER OATH

would know it for a command to bear it as became us and her, and so submit to the will of God ; and in thus obeying, find assuagement of our grief. It was like her, for she was always thinking of others, not of herself. Yes, her heart was sore for us ; she could find time to think of us, the humblest of her servants, and try to soften our pain, lighten the burden of our troubles,—she that was drinking of the bitter waters ; she that was walking in the Valley of the Shadow of Death.

I wrote the letter. You will know what it cost me, without my telling you. I wrote it with the same wooden stylus which had put upon parchment the first words ever dictated by Joan of Arc—that high summons to the English to vacate France, two years past, when she was a lass of seventeen ; it had now set down the last ones which she was ever to dictate. Then I broke it. For the pen that had served Joan of Arc could not serve any that would come after her in this earth without abasement.

The next day, May 29th, Cauchon summoned his serfs, and forty-two responded. It is charitable to believe that the other twenty were ashamed to come. The forty-two pronounced her a relapsed heretic, and condemned her to be delivered over to the secular arm. Cauchon thanked them. Then he sent orders that Joan be conveyed the next morning to the place known as the Old Market ; and that she be then delivered to the civil judge, and by the civil judge to the executioner. That meant that she would be burnt.

All the afternoon and evening of Tuesday the 29th the news was flying, and the people of the country-side flocking to Rouen to see the tragedy—all, at least, who could prove their English sympathies and count upon admission. The press grew thicker and thicker in the streets, the excitement grew higher and higher. And now a thing was noticeable again which had been noticeable more than once before— that there was pity for Joan in the hearts of many of these people. Whenever she had been in great danger it had manifested itself, and now it was apparent again—manifest in a pathetic dumb sorrow which was visible in many faces.

Early the next morning, Wednesday, Martin Ladvenu and another friar were sent to Joan to prepare her for death; and Manchon and I went with them—a hard service for me. We tramped through the dim corridors, winding this way and that, and piercing ever deeper and deeper into that vast heart of stone, and at last we stood before Joan. But she did not know it. She sat with her hands in her lap and her head bowed, thinking, and her face was very sad. One might not know what she was thinking of. Of her home, and the peaceful pastures, and the friends she was no more to see? Of her wrongs, and her forsaken estate, and the cruelties which had been put upon her? Or was it of death—the death which she had longed for, and which was now so close? Or was it of the *kind* of death she must suffer? I hoped not; for she feared only one kind, and that one had for her unspeakable terrors. I believed she so feared that one that with her strong will she would shut the thought of it wholly out of her mind, and hope and believe that God would take pity on her and grant her an easier one; and so it might chance that the awful news which we were bringing might come as a surprise to her, at last.

We stood silent awhile, but she was still unconscious of us, still deep in her sad musings and far away. Then Martin Ladvenu said, softly—

"Joan."

She looked up then, with a little start, and a wan smile, and said—

"Speak. Have you a message for me?"

"Yes, my poor child. Try to bear it. Do you think you can bear it?"

"Yes"—very softly, and her head drooped again.

"I am come to prepare you for death."

A faint shiver trembled through her wasted body. There was a pause. In the stillness we could hear our breathings. Then she said, still in that low voice—

"When will it be?"

The muffled notes of a tolling bell floated to our ears out of the distance.

"Now. The time is at hand."

That slight shiver passed again.

"It is so soon—ah, it is so soon!"

There was a long silence. The distant throbbings of the bell pulsed through it, and we stood motionless and listening. But it was broken at last—

"What death is it?"

"By fire!"

"Oh, I knew it, I knew it!" She sprang wildly to her feet, and wound her hands in her hair, and began to writhe and sob, oh, so piteously, and mourn and grieve and lament, and turn to first one and then another of us, and search our faces beseechingly, as hoping she might find help and friendliness there, poor thing—she that had never denied these to any creature, even her wounded enemy on the battle-field.

"Oh, cruel, cruel, to treat me so! And must my body, that has never been defiled, be consumed to-day and turned to ashes? Ah, sooner would I that my head were cut off seven times than suffer this woful death. I had the promise of the Church's prison when I submitted, and if I had but been there, and not left here in the hands of my enemies, this miserable fate had not befallen me. Oh, I appeal to God the Great Judge, against the injustice which has been done me."

There was none there that could endure it. They turned away, with the tears running down their faces. In a moment I was on my knees at her feet. At once she thought only of my danger, and bent and whispered in my ear: "Up!—do not peril yourself, good heart. There—God bless you always!" and I felt the quick clasp of her hand. Mine was the last hand she touched with hers in life. None saw it; history does not know of it or tell of it, yet it is true, just as I have told it. The next moment she saw Cauchon coming, and she went and stood before him and reproached him, saying—

"Bishop, it is by you that I die!"

He was not shamed, not touched; but said, smoothly—

"Ah, be patient, Joan. You die because you have not kept your promise, but have returned to your sins."

29

"Alas," she said, "if you had put me in the Church's prison, and given me right and proper keepers, as you promised, this would not have happened. And for this I summon you to answer before God!"

Then Cauchon winced, and looked less placidly content than before, and he turned him about and went away.

Joan stood awhile musing. She grew calmer, but occasionally she wiped her eyes, and now and then sobs shook her body; but their violence was modifying now, and the intervals between them were growing longer. Finally she looked up and saw Pierre Maurice, who had come in with the Bishop, and she said to him—

"Master Peter, where shall I be this night?"

"Have you not good hope in God?"

"Yes—and by His grace I shall be in Paradise."

Now Martin Ladvenu heard her in confession; then she begged for the sacrament. But how grant the communion to one who had been publicly cut off from the Church, and was now no more entitled to its privileges than an unbaptized pagan? The brother could not do this, but he sent to Cauchon to inquire what he must do. All laws, human and divine, were alike to that man—he respected none of them. He sent back orders to grant Joan whatever she wished. Her last speech to him had reached his fears, perhaps: it could not reach his heart, for he had none.

The Eucharist was brought now to that poor soul that had yearned for it with such unutterable longing all these desolate months. It was a solemn moment. While we had been in the deeps of the prison, the public courts of the castle had been filling up with crowds of the humbler sort of men and women, who had learned what was going on in Joan's cell, and had come with softened hearts to do — they knew not what; to hear—they knew not what. We knew nothing of this, for they were out of our view. And there were other great crowds of the like caste gathered in masses outside the castle gates. And when the lights and the other accompaniments of the Sacrament passed by, coming to Joan in the prison, all those multitudes kneeled down and began to pray for her, and many wept; and when the solemn ceremony of the communion began in Joan's cell, out of the distance a moving sound was borne moaning to our ears—it was those invisible multitudes chanting the litany for a departing soul.

The fear of the fiery death was gone from Joan of Arc now, to come again no more, except for one fleeting instant—then it would pass, and serenity and courage would take its place and abide till the end.

CHAPTER XXIV

AT nine o'clock the Maid of Orleans, Deliverer of France, went forth in the grace of her innocence and her youth to lay down her life for the country she loved with such devotion, and for the King that had abandoned her. She sat in the cart that is used only for felons. In one respect she was treated worse than a felon; for whereas she was on her way to be sentenced by the civil arm, she already bore her judgment inscribed in advance upon a mitre-shaped cap which she wore:

HERETIC, RELAPSED, APOSTATE, IDOLATER.

In the cart with her sat the friar Martin Ladvenu and Maître Jean Massieu. She looked girlishly fair and sweet and saintly in her long white robe, and when a gush of sunlight flooded her as she emerged from the gloom of the prison and was yet for a moment still framed in the arch of the sombre gate, the massed multitudes of poor folk murmured "A vision! a vision!" and sank to their knees praying, and many of the women weeping; and the moving invocation for the dying rose again, and was taken up and borne along, a majestic wave of sound, which accompanied the doomed, solacing and blessing her, all the sorrowful way to the place of death. "Christ have pity! Saint Margaret have pity! Pray for her, all ye saints, archangels, and blessed martyrs, pray for her! Saints and angels intercede for her! From thy wrath, good Lord, deliver her! O Lord God, save her! Have mercy on her, we beseech Thee, good Lord!"

It is just and true, what one of the histories has said:

" The poor and the helpless had nothing but their prayers to give Joan of Arc; but these we may believe were not unavailing. There are few more pathetic events recorded in history than this weeping, helpless, praying crowd, holding their lighted candles and kneeling on the pavement beneath the prison walls of the old fortress."

And it was so all the way: thousands upon thousands massed upon their knees and stretching far down the distances, thick-sown with the faint yellow candle-flames, like a field starred with golden flowers.

But there were some that did not kneel; these were the English soldiers. They stood elbow to elbow, on each side of Joan's road, and walled it in, all the way; and behind these living walls knelt the multitudes.

By-and-by a frantic man in priest's garb came wailing and lamenting, and tore through the crowd and the barrier of soldiers and flung himself on his knees by Joan's cart and put up his hands in supplication, crying out—

" O, forgive, forgive !"

It was Loyseleur !

And Joan forgave him; forgave him out of a heart that knew nothing but forgiveness, nothing but compassion, nothing but pity for all that suffer, let their offence be what it might. And she had no word of reproach for this poor wretch who had wrought day and night with deceits and treacheries and hypocrisies to betray her to her death.

The soldiers would have killed him, but the Earl of Warwick saved his life. What became of him is not known. He hid himself from the world somewhere, to endure his remorse as he might.

In the square of the Old Market stood the two platforms and the stake that had stood before in the church-yard of St. Ouen. The platforms were occupied as before, the one by Joan and her judges, the other by great dignitaries, the principal being Cauchon and the English Cardinal—Winchester. The square was packed with people, the windows and roofs of the blocks of buildings surrounding it were black with them.

When the preparations had been finished, all noise and movement gradually ceased, and a waiting stillness followed which was solemn and impressive.

And now, by order of Cauchon, an ecclesiastic named Nicholas Midi preached a sermon, wherein he explained that when a branch of the vine—which is the Church—becomes diseased and corrupt, it must be cut away or it will corrupt and destroy the whole vine. He made it appear that Joan, through her wickedness, was a menace and a peril to the Church's purity and holiness, and her death therefore necessary. When he was come to the end of his discourse he turned toward her and paused a moment, then he said—

"Joan, the Church can no longer protect you. Go in peace !"

Joan had been placed wholly apart and conspicuous, to signify the Church's abandonment of her, and she sat there in her loneliness, waiting in patience and resignation for the end. Cauchon addressed her now. He had been advised to read the form of her abjuration to her, and had brought it with him ; but he changed his mind, fearing that she would proclaim the truth—that she had never knowingly abjured—and so bring shame upon him and eternal infamy. He contented himself with admonishing her to keep in mind her wickednesses, and repent of them, and think of her salvation. Then he solemnly pronounced her excommunicate and cut off from the body of the Church. With a final word he delivered her over to the secular arm for judgment and sentence.

Joan, weeping, knelt and began to pray. For whom ? Herself ? Oh no—for the King of France. Her voice rose sweet and clear, and penetrated all hearts with its passionate pathos. She never thought of his treacheries to her, she never thought of his desertion of her, she never remembered that it was because he was an ingrate that she was here to die a miserable death ; she remembered only that he was her King, that she was his loyal and loving subject, and that his enemies had undermined his cause with evil reports and false charges, and he not by to defend himself. And so, in the very presence of death, she forgot her own troubles to implore all in her hearing to be just to him; to believe that he was good and noble and sincere, and not in any way to blame for any acts of hers, neither advising them nor urging them, but being wholly clear and free of all responsibility for them. Then, closing, she begged in humble and touching words that all here present would pray for her and would pardon her, both her enemies and such as might look friendly upon her and feel pity for her in their hearts.

There was hardly one heart there that was not touched—even the English, even the judges showed it, and there was many a lip that trembled and many an eye that was blurred with tears; yes, even the English Cardinal's—that man with a political heart of stone but a human heart of flesh.

The secular judge who should have delivered judgment and pronounced sentence was himself so disturbed that he forgot his duty, and Joan went to her death unsentenced—thus completing with an illegality what had begun illegally and had so continued to the end. He only said—to the guards—

"Take her"; and to the executioner, "Do your duty."

Joan asked for a cross. None was able to furnish one. But an English soldier broke a stick in two and crossed the pieces and tied them together, and this cross he gave her, moved to it by the good heart that was in him; and she kissed it and put it in her bosom. Then Isambard de la Pierre went to the church near by and brought her a consecrated one; and this one also she kissed, and pressed it to her bosom with rapture, and then kissed it again and again, covering it with tears and pouring out her gratitude to God and the saints.

And so, weeping, and with her cross to her lips, she climbed up the cruel steps to the face of the stake, with the friar Isambard at her side. Then she was helped up to the top of the pile of wood that was built around the lower third of the stake, and stood upon it with her back against the stake, and the world gazing up at her breathless. The executioner ascended to her side and wound chains about her slender body, and so fastened her to the stake. Then he descended to finish his dreadful office; and there she remained alone—she that had had so many friends in the days when she was free, and had been so loved and so dear.

All these things I saw, albeit dimly and blurred with tears; but I could bear no more. I continued in my place, but what I shall deliver to you now I got by others' eyes and others' mouths. Tragic sounds there were that pierced my ears and wounded my heart as I sat there, but it is as I tell you: the latest image recorded by my eyes in that desolating hour was Joan of Arc with the grace of her comely youth still unmarred; and that image, untouched by time or decay, has remained with me all my days. Now I will go on.

If any thought that now, in that solemn hour when all transgressors repent and confess, she would revoke her revocation and say her great deeds had been evil deeds and Satan and his fiends their source, they erred. No such thought was in her blameless mind. She was not thinking of herself and her troubles, but of others, and of woes that might befall them. And so, turning her grieving eyes about her, where rose the towers and spires of that fair city, she said—

"Oh, Rouen, Rouen, must I die here, and must you be my tomb? Ah, Rouen, Rouen, I have great fear that you will suffer for my death."

A whiff of smoke swept upward past her face, and for one moment terror seized her and she cried out, "Water! Give me holy water!" but the next moment her fears were gone, and they came no more to torture her.

She heard the flames crackling below her, and immediately distress for a fellow-creature who was in danger took possession of her. It was the friar Isambard. She had given him her cross and begged him to raise it toward her face and let her eyes rest in hope and consolation upon it till she was entered into the peace of God. She made him go out from the danger of the fire. Then she was satisfied, and said—

"Now keep it always in my sight until the end."

Not even yet could Cauchon, that man without shame, en-

THE MARTYRDOM OF THE MAID OF ORLEANS

dure to let her die in peace, but went toward her, all black with crimes and sins as he was, and cried out—

"I am come, Joan, to exhort you for the last time to repent and seek the pardon of God."

"I die through you," she said, and these were the last words she spoke to any upon earth.

Then the pitchy smoke, shot through with red flashes of flame, rolled up in a thick volume and hid her from sight; and from the heart of this darkness her voice rose strong and eloquent in prayer, and when by moments the wind shredded somewhat of the smoke aside, there were veiled glimpses of an upturned face and moving lips. At last a mercifully swift tide of flame burst upward, and none saw that face any more nor that form, and the voice was still.

Yes, she was gone from us: JOAN OF ARC! What little words they are, to tell of a rich world made empty and poor!

CONCLUSION

JOAN's brother Jacques died in Domremy during the Great Trial at Rouen. This was according to the prophecy which Joan made that day in the pastures the time that she said the rest of us would go to the great wars.

When her poor old father heard of the martyrdom it broke his heart and he died.

The mother was granted a pension by the City of Orleans, and upon this she lived out her days, which were many. Twenty-four years after her illustrious child's death she travelled all the way to Paris in the winter time and was present at the opening of the discussion in the Cathedral of Nôtre Dame which was the first step in the Rehabilitation. Paris was crowded with people, from all about France, who came to get sight of the venerable dame, and it was a touching spectacle when she moved through these reverend wet-eyed multitudes on her way to the grand honors awaiting her at the cathedral. With her were Jean and Pierre, no longer the light-hearted youths who marched with us from Vaucouleurs, but war-worn veterans with hair beginning to show frost.

After the martyrdom Noël and I went back to Domremy, but presently when the Constable Richemont superseded La Tremouille as the King's chief adviser and began the completion of Joan's great work, we put on our harness and returned to the field and fought for the King all through the wars and skirmishes until France was freed of the English. It was what Joan would have desired of us; and, dead or alive, her desire was law for us. All the survivors of the personal staff were faithful to her memory and fought for the

King to the end. Mainly we were well scattered, but when Paris fell we happened to be together. It was a great day and a joyous; but it was a sad one at the same time, because Joan was not there to march into the captured capital with us.

Noël and I remained always together, and I was by his side when death claimed him. It was in the last great battle of the war. In that battle fell also Joan's sturdy old enemy, Talbot. He was eighty-five years old, and had spent his whole life in battle. A fine old lion he was, with his flowing white mane and his tameless spirit; yes, and his indestructible energy as well; for he fought as knightly and vigorous a fight that day as the best man there.

La Hire survived the martyrdom thirteen years; and always fighting, of course, for that was all he enjoyed in life. I did not see him in all that time, for we were far apart, but one was always hearing of him.

The Bastard of Orleans and D'Alençon and D'Aulon lived to see France free, and to testify with Jean and Pierre d'Arc and Pasquerel and me at the Rehabilitation. But they are all at rest now, these many years. I alone am left of those who fought at the side of Joan of Arc in the great wars. She said I would live until these wars were forgotten—a prophecy which failed. If I should live a thousand years it would still fail. For whatsoever had touch with Joan of Arc, that thing is immortal.

Members of Joan's family married, and they have left descendants. Their descendants are of the nobility, but their family name and blood bring them honors which no other nobles receive or may hope for. You have seen how everybody along the way uncovered when those children came yesterday to pay their duty to me. It was not because they are noble, it is because they are grandchildren of the brothers of Joan of Arc.

Now as to the Rehabilitation. Joan crowned the King at Rheims. For reward he allowed her to be hunted to her death without making one effort to save her. During the next twenty-three years he remained indifferent to her memory; in-

different to the fact that her good name was under a damning
blot put there by the priests because of the deeds which she
had done in saving him and his sceptre; indifferent to the
fact that France was ashamed, and longed to have the Deliverer's fair fame restored. Indifferent all that time. Then he
suddenly changed and was anxious to have justice for poor
Joan, himself. Why? Had he become grateful at last? Had
remorse attacked his hard heart? No, he had a better reason—a better one for his sort of man. This better reason
was that, now that the English had been finally expelled from
the country, they were beginning to call attention to the fact
that this King had gotten his crown by the hands of a person
proven by the priests to have been in league with Satan and
burnt for it by them as a sorceress—therefore, of what value
or authority was such a Kingship as that? Of no value at all;
no nation could afford to allow such a king to remain on the
throne.

It was high time to stir, now, and the King did it. That is
how Charles VII. came to be smitten with anxiety to have
justice done the memory of his benefactress.

He appealed to the Pope, and the Pope appointed a great
commission of churchmen to examine into the facts of Joan's
life and award judgment. The Commission sat at Paris, at
Domremy, at Rouen, at Orleans, and at several other places,
and continued its work during several months. It examined
the records of Joan's trials, it examined the Bastard of Orleans, and the Duke d'Alençon, and D'Aulon, and Pasquerel,
and Courcelles, and Isambard de la Pierre, and Manchon, and
me, and many others whose names I have made familiar to
you; also they examined more than a hundred witnesses
whose names are less familiar to you—friends of Joan in
Domremy, Vaucouleurs, Orleans, and other places, and a number of judges and other people who had assisted at the Rouen
trials, the abjuration, and the martyrdom. And out of this
exhaustive examination Joan's character and history came
spotless and perfect, and this verdict was placed upon record,
to remain forever.

I was present upon most of these occasions, and saw again many faces which I have not seen for a quarter of a century; among them some well-beloved faces—those of our generals and that of Catherine Boucher (married, alas!), and also among them certain other faces that filled me with bitterness—those of Beaupère and Courcelles and a number of their fellow-fiends. I saw Haumette and Little Mengette—edging along toward fifty, now, and mothers of many children. I saw Noël's father, and the parents of the Paladin and the Sun-flower.

It was beautiful to hear the Duke d'Alençon praise Joan's splendid capacities as a general, and to hear the Bastard in-dorse these praises with his eloquent tongue and then go on and tell how sweet and good Joan was, and how full of pluck and fire and impetuosity, and mischief, and mirthfulness, and tenderness, and compassion, and everything that was pure and fine and noble and lovely. He made her live again be-fore me, and wrung my heart.

I have finished my story of Joan of Arc, that wonderful child, that sublime personality, that spirit which in one re-gard has had no peer and will have none—this: its purity from all alloy of self-seeking, self-interest, personal ambition. In it no trace of these motives can be found, search as you may, and this cannot be said of any other person whose name appears in profane history.

With Joan of Arc love of country was more than a senti-ment—it was a passion. She was the Genius of Patriotism—she was Patriotism embodied, concreted, made flesh, and pal-pable to the touch and visible to the eye.

Love, Mercy, Charity, Fortitude, War, Peace, Poetry, Mu-sic—these may be symbolized as any shall prefer: by figures of either sex and of any age; but a slender girl in her first young bloom, with the martyr's crown upon her head, and in her hand the sword that severed her country's bonds—shall not this, and no other, stand for PATRIOTISM through all the ages until time shall end?

SOME BOOKS FOR THE LIBRARY

THE ABBEY SHAKESPEARE.
The Comedies of William Shakespeare. With Many Drawings by EDWIN A. ABBEY, Reproduced by Photogravure. Four Volumes. Large 8vo, Half Cloth, Deckel Edges and Gilt Tops, $30 00. Net. (*In a Box.*)

MARK TWAIN.
The following volumes will be issued shortly:
THE ADVENTURES OF HUCKLEBERRY FINN. Illustrated.—A CONNECTICUT YANKEE IN KING ARTHUR'S COURT. Illustrated. Other volumes to follow.

ON SNOW-SHOES TO THE BARREN GROUNDS.
Twenty-eight Hundred Miles after Musk-Oxen and Wood-Bison. By CASPAR WHITNEY. Illustrated. 8vo, Cloth, Ornamental. (*In Press.*)

MEMOIRS OF BARRAS,
Member of the Directorate. Edited, with a General Introduction, Prefaces, and Appendices, by GEORGE DURUY. Translated. With Seven Portraits in Photogravure, Two Fac-similes, and Two Plans. Complete in Four Volumes. 8vo, Cloth, Uncut Edges and Gilt Tops. Vols. I. and II., $3 75 each. Vols. III. and IV., *just ready.*

FROM THE BLACK SEA
Through Persia and India. Written and Illustrated by EDWIN LORD WEEKS. With Photogravure Portrait. 8vo, Cloth, Ornamental, Uncut Edges and Gilt Top, $3 50.

A FEW MEMORIES.
By MARY ANDERSON (MADAME DE NAVARRO). With Six Portraits, of which Five are Photogravures. 8vo, Cloth, Deckel Edges and Gilt Top, $2 50.

NOTES IN JAPAN.
Written and Illustrated by ALFRED PARSONS. Crown 8vo, Cloth, Ornamental, Uncut Edges and Gilt Top, $3 00.

PONY TRACKS.
Written and Illustrated by FREDERIC REMINGTON. 8vo, Cloth, Ornamental, $3 00.

STOPS OF VARIOUS QUILLS.

Poems. By W. D. HOWELLS. With Illustrations by HOWARD PYLE. 4to, Cloth, Ornamental, Uncut Edges and Gilt Top, $2 50.

Limited Edition of Fifty Numbered Copies on Hand-made Paper, signed by Mr. HOWELLS and Mr. PYLE. 4to, Deckel Edges, Half Cloth, Gilt Top, $15 00. (*In a Box.*)

OUR EDIBLE TOADSTOOLS AND MUSHROOMS,

And How to Distinguish Them. By W. HAMILTON GIBSON. With Thirty Colored Plates, and Fifty-seven other Illustrations by the Author. 8vo, Cloth, Ornamental, Uncut Edges and Gilt Top, $7 50. (*In a Box.*)

STUDIES OF MEN.

By GEORGE W. SMALLEY, Author of "London Letters, and Some Others." Crown 8vo, Cloth, Uncut Edges and Gilt Top, $2 50.

RHODES'S UNITED STATES.

History of the United States from the Compromise of 1850. By JAMES FORD RHODES. With Maps. 8vo, Cloth, Uncut Edges and Gilt Top. Vols. I. and II., 1850–60, $5 00 ; Vol. III., 1860–62, $2 50.

GREEN'S SHORT HISTORY. *Illustrated.*

A Short History of the English People. By JOHN RICHARD GREEN. Illustrated Edition. Edited by Mrs. J. R. GREEN and Miss KATE NORGATE. In Four Volumes. With Colored Plates, Maps, and Numerous other Illustrations. Royal 8vo, Illuminated Cloth, Uncut Edges and Gilt Tops, $5 00 per volume.

FOUR AMERICAN UNIVERSITIES.

Harvard, Yale, Princeton, and Columbia. With Many Illustrations. 4to, Cloth, Ornamental, $3 50.

CHAPTERS FROM SOME UNWRITTEN MEMOIRS.

By ANNE THACKERAY RITCHIE. Author of " Records of Tennyson, Ruskin, Browning," etc. Crown 8vo, Cloth, Ornamental, Uncut Edges and Gilt Top, $2 00.

PUBLISHED BY HARPER & BROTHERS, NEW YORK

☞ *The above works are for sale by all booksellers, or will be mailed by the publishers, postage prepaid, on receipt of the price.*

AFTERWORD

Susan K. Harris

"I dare say there are a good many faults in the book," William Dean Howells wrote of Mark Twain's *Personal Recollections of Joan of Arc* in 1896. "It is unequal; its archaism is often superficially a failure; if you look at it merely on the technical side, the outbursts of the nineteenth-century American in the armor of the fifteenth-century Frenchman are solecisms. But in spite of all this, the book has a vitalizing force. Joan lives in it again, and dies, and then lives on in the love and pity and wonder of the reader."[1]

For Howells, Twain's old friend and staunch supporter, *Joan of Arc*'s pathos transcended its technical faults. Few other critics agreed with him. The most stringent criticism came from James Westfall Thompson, in a long review essay in the *Dial* that surveyed Francis C. Lowell's historical *Joan of Arc*, Mrs. Oliphant's biographical *Jeanne d'Arc, Her Life and Death*, and Mark Twain's novel, all published in 1896. Exhibiting values antithetical to Howells', Thompson insisted that a life as remarkable as Joan's did not need fictionalizing; for him, Twain's book was neither good history nor good literature. Condemning its "labored spontaneity" and artificial style, Thompson targeted *Joan*'s anachronisms as its worst fault, taking special offense at the populist mentality displayed by its narrator, the Sieur Louis de Conte. Mark Twain's own politics, this reviewer perceived, played havoc with his desire to write historical fiction.[2]

Most subsequent readers have sided with Thompson. Though few have so vehemently insisted on historical accuracy, nearly all have expressed concern about Twain's controlling political presence in the text, and about his narrator's stylistic inconsistencies. Twain's attitude toward Joan, too, has constituted a problem for twentieth-century readers, who often interpret his veneration of her either as mere sentiment or as evidence of his own sexual immaturity. Finally, readers are often troubled by the novel's balance between history and fiction.

But it is helpful to put *Personal Recollections of Joan of Arc* into the contexts from which it sprang. The most telling may be the fact that, as Thompson's review indicates, the *fin de siècle* saw an outpouring of books, articles, paintings, plays, and other material focusing on Joan of Arc. Fascination with her had been building for half a century, ever since she was "rediscovered" as a result of the publication, for the first time, of the manuscripts of her trial. It is sometimes difficult for us to remember that she was not canonized until 1920. Prior to the turn of the century she primarily existed as a legend, generally portrayed in France as a popular hero (although her countryman Voltaire burlesqued her), and in England, not surprisingly, as an impostor. In 1840 Jules Quicherat, a French historian, discovered the Latin manuscripts of her trial, translated them into French, and published them between 1841 and 1849. Other historians flocked to the new materials, the best known being Jules Michelet. Most were awed by the Maid's accomplishments and by her religious sincerity. Since the concept of historical objectivity had not yet been invented, the early histories of Joan tended to be extremely adulatory.

Quicherat's discovery did invite awe. The record of Joan's trial still makes sensational reading. All commentators remark on Joan's extraordinary poise, wit, and political savvy; none can explain how an illiterate peasant girl of seventeen (nineteen when tried and executed) could have accomplished her military and political feats. As Howells noted, explanations are easiest for the faithful: for them Joan was, as she claimed, inspired by God. For rationalists, explaining her is more difficult; many, like George Bernard Shaw, have opted for genius (itself a nineteenth-century concept developed to secularize the idea of divine inspiration). Complicating interpretations of her acts, for these rediscoverers as for her original prosecutors, was the fact of her sex: while part of the Inquisition's case against her focused on her refusal to wear women's clothes, part of her rediscoverers' problem lay in their own discomfort with her masculine behavior. Early hostile historians had solved this problem easily by labeling her a whore; late-nineteenth- and early-twentieth-century writers handled it by transferring their own ambivalences to her, portraying her as torn between her nationalist commitments and her simple love of home, stressing her affection for her family, and recording her laments that the life of a normal woman was denied her. For Michelet, on whom Twain depended most heavily, Joan's womanliness was central to her power, enabling her, for instance, to persuade the wild Armagnac soldiers to come to confession and to dismiss their prostitutes. Even Shaw, a leading feminist for his age, portrayed Joan within a decidedly womanly, if not feminine, frame. Themselves confused about the relationship between biology and culture, sexuality and behavior, few of Joan's sympathetic rediscoverers were able to reconcile her gender and her deeds.

Certainly this cultural ambivalence lies at the heart of *Personal Recollections of Joan of Arc*. Like Michelet, Twain insists that beneath her military facade, donned only to expedite her mission, Joan was a model woman: gentle, sensitive, self-effacing, feminine. In Twain's reading, Joan's most womanly act lies precisely in the self-sacrifice she makes when she assumes masculine prerogatives, from dress to military power; in her heart, he claims, she would rather be home with her mother. In addition to stressing her youth, emotional vulnerability, and selflessness, Twain partially resolves Joan's gender contradictions by suggesting her maternalism; in this text Charles VII and France are Joan's children, whom her nurturing instincts compel her to protect. Emotionally the devoted daughter and mother, Twain's Joan balances her masculine behavior with her feminine sensibility.

Twain's and his contemporaries' ambivalence is especially interesting within the context of the feminist movement of the late nineteenth and early twentieth centuries. The years between 1890 and the First World War constituted a watershed for white middle-class western women, who suddenly began to achieve some of the goals they had been working toward for at least fifty years. During this period institutions of higher learning hitherto closed to women became accessible, and white women became a noticeable presence in the workforce. On the domestic front, many women realized that they could be happy living alone, or with another woman, thus undermining the demand that all women be enmeshed in heterosexual families. Additionally, sexuality began to be discussed, which helped middle-class women understand their own drives, and contraceptive information began to become available, especially in England, facilitating women's control over their own reproduction. On the trivial front, women cut their hair, shortened their skirts, and learned to ride bicycles. In the United States the movement culminated in women's gaining the vote in 1920.

But freedom for these "New Women" did not come easily. The psychological battle was as difficult as the political and economic ones. Olive Schreiner, a South African feminist, listed numerous cases of "asthma," as she called it, among her friends and acquaintances,[3] and both sympathetic and hostile observers recorded other manifestations of stress and of the deep, almost pathological unhappiness many women experienced concomitantly with their new lifestyles. The constant carping on the subject by the popular press meant that New Women were surrounded by caricatures of themselves, and conservative commentators, like the anti-abortion forces of today, never let them forget that they were living in a state of sin. In the midst of their hard-won freedoms, the New Women were often haunted by doubt and by the open disapproval of their contemporaries. On the subject of women's roles, as on so many other subjects, turn-of-the-century western culture was deeply and emotionally conflicted, and the women themselves bore the brunt of it.

These conflicts are reflected in contemporary portrayals of Joan of Arc. Coming to attention during the latter part of the nineteenth century, the figure who could have been a symbol for the New Woman was as often coopted by the conservative forces as by the liberal, becoming an anomaly used to highlight women's "essential" femininity and, most significantly, to support traditional sexual values. Cultural anxieties about women's sexuality were transformed into a celebration of Joan's virginity; cultural anxieties about women's domestic roles were transformed into the insistence that Joan's personal preference would have been to stay home spinning instead of leading the French armies to victory. At heart a nineteenth-century "True" woman rather than a twentieth-century "New" one, Joan could be heralded as a leader whose heroism lay, paradoxically, in her femininity, especially in her feminine sacrifice of self for her country, her God, and her king.

Certainly this ideology informs Mark Twain's text. Although the evidence suggests that he and his wife, Olivia Langdon Clemens, had a sexually active and fulfilling marriage, on paper at least Twain was always a sexual conservative, especially (though not exclusively) in regard to women. His dislike of female sexual display of any kind amounted to a neurosis, and the contemplation of rape seems to have paralyzed him. But he also recognized, increasingly as he aged and experienced his family of growing daughters, that women were capable of far more physical and intellectual activity than they were generally permitted. At the same time, he was deeply interested — some have said obsessed — with young girls. Hence Joan of Arc was for him a fascinating female, both virginal and intellectually gifted, feminine and physically active. The ambivalences of the age, which entered into Twain's own sexual codes, are manifest in the framing and handling of gender issues in his novel about the Maid of Orléans.

But gender roles are not the only problematic issue in this text. As the reviewer for the *Dial* was the first — but certainly not the last — to note, it was extremely odd for Mark Twain to choose to celebrate a Catholic heroine, since he himself was virulently, almost viscerally anti-Catholic. Growing up in the Protestant sector of the Mississippi Valley, with its longstanding hostility to the Catholics who traded up and down the river — from the French in the North to the Spanish in the South — Samuel Clemens imbibed anti-Catholicism early, and thoroughly. Unlike Harriet Beecher Stowe, who grew up in an equally anti-Catholic environment at an even earlier period but became interested in Catholicism and its cultures and eventually attended the Episcopal Church, Clemens was never able to shake his intense distrust of Catholic theology, church personnel, and the medieval culture that the church produced. Even in novels such as *Joan of Arc* and *No. 44, The Mysterious Stranger*, which take place in medieval Europe and feature Catholic characters, Twain's dislike for and distrust of the church is paramount. In his writings, a good priest is an anomaly, a character to be remarked — not a promising framework for works proposing to explore people and events of the medieval world. Despite his occasional efforts to overcome his prejudices, Twain, like many of his Protestant contemporaries, was convinced that the church was the anti-Christ.

This ideology is painfully evident in *Personal Recollections of Joan of Arc*. As William Searle notes, in writing a "reverent biography" of Joan, Twain was working in opposition to his own beliefs, especially the philosophical determinism that hardened as he aged.[4] The result could not be other than strained. Structurally, the greatest strain falls on the narrator, the Sieur Louis de Conte, who becomes the mouthpiece for Twain's mixed intentions. On the one hand, de Conte, like August Feldner in *No. 44, The Mysterious Stranger*, is (consciously at least) a devout Catholic. On the other hand, he harbors reservations about Catholic doctrine and resentments toward church authority that no fifteenth-century Frenchman could have conceived of. Even when Twain is trying to present de Conte as double-voiced, speaking truths he wots not of, à la Huck Finn, he is unable to sustain the fiction of a naive narrator.

In part this is the result of de Conte's retrospective stance; his putative "recollections" are told from his old age. But it is also the result of Mark Twain's inability to believe in Catholic doctrine, even in the guise of a fictional character, and of his conviction that those who do believe are gullible. By his logic, sophisticated Catholics must be evil, hypocrites who prate church doctrine and manipulate believers for their own ends. Since he does not intend de Conte as an evil character, Twain tries to make him a naif who entertains some doubts. At the same time, however, he makes de Conte a fledgling democrat, a literate nobleman, and an alienated observer. In the end, de Conte's Catholicism is manifested more as a target of Twain's satire than as an attempt at reproducing the mind-set of a fifteenth-century memoirist.

If our standards for a good historical novel require characters to reflect the ideologies of their own eras, *Joan* certainly fails in this respect, as early reviewers noted. But for historiographically minded readers, this aesthetic flaw may be offset by the book's representation of late-nineteenth-century controversies. Mark Twain worked well with historical materials, and the trial records on which this novel is based are skillfully handled. Twain's sequencing of events, his cast of characters (many of them historically verifiable), and his use of the primary documents reprinted by Michelet are impressive. The novel's problem, for most readers, lies not in the facts presented but in their interpretation through a narrator who speaks for Mark Twain's time rather than for his own. If, however, we approach this novel not through a requirement that it display historically appropriate ideologies but through an interest in its display of its own era's concerns, we can see that its celebration of Joan generally, its portrayal of her as a "True Woman" despite her transgressions into masculine territories, and its peculiarly American mistrust of Catholic sensibilities all mark Twain's Joan of Arc as a text richly reflective of surrounding cultural discourses.

But *Joan of Arc* is not only a bundle of contemporary debates. It is also distinctively a product of Mark Twain's own sensibility. This is particularly evident in its narrative structure, which, as most early reviewers remarked, is extremely odd. Masking as a translation from the memoirs of Joan's secretary and old friend, the text distances itself from Joan as much as possible. De Conte's reverence toward Joan only serves to increase this distance. The result, for many, is hagiography rather than either history or biography; some also note that in fact — as the text actually indicates quite openly — *Personal Recollections* is far more a record of de Conte's life than of Joan's. Perhaps it is most useful to see it as a collage of cultural discourses ordered by Mark Twain's sensibility. The text's portrayal of Joan, as we have seen, reflects turn-of-the-century conflicts about the status, and nature, of women; its portrayal of Catholic doctrine and world view reflects North American paranoia about the church in general and Twain's own distaste in particular; and de Conte's incipient populism reflects a combination of American self-confidence and Clemens' own ardent republicanism. But while these major discourses in the novel act at cross-purposes, they are united and made coherent through de Conte's narrative alienation, which as an element of *Joan of Arc* holds the text together, and as an element of Mark Twain's writing echoes many of his other first-person narrators, from Huck Finn to August Feldner and Theodor Fischer of the late, unfinished "Mysterious Stranger" manuscripts.

Though not the focal character, the Sieur Louis de Conte is the most powerful character in *Personal Recollections of Joan of Arc*. Both narrator and actor, at once intimately engaged in the events of his age and agonizingly disaffected from the zeitgeist, he controls readers' access to every element of the text. Consequently we cannot examine any part of *Joan* without examining de Conte's relation to it. In gender terms, this has led one reader to argue that Twain's work is in fact feminist in outlook; in "Woman as Force in Twain's *Joan of Arc*: The Unwordable Fascination," Christina Zwarg suggests that Twain parodies prevailing historiographical modes, especially as they wrote women out of the story of the past. De Conte's narration, which tells Joan's story but rarely permits her to speak, is for Zwarg far more "about" the writing of history than about the complexities of Joan herself.[5]

Less determinedly deconstructive analyses of de Conte's relation to the novel tend to be less convinced of Twain's historiographical acumen. While he was probably *the* master deconstructionist in nineteenth-century America, Twain rarely brought this talent to bear on issues concerning women or Catholicism. Nevertheless, the point with which Zwarg begins her argument, a point central, as she notes, to the feminist critique, is absolutely true: the layers of tellers of this story, from Clemens to Twain to "Jean François Alden" (the putative translator of de Conte's text) to de Conte himself, make it almost impossible for readers to confront whatever vestiges of the historical Joan might exist. Only by isolating her responses at the trial from de Conte's incessant interpretations do we have any chance to "hear" Joan speak.

In religious terms, de Conte's relation to the story he tells is equally complicated. Awkwardly and anachronistically positioned as a son of the church (an orphan, he was reared and educated by a "good" priest), he is finally completely estranged from both the religion and its culture — which means he is finally estranged from his time. De Conte does project another possibility for faith, but this spiritual "Other" is itself an anachronism in de Conte's era, a throwback, as some critics have seen it, to European nature religions predating (and, some argue, surviving as an underground current along with) Christianity. In the novel this spiritual alternative is figured by the "Arbre Fée de Bourlemont," the Fairy Tree, which grows near Domrémy, Joan's and de

Conte's childhood home. The historical record does suggest that some kind of tree played a part in a minor nature cult in the lives of Domrémy villagers during Joan's childhood; her interlocutors tried to make something of it, and failed.[6] Twain, however, has the tree assume major significance, representing uncorrupted nature and spiritual purity. Inhabited by fairies who are friends to the village children, the tree in Twain's novel has been a beneficent force in Domrémy life for five hundred years prior to de Conte's story, but the priest banishes the fairies during de Conte's adolescence. De Conte never recovers from the blow; his guardian's act becomes, within the novel's structure, the first of his experiences of disillusionment with the church. Despite the fairies' banishment, for de Conte the tree stands as an alternative to the church that has failed him. The Fairy Tree also rounds out his life, appearing in his dreams to signal his impending, and welcome, death: "I myself, old and broken, wait with serenity; for I have seen the vision of the Tree," he tells us at the opening of his tale. "I have seen it, and am content" (10).

The alienation initiated by the incident of the tree becomes the hallmark of de Conte's life, and in this he is a quintessentially Mark Twain character. Never totally a part of Domrémy life — his parents, minor nobility from Paris, were killed when he was six years old, which means he is neither rooted in Domrémy nor a peasant like his companions, and he and the priest are the only villagers who can read and write — de Conte expresses an almost pre-modernist sense of unease. While his rank and education make him an anomaly in Domrémy, outside the village his loyalty to Joan and his disaffection from the church alienate him from the noblemen and churchmen among whom he should feel at home. Additionally, he claims to be the only one of Joan's acquaintances who can "hear" her muttered, apparently only semiconscious prophecies — a dubious privilege, since it means de Conte alone knows that she has prophesied her own death. Like all autobiographical narrators, he is privileged to know the end of the story before he begins telling it, and this privileged position sets him apart. Like Hank Morgan, or like August Feldner of *No. 44, The Mysterious Stranger*, he is constantly aware of his difference from his fellows. Like the protagonists of all of the unfinished stories collected under the umbrella title *Which Was the Dream?*,[7] he feels separated from his time, adrift in an uncertain universe. The strain of alien-ation that marks Twain's writing almost from its inception is intrinsic to de Conte's sense of self and to the narrative structuring of *Joan of Arc*.

So, too, is the collection of images that provide fragile respite from that alienation. The Fairy Tree, stripped of its inhabitants but still retaining sym-bolic power, is intimately associated in de Conte's account with both the image of Joan and the notion of home. Many of the most lyric passages in *Joan of Arc* occur in de Conte's memories of his and Joan's Domrémy child-hood, and these passages signal a fusion of images representing what he can imagine of peace and contentment. "I know that when the Children of the Tree die in a far land, . . . they turn their longing eyes toward [the Fairy Tree and] the bloomy mead sloping away to the river, and to their perishing nos-trils is blown faint and sweet the fragrance of the flowers of home," he tells us in his opening description of Domrémy (11). This peacefulness is not unrelat-ed to the tranquility Huck and Jim experience on the river, or the peace Mark Twain "remembers" about summers spent on his Uncle John Quarles' Missouri farm.[8] This is where the figuring of Joan as the essential female is most operative; in de Conte's imagination she is part of nature, of uncorrupt-ed childhood, of safety and security. For him, beneath her military mask lies his only hope of spiritual salvation.

As almost every reader has perceived, Mark Twain was not particularly interested in women in his fiction, and most of his female characters are not very complex or in the end very interesting. His male characters, however, are pivotal figures in American literature, looking back toward oral traditions and various forms of the folktale, and forward to the anxious, unsettled protago-nists of the modernist era. In *Personal Recollections of Joan of Arc by the Sieur Louis de Conte*, Twain's male narrator is far more central to his concerns, espe-cially his sense of alienation, than is the figure that gives the narrator reason for being. One of the last of his completed novels, *Joan of Arc* is marked by grief — not the grief de Conte consciously expresses about Joan's martyrdom, but the unconscious grief he represents. This is the grief of loss, of disloca-tion, of disaffection. It is the grief that accompanied the euphoria of late-nine-teenth-century western progress, and it is the grief that highlighted Mark

Twain's own progress through life. It is the grief of the age, and like the other cultural discourses informing *Joan of Arc*, it reflects a communal experience refracted through one particular author's sensibility. In this, *Personal Recollections of Joan of Arc* is an important book. Carefully researched, written with passion, it is one of those texts that tells us more about its writer and his period than about the events it purports to record. It is especially important for students of Mark Twain, but it is also significant for students of late-nineteenth-century American culture. Both as a testament to the cult of Joan of Arc and as a document of *fin-de-siécle* loss and alienation, Twain's novel points to some of the major currents, and contradictions, of turn-of-the-century life and thought.

NOTES

1. William Dean Howells, review of *Personal Recollections of Joan of Arc* by Mark Twain, *Harper's Weekly*, vol. 41 (May 30, 1896), 535–36.

2. James Westfall Thompson, "The Maid of Orléans," New Books column, *Dial*, vol. 20 (June 16, 1896), 351–56.

3. Lloyd Fernando, *"New Women" in the Late Victorian Novel* (University Park: Pennsylvania State Univ. Press, 1977), 21–23.

4. William Searle, *The Saint and the Skeptics: Joan of Arc in the Work of Mark Twain, Anatole France, and Bernard Shaw.* (Detroit: Wayne State Univ. Press, 1976), 15.

5. Christina Zwarg, "Woman as Force in Twain's *Joan of Arc*: The Unwordable Fascination," *Criticism* 27 (Winter 1985), 57–72.

6. William P. Barrett, *The Trial of Jeanne d'Arc: Translated into English from the Original Latin and French Documents* (n.p.: Gotham House, 1932), 54–55. Most commentators on *Personal Recollections of Joan of Arc* have discussed the Fairy Tree in terms of its relation to nature cults. See especially Albert E. Stone, *The Innocent Eye: Childhood in Mark Twain's Imagination* (New Haven: Yale Univ. Press, 1961), 223–24; James D. Wilson, "In Quest of Redemptive Vision: Mark Twain's *Joan of Arc*," *Texas Studies in Language and Literature* 20, no. 1 (Spring 1978), 181–98; and Roger B. Salomon, *Twain and the Image of History* (New Haven: Yale Univ. Press, 1961), 174, 180–84.

7. John S. Tuckey, ed., *Mark Twain's Which Was the Dream? and Other Symbolic Writings of the Later Years* (Berkeley: Univ. of California Press, 1968).

8. Clemens' memories of John Quarles' farm are scattered throughout his writings. See especially *Mark Twain's Autobiography*, ed. Albert B. Paine, 2 vols. (New York: Harper, 1924).

FOR FURTHER READING

Susan K. Harris

Readers interested in histories of Joan of Arc should look at William P. Barrett's *The Trial of Jeanne d'Arc: Translated into English from the Original Latin and French Documents* (n.p.: Gotham House, 1932) and Vita Sackville-West's *Saint Joan of Arc* (New York: Literary Guild, 1936). Evaluations of Twain's handling of the story can be found in Roger B. Salomon's *Twain and the Image of History* (New Haven: Yale Univ. Press, 1961), William Searle's *The Saint and the Skeptics: Joan of Arc in the Work of Mark Twain, Anatole France, and Bernard Shaw* (Detroit: Wayne State Univ. Press, 1976), Susan K. Harris's *Mark Twain's Escape from Time: A Study of Patterns and Images* (Columbia: Univ. of Missouri Press, 1982), Rolande Ballorain's "Mark Twain's Capers: A Chameleon in King Carnival's Court," in *American Novelists Revisited: Essays in Feminist Criticism*, ed. Fritz Fleischmann (Boston: G. K. Hall, 1982), and J. D. Stahl's *Mark Twain, Culture and Gender: Envisioning America Through Europe* (Athens: Univ. of Georgia Press, 1994). Discussions of the New Woman in history and literature can be found in Lloyd Fernando's *"New Women" in the Late Victorian Novel* (University Park: Pennsylvania State Univ. Press, 1977), Gail Cunningham's *The New Woman and the Victorian Novel* (New York: Macmillan, 1978), Patricia Marks's *Bicycles, Bangs, and Bloomers: The New Woman in the Popular Press* (Lexington: Univ. Press of Kentucky, 1990), and Adele Heller and Lois Rudnick, eds., *1915, The Cultural Moment: The New Politics, the New Woman, the New Psychology, the New Art and the New Theatre in America* (New Brunswick, N.J.: Rutgers Univ. Press, 1991).

ILLUSTRATORS AND ILLUSTRATIONS
IN MARK TWAIN'S FIRST AMERICAN EDITIONS

Beverly R. David & Ray Sapirstein

From the "gorgeous gold frog" stamped into the cover of *The Celebrated Jumping Frog of Calaveras County* in 1867 to the comet-riding captain on the frontispiece of *Extract from Captain Stormfield's Visit to Heaven* in 1909, illustrators and illustrations were an integral part of Mark Twain's first editions.

Twain marketed most of his major works by subscription, and illustration functioned as an important sales tool. Subscription books were packed with pictures of every type and size and were bound in brassy gold-stamped covers. The books were sold by agents who flipped through a prospectus filled with lively illustrations, selected text, and binding samples. Illustrations quickly conveyed a sense of the story, condensing the proverbial "thousand words" and outlining the scope and tone of the work, making an impression on the potential purchaser even before the full text had been printed. Book canvassers were rewarded with up to 50 percent of the selling price, which started at $3.50 and ranged as high as $7.00 for more ornate bindings. The books themselves were seldom produced until a substantial number of customers had placed orders. To justify the relatively high price and to reassure buyers that they were getting their money's worth, books published by subscription had to offer sensational volume and apparent substance. As Frank Bliss of the American Publishing Company observed, these consumers "would not pay for blank paper and wide margins. They wanted everything filled up with type or pictures." While authors of trade books generally tolerated lighter sales, gratified by attracting a "better class of readers," as Hamlin Hill put it, authors of subscription books sacrificed literary respectability for popular appeal and considerable profit.[1]

The humorist George Ade remembered Twain's books vividly, offering us a child's-eye view of the nineteenth-century subscription book market.

Just when front-room literature seemed at its lowest ebb, so far as the American boy was concerned, along came Mark Twain. His books looked at a distance, just like the other distended, diluted, and altogether tasteless volumes that had been used for several decades to balance the ends of the center table . . . so thick and heavy and emblazoned with gold that [they] could keep company with the bulky and high-priced Bible. . . . The publisher knew his public, so he gave a pound of book for every fifty cents, and crowded in plenty of wood-cuts and stamped the outside with golden bouquets and put in a steel engraving of the author, with a tissue paper veil over it, and "sicked" his multitude of broken-down clergymen, maiden ladies, grass widows, and college students on the great American public.

Can you see the boy, Sunday morning prisoner, approach the book with a dull sense of foreboding, expecting a dose of Tupper's *Proverbial Philosophy?* Can you see him a few minutes later when he finds himself linked arm-in-arm with Mulberry Sellers or Buck Fanshaw or the convulsing idiot who wanted to know if Christopher Columbus was sure-enough dead? No wonder he curled up on the hair-cloth sofa and hugged the thing to his bosom and lost all interest in Sunday school. *Innocents Abroad* was the most enthralling book ever printed until *Roughing It* appeared. Then along came *The Gilded Age, Life on the Mississippi,* and *Tom Sawyer. . . .* While waiting for a new one we read the old ones all over again.[2]

Publishers, editors, and Twain himself spent a good deal of time on design — choosing the most talented artists, directing their interpretations of text, selecting from the final prints, and at times removing material they deemed unfit for illustration.[3]

With the exception of *Following the Equator* (1897), books released in the twilight of Twain's career were not sold by subscription. Twain's later books, published for the trade market by Harper and Brothers, seldom contained more than a frontispiece and a dozen or so tasteful illustrations, rather than the hundreds of illustrations per volume that subscription publishing demanded. Illustration, however, remained a major component of Twain's later work in two important cases: *Extracts from Adam's Diary,* illustrated by Fred

Strothmann in 1904, and *Eve's Diary*, illustrated by Lester Ralph in 1906.

The stories behind the illustrators and illustrations of Mark Twain's first editions abound in back-room intrigue. The besotted or negligent lapses of some of the artists and the procrastinations of the engravers are legendary. The consequent production delays, mistimed releases, and copyright infringements all implied a lack of competent supervision that frequently infuriated Twain and ultimately encouraged him to launch his own publishing company.

In many cases, Twain took illustrations into account as he wrote and edited his text, using them as counterpoint and accompaniment to his words, often allowing them to inform his general narrative strategy and to influence the amount of detail he felt necessary to include in his written descriptions. In the most artful and carefully considered illustrated works, an analysis of the relationships between author and illustrator and between text and pictures illuminates key dimensions of Twain's writings and the responses they have elicited from readers. Examinations of even the most straightforward examples of decorative imagery yield insights into the publishing history of Twain's books and his attitudes toward the production process.

The original illustrations in Twain's works have often been replaced in the twentieth century by subsequent visual interpretations. But while Norman Rockwell's well-known nostalgic renderings of *Tom Sawyer* and *Huckleberry Finn* may tell us much about 1930s sensibilities, we would do well to reacquaint ourselves with the first American editions and the artwork they contained if we want to understand the books Twain wrote and the world they affected.

Illustrated books, like the illustrated weekly magazines that first appeared in the 1860s, were a significant source of visual images entering nineteenth-century homes. Because of their widespread popularity and the relative paucity of other sources of visual information, Twain's books helped to define America's images of remote people, exotic scenes, and historic events. In addition to being an essential element of Mark Twain's body of work, illustrations are a documentary source in their own right, a window into Twain's world and our own.

NOTES

1. For background on subscription book publishing, see Hamlin Hill, *Mark Twain and Elisha Bliss* (Columbia: University of Missouri Press, 1964), chapter 1. See also R. Kent Rasmussen, "Subscription-book publishing" entry, *Mark Twain A to Z: The Essential Reference to His Life and Writings* (New York: Facts on File, 1995), p. 448.

2. George Ade, "Mark Twain and the Old-Time Subscription Book," *Review of Reviews* 61 (June 10, 1910): 703–4; reprinted in Frederick Anderson, ed., *Mark Twain: The Critical Heritage* (London: Routledge and Kegan Paul, 1971), pp. 337–39.

3. Beverly R. David, *Mark Twain and His Illustrators, Volume 1 (1869–1875)* (Troy, N.Y.: Whitston Publishing Company, 1986), discusses in detail Twain's involvement in the production of his early books.

READING THE ILLUSTRATIONS IN *JOAN OF ARC*

Ray Sapirstein

Unlike many of Mark Twain's previous illustrators, Frank Vincent Du Mond (1865–1951) was more an artist than an illustrator. His academic style and artistic credentials lent *Personal Recollections of Joan of Arc* (1896) an aura of seriousness — an indication of Twain's retreat into conventional respectability and sentimentality. Born in 1865 in Rochester, New York, Du Mond was studying in France when Twain met him in 1894 and later hired him.[1] His name is the only authentic attribution on a title page packed with fictions. With its French sound — perhaps a factor in Twain's choice of the artist — the name tacitly validates the work and maintains the impression of French authorship that Twain sought to foster in hopes of distancing *Joan of Arc* from his reputation as a humorist.

A painter and teacher at the Art Students League in New York at the time the work was published, Du Mond had been trained in Paris and had earned a medal in the Paris Salon of 1890. Firmly entrenched in the academic art establishment, he directed the Department of Fine Arts at the Lewis and Clark Exposition of 1905, and received several prizes at various international expositions from 1901 through 1915, well after modern and impressionist art migrated from France to the United States.[2]

While Du Mond executed the majority of the images that appear in *Joan of Arc*, the book also features several "reproductions from old paintings and statues" duplicated photographically and printed in ink, a process still being perfected in 1896. Du Mond's illustrations are typically signed with a stylized monogram at the lower left, and the reproductions are attributed and captioned. Otherwise Du Mond's work is not easily distinguishable from the "old paintings." It suggests a Franco-American variant of the neo-medievalist, Pre-Raphaelite painting style popular during the nineteenth century and promoted by such English artist-theorists as John Ruskin, William Morris, Edward Burne-Jones, and Dante Gabriel Rossetti. A few years earlier and in a different turn of mind, Twain might have found the style abhorrent.

Melodramatic and nationalistic idealizations of the medieval and chivalric traced their origin to Sir Walter Scott, Twain's literary arch-nemesis, the frequent target of his vituperation.

Du Mond's illustrations appear to have been fastidiously researched, in the sense that the props, costumes, and insignia seem historically accurate and faithfully rendered. Yet history painting, like much historical writing at the time, often indulged in crude subjectivity, adulation, and melodrama, using minutely researched costumes and settings to grant interpretations the appearance of established truth. Expert rendering and accurate detail tend to give the illusion of reality to what is often no more than a rough representation based upon little concrete information. While Du Mond's attention to detail made his drawings seem more real, they also conferred upon Twain's words a semblance of historical accuracy and actuality.

Du Mond might have undertaken to counterpoint Twain's voice, but instead the illustrations generally offer a prosaic, umabiguous reading, minimally challenging and modifying the text. Du Mond was a deferential accompanist, content to provide decorative interludes rather than substantive collaboration. Taking his cue from Twain, he depicted Joan of Arc as a maternal figure costumed in a decidedly matronly and conservatively feminine fashion, a contradiction of the historical record, as Susan Harris points out in the afterword to this volume. In the bulk of the illustrations, she wears a hybrid between petticoated skirts and a suit of armor, and her bearing recalls the reformist militarism of the Salvation Army of the late nineteenth century rather than any costume of the fifteenth. Joan appears decidedly older than her nineteen years, more the "true" woman in Du Mond's work than in the "old paintings," which more ambiguously depict her in a full suit of armor. Throughout the book, Du Mond's Joan wears male garb as little as possible, dressing in a full skirt on horseback and in battle, and only rarely donning tunic and tights. Twain actually seems to have grown to resent Du Mond's interpretation of Joan, implying that he made her too much like a common peasant girl with "the figure of a cotton-bale."[3] Elsewhere Twain suggested that Joan was too often painted "as a strapping middle-aged fishwoman, with costume to match," and further noted that "supremely great souls are never lodged in gross bodies."[4]

Although several of Du Mond's illustrations do challenge nineteenth-century notions of feminine propriety — including a few images that show Joan riding a horse astride, rather than in the more "chaste" sidesaddle position — the artist primarily conveys Joan's traditional femininity, deference, and devotion. Among the devices he uses to suggest Joan's passivity, he frequently depicts her in stiff, melodramatic poses, mawkishly nonconfrontational, eyes uplifted in piety or downcast in modesty, with "the fixed expression of a ham," Twain later complained.[5] However, Du Mond was occasionally compelled to show us Joan's strength, to represent the several active, convincingly confrontational postures described in the text: "Joan Before the Governor" facing 58), for instance, and "Joan Reprimands the Conspirators" (facing 92). But if he portrayed Joan dynamically in such cases, he did so, for the most part, halfheartedly and awkwardly, perhaps betraying a distinctly more hostile attitude toward self-possessed women than Twain expressed.

The unsigned frontispiece is the most evocative and transcendent of Du Mond's images in the book, surely his own design rather than a reproduction (Joan's breastplate matches the one that appears in many of his signed illustrations). Joan wears a graceful and suggestive form-fitting full suit of armor, her eyes fixed upon the viewer, her arms and weapons raised in a paradoxical gesture indicating both victory and surrender. She appears both passive and resolute, her posture and figure inviting, her gaze and armor impervious. The image captures the complex attitudes of the nineteenth-century social elite toward women's social participation and sexuality. It is both threatening and tempting, exciting and dangerous, a further testament to the power of Joan of Arc as a figure of the insecurity generated by the loosening of traditional gender roles.

Examined closely, the frontispiece is almost Wagnerian in its heroic and iconic medievalism and nationalism. Displaying her sword while triumphantly perched atop the head of the English king, Henry V, and the severed heads of two male demons, Joan fundamentally and deeply threatens the ideology of male dominance. In the reproduction of the painting *The Maid of Orleans* (facing 370), the French text explicitly likens Joan, who "cut off the head of the English Holofernes," to the biblical Judith, an allusion no doubt intended in Du Mond's frontispiece as well. Along with the decorative quality of Du Mond's background, this theme calls to mind the work of the fin-de-siècle Viennese artist Gustav Klimt, who painted several vivid images of Judith holding the head of the Assyrian general Holofernes. Like Klimt's much analyzed portraits of women during the same era, Du Mond's Joan is menacing, her independence, indeterminacy, and sexuality the source of anxiety for men in an era that rigidly prescribed the behaviors expected of both sexes.

Joan, like Judith, was both a martial heroine who thwarted a national enemy and an archetype to be invoked by traditionalists seeking essential divisions between the sexes. Both author and illustrator chose to emphasize Joan's passivity and femininity, although Du Mond's frontispiece goes beyond Twain's text in portraying her self-sufficiency and command as a dangerous challenge to social stability. Whether by Du Mond's prerogative or editorial decision, it is clear the illustrations were carefully assembled and considered. As Twain often stipulated, the illustrations were subject to his approval, and ultimately the appearance and content of the work rest within Twain's authority, whether by intent or default.

NOTES

1. R. Kent Rasmussen, "Du Mond, Frank Vincent" and "Joan of Arc, Personal Recollections of" entries, *Mark Twain A to Z: The Essential Reference to His Life and Writings* (New York, Facts on File, 1995), pp. 121, 262.

2. "Du Mond, Frank V." entry, *Who Was Who in American Art*, Peter Hastings Falk, ed. (Madison, Conn: Soundview Press, 1985), p. 175.

3. Society of Illustrators speech, New York, December 21, 1905, *Mark Twain's Speeches*, The Oxford Mark Twain (New York: Oxford University Press, 1996), p. 242.

4. Cited in Rasmussen, "Joan of Arc" entry, p. 263.

5. Society of Illustrators speech, p. 243.

A NOTE ON THE TEXT

Robert H. Hirst

This text of *Personal Recollections of Joan of Arc* is a photographic facsimile of a copy of the first American edition dated 1896 on the title page. Although books printed from the first edition plates were manufactured until at least 1915, the earliest copies of the first edition were published in early May 1896. Two copies were deposited with the Copyright Office on May 1. The copy reproduced here is an example of Jacob Blanck's "first state," judging by the advertisements at the back (*BAL* 3446). The original volume is in the collection of the Mark Twain House in Hartford, Connecticut (810/C625per/1896/c. 3).

THE MARK TWAIN HOUSE

The Mark Twain House is a museum and research center dedicated to the study of Mark Twain, his works, and his times. The museum is located in the nineteen-room mansion in Hartford, Connecticut, built for and lived in by Samuel L. Clemens, his wife, and their three children, from 1874 to 1891. The Picturesque Gothic-style residence, with interior design by the firm of Louis Comfort Tiffany and Associated Artists, is one of the premier examples of domestic Victorian architecture in America. Clemens wrote *Adventures of Huckleberry Finn*, *The Adventures of Tom Sawyer*, *A Connecticut Yankee in King Arthur's Court*, *The Prince and the Pauper*, and *Life on the Mississippi* while living in Hartford.

The Mark Twain House is open year-round. In addition to tours of the house, the educational programs of the Mark Twain House include symposia, lectures, and teacher training seminars that focus on the contemporary relevance of Twain's legacy. Past programs have featured discussions of literary censorship with playwright Arthur Miller and writer William Styron; of the power of language with journalist Clarence Page, comedian Dick Gregory, and writer Gloria Naylor; and of the challenges of teaching *Adventures of Huckleberry Finn* amidst charges of racism.

CONTRIBUTORS

Beverly R. David is professor emerita of humanities and theater at Western Michigan University in Kalamazoo. She is currently working on volume 2 of *Mark Twain and His Illustrators*, and on a Mark Twain mystery entitled *Murder at the Matterhorn*. She has written a number of sections on illustration for the *Mark Twain Encyclopedia* and her *Mark Twain and His Illustrators, Volume 1 (1869–1875)* was published in 1989. Dr. David resides in Allegan, Michigan, in the summer and Green Valley, Arizona, in the winter.

Shelley Fisher Fishkin, professor of American Studies and English at the University of Texas at Austin, is the author of the award-winning books *Was Huck Black? Mark Twain and African-American Voices* (1993) and *From Fact to Fiction: Journalism and Imaginative Writing in America* (1985). Her most recent book is *Lighting Out for the Territory: Reflections on Mark Twain and American Culture* (1996). She holds a Ph.D. in American Studies from Yale University, has lectured on Mark Twain in Belgium, England, France, Israel, Italy, Mexico, the Netherlands, and Turkey, as well as throughout the United States, and is president-elect of the Mark Twain Circle of America.

Susan K. Harris is professor of American literature at Pennsylvania State University in State College. She is co-editor of *Legacy: A Journal of American Women Writers,* and author of the notes for the Library of America's volume *Mark Twain, Historical Romances: The Prince and the Pauper, A Connecticut Yankee, Joan of Arc* (1994). Her other works include *Mark Twain's Escape from Time* (1982), *Nineteenth-Century American Women's Novels: Interpretive Strategies* (1990), *The Courtship of Olivia Langdon and Mark Twain* (1996), and related articles and essays.

Robert H. Hirst is the General Editor of the Mark Twain Project at The Bancroft Library, University of California in Berkeley. Apart from that, he has no other known eccentricities.

Justin Kaplan is the author of *Mr. Clemens and Mark Twain* (1966; winner of the Pulitzer Prize and National Book Award), *Lincoln Steffens: A Biography* (1974), *Mark Twain and His World* (1974), and *Walt Whitman: A Life* (1980), and general editor of the sixteenth edition of *Bartlett's Familiar Quotations* (1992). From 1992 to 1995 he was Jenks Professor of Contemporary Letters at the College of the Holy Cross, Worcester, Massachusetts. He and his wife, the novelist Anne Bernays, are co-authors of a work of nonfiction, *The Language of Names*, to be published in the spring of 1997.

Ray Sapirstein is a doctoral student in the American Civilization Program at the University of Texas at Austin. He curated the 1993 exhibition *Another Side of Huckleberry Finn: Mark Twain and Images of African Americans* at the Harry Ransom Humanities Research Center at the University of Texas at Austin. He is currently completing a dissertation on the photographic illustrations in several volumes of Paul Laurence Dunbar's poetry.

ACKNOWLEDGMENTS

There are a number of people without whom The Oxford Mark Twain would not have happened. I am indebted to Laura Brown, senior vice president and trade publisher, Oxford University Press, for suggesting that I edit an "Oxford Mark Twain," and for being so enthusiastic when I proposed that it take the present form. Her guidance and vision have informed the entire undertaking.

Crucial as well, from the earliest to the final stages, was the help of John Boyer, executive director of the Mark Twain House, who recognized the importance of the project and gave it his wholehearted support.

My father, Milton Fisher, believed in this project from the start and helped nurture it every step of the way, as did my stepmother, Carol Plaine Fisher. Their encouragement and support made it all possible. The memory of my mother, Renée B. Fisher, sustained me throughout.

I am enormously grateful to all the contributors to The Oxford Mark Twain for the effort they put into their essays, and for having been such fine, collegial collaborators. Each came through, just as I'd hoped, with fresh insights and lively prose. It was a privilege and a pleasure to work with them, and I value the friendships that we forged in the process.

In addition to writing his fine afterword, Louis J. Budd provided invaluable advice and support, even going so far as to read each of the essays for accuracy. All of us involved in this project are greatly in his debt. Both his knowledge of Mark Twain's work and his generosity as a colleague are legendary and unsurpassed.

Elizabeth Maguire's commitment to The Oxford Mark Twain during her time as senior editor at Oxford was exemplary. When the project proved to be more ambitious and complicated than any of us had expected, Liz helped make it not only manageable, but fun. Assistant editor Elda Rotor's wonderful help in coordinating all aspects of The Oxford Mark Twain, along with literature editor T. Susan Chang's enthusiastic involvement with the project in its final stages, helped bring it all to fruition.

I am extremely grateful to Joy Johannessen for her astute and sensitive copyediting, and for having been such a pleasure to work with. And I appreciate the conscientiousness and good humor with which Kathy Kuhtz Campbell heroically supervised all aspects of the set's production. Oxford president Edward Barry, vice president and editorial director Helen McInnis, marketing director Amy Roberts, publicity director Susan Rotermund, art director David Tran, trade editorial, design and production manager Adam Bohannon, trade advertising and promotion manager Woody Gilmartin, director of manufacturing Benjamin Lee, and the entire staff at Oxford were as supportive a team as any editor could desire.

The staff of the Mark Twain House provided superb assistance as well. I would like to thank Marianne Curling, curator, Debra Petke, education director, Beverly Zell, curator of photography, Britt Gustafson, assistant director of education, Beth Ann McPherson, assistant curator, and Pam Collins, administrative assistant, for all their generous help, and for allowing us to reproduce books and photographs from the Mark Twain House collection. One could not ask for more congenial or helpful partners in publishing.

G. Thomas Tanselle, vice president of the John Simon Guggenheim Memorial Foundation, and an expert on the history of the book, offered essential advice about how to create as responsible a facsimile edition as possible. I appreciate his very knowledgeable counsel.

I am deeply indebted to Robert H. Hirst, general editor of the Mark Twain Project at The Bancroft Library in Berkeley, for bringing his outstanding knowledge of Twain editions to bear on the selection of the books photographed for the facsimiles, for giving generous assistance all along the way, and for providing his meticulous notes on the text. The set is the richer for his advice. I would also like to express my gratitude to the Mark Twain Project, not only for making texts and photographs from their collection available to us, but also for nurturing Mark Twain studies with a steady infusion of matchless, important publications.

I would like to thank Jeffrey Kaimowitz, curator of the Watkinson Library at Trinity College, Hartford (where the Mark Twain House collection is kept), along with his colleagues Peter Knapp and Alesandra M. Schmidt, for having been instrumental in Robert Hirst's search for first editions that could be safely reproduced. Victor Fischer, Harriet Elinor Smith, and especially Kenneth M. Sanderson, associate editors with the Mark Twain Project, reviewed the note on the text in each volume with cheerful vigilance. Thanks are also due to Mark Twain Project associate editor Michael Frank and administrative assistant Brenda J. Bailey for their help at various stages.

I am grateful to Helen K. Copley for granting permission to publish photographs in the Mark Twain Collection of the James S. Copley Library in La Jolla, California, and to Carol Beales and Ron Vanderhye of the Copley Library for making my research trip to their institution so productive and enjoyable.

Several contributors — David Bradley, Louis J. Budd, Beverly R. David, Robert Hirst, Fred Kaplan, James S. Leonard, Toni Morrison, Lillian S. Robinson, Jeffrey Rubin-Dorsky, Ray Sapirstein, and David L. Smith — were particularly helpful in the early stages of the project, brainstorming about the cast of writers and scholars who could make it work. Others who participated in that process were John Boyer, James Cox, Robert Crunden, Joel Dinerstein, William Goetzmann, Calvin and Maria Johnson, Jim Magnuson, Arnold Rampersad, Siva Vaidhyanathan, Steve and Louise Weinberg, and Richard Yarborough.

Kevin Bochynski, famous among Twain scholars as an "angel" who is gifted at finding methods of making their research run more smoothly, was helpful in more ways than I can count. He did an outstanding job in his official capacity as production consultant to The Oxford Mark Twain, supervising the photography of the facsimiles. I am also grateful to him for having put me in touch via e-mail with Kent Rasmussen, author of the magisterial *Mark Twain A to Z*, who was tremendously helpful as the project proceeded, sharing insights on obscure illustrators and other points, and generously being "on call" for all sorts of unforeseen contingencies.

I am indebted to Siva Vaidhyanathan of the American Studies Program of the University of Texas at Austin for having been such a superb research assistant. It would be hard to imagine The Oxford Mark Twain without the benefit of his insights and energy. A fine scholar and writer in his own right, he was crucial to making this project happen.

Georgia Barnhill, the Andrew W. Mellon Curator of Graphic Arts at the American Antiquarian Society in Worcester, Massachusetts, Tom Staley, director of the Harry Ransom Humanities Research Center at the University of Texas at Austin, and Joan Grant, director of collection services at the Elmer Holmes Bobst Library of New York University, granted us access to their collections and assisted us in the reproduction of several volumes of The Oxford Mark Twain. I would also like to thank Kenneth Craven, Sally Leach, and Richard Oram of the Harry Ransom Humanities Research Center for their help in making HRC materials available, and Jay and John Crowley, of Jay's Publishers Services in Rockland, Massachusetts, for their efforts to photograph the books carefully and attentively.

I would like to express my gratitude for the grant I was awarded by the University Research Institute of the University of Texas at Austin to defray some of the costs of researching The Oxford Mark Twain. I am also grateful to American Studies director Robert Abzug and the University of Texas for the computer that facilitated my work on this project (and to UT systems analyst Steve Alemán, who tried his best to repair the damage when it crashed). Thanks also to American Studies administrative assistant Janice Bradley and graduate coordinator Melanie Livingston for their always generous and thoughtful help.

The Oxford Mark Twain would not have happened without the unstinting, wholehearted support of my husband, Jim Fishkin, who went way beyond the proverbial call of duty more times than I'm sure he cares to remember as he shared me unselfishly with that other man in my life, Mark Twain. I am also grateful to my family — to my sons Joey and Bobby, who cheered me on all along the way, as did Fannie Fishkin, David Fishkin, Gennie Gordon, Mildred Hope Witkin, and Leonard, Gillis, and Moss

Plaine — and to honorary family member Margaret Osborne, who did the same.

My greatest debt is to the man who set all this in motion. Only a figure as rich and complicated as Mark Twain could have sustained such energy and interest on the part of so many people for so long. Never boring, never dull, Mark Twain repays our attention again and again and again. It is a privilege to be able to honor his memory with The Oxford Mark Twain.

Shelley Fisher Fishkin
Austin, Texas
April 1996